Biophysics of Photoreception

Molecular and Phototransductive Events

Istituto Italiano per gli Studi Filosofici

Series on Biophysics and Biocybernetics

Coordinating Editor: C. Taddei-Ferretti

Vol. 2: Biocybernetics of Vision: Integrative Mechanisms and Cognitive Processes
edited by: C. Taddei-Ferretti

Forthcoming volumes:

Vol. 3: High Dilution Effects on Cells and Integrated Systems
edited by: C. Taddei-Ferretti and P. Marotta

Vol. 4: Macromolecular Interplay in Brain Associative Mechanisms
edited by: A. Neugebauer

ISTITUTO ITALIANO PER GLI STUDI FILOSOFICI
SERIES ON BIOPHYSICS AND BIOCYBERNETICS
Vol. 1 – Biophysics

Istituto Italiano per gli Studi Filosofici
Gruppo Nazionale di Cibernetica e Biofisica, CNR

Biophysics of Photoreception

Molecular and Phototransductive Events

Proceedings of the International School of Biophysics
Casamicciola, Napoli, Italy, 10–16 October 1994

Edited by

C. Taddei-Ferretti
Istituto di Cibernetica, CNR, Arco Felice, Napoli, Italy

World Scientific
Singapore • New Jersey • London • Hong Kong

Published by

World Scientific Publishing Co. Pte. Ltd.

P O Box 128, Farrer Road, Singapore 912805

USA office: Suite 1B, 1060 Main Street, River Edge, NJ 07661

UK office: 57 Shelton Street, Covent Garden, London WC2H 9HE

British Library Cataloguing-in-Publication Data
A catalogue record for this book is available from the British Library.

BIOPHYSICS OF PHOTORECEPTION: MOLECULAR AND PHOTOTRANSDUCTIVE EVENTS

ISBN 981-02-3228-4

This book is printed on acid-free paper.

Printed in Singapore by Uto-Print

The meeting of **Hennig Stieve** (left), who is leaving the Hotel Gran Paradiso (Casamicciola at Isle of Ischia, Naples, Italy) at the end of the "*International School A. Borsellino and W. Reichardt on Biophysics of Photoreception: Molecular and Phototransductive Events*", and of **Miguel García Pérez** (right), who is just arrived to the Hotel at the beginning of the "*International School E. Caianiello on Biocybernetics of Vision: Integrative Mechanisms and Cognitive Processes*", simbolyzes the distinctions and interconnections of the two fields that will periodically alternate in the publication of the "Series on Biophysics and Biocybernetics".

THE "SERIES ON BIOPHYSICS AND BIOCYBERNETICS"

With this volume a new series of works on Biophysics and on Biocybernetics begins.

The "*Series on Biophysics and Biocybernetics*" is brought to life thanks to the activity of the Istituto Italiano per gli Studi Filosofici, a Foundation receptive to the scientific enterprise and the continually evolving cutting edge of science, and thanks to the inspiration of its President, Gerardo Marotta.

The series is a periodic publication: two volumes are foreseen for each issue, one concerning Biophysics and the other one Biocybernetics.

Each volume will contain a collection of lectures by specialists and selected interventions by young scholars held at the "International School on Biophysics" and at the "International School on Biocybernetics", organized by the Istituto Italiano per gli Studi Filosofici in order to provide the opportunity of discussions among all the participants.

The Istituto Italiano per gli Studi Filosofici intends the publication of this series to be a means to further the integration between philosophic thought and the advances of science and technology in a critical approach.

Cloe Taddei-Ferretti

Series Coordinator

PREFACE

This is the first volume of the Series on Biophysics and Biocybernetics promoted by the Istituto Italiano per gli Studi Filosofici.

It appears as the Proceedings of the "International School A. Borsellino and W. Reichardt on Biophysics of Photoreception: Molecular and Phototransductive Events", which was inaugurated at the site of the Istituto, Palazzo Serra di Cassano, Naples, Italy, and was held at Casamicciola on the isle of Ischia, Italy, on October 10-16, 1994, under the direction of this volume's Editor.

This School was promoted and supported jointly by the Istituto Italiano per gli Studi Filosofici and by the Gruppo Nazionale di Cibernetica e Biofisica of the Italian Consiglio Nazionale delle Ricerche (CNR).

The School has been organized by the Istituto di Cibernetica, CNR, in the 25th year of its foundation: for this reason it has been dedicated to the memory of Prof. A. Borsellino and of Prof. W. Reichardt, who were in the Advisory Board of the Institute.

The topics covered ranged from "Photoreception and Phototransduction in Microorganisms", to "Photoreception and Phototransduction in Invertebrates" and to "Photoreception and Phototransduction in Vertebrates".

A main problem in biophysics is the interaction of light information with functional living structures, in order to shape the organism's behaviour.

Although the processes of photoreception and phototransduction are variously articulated in different living beings, as it appears in the subdivision of the School topics in microorganisms, invertebrates and vertebrates, general ways of light signal reception and transduction, through light energy - photosensitive molecule interactions, could be observed. Highly sophisticated advanced techniques are employed in this research field.

I would like to thank the members of the School Advisory Board for their fruitful advice and suggestions, as well as for their helpful cooperation also as reviewers of the participants' communications: D.-P. Häder (D), R.C. Hardie (UK), U.B. Kaupp (D), F. Lenci (I), B. Minke (IL), H. Stieve (D), A. Packard (I), I. Pepe (I).

I would also acknowledge the partial financial support of the Comitato Nazionale per la Fisica, CNR, the Istituto di Cibernetica, CNR, the Società Italiana di Biofisica Pura ed Applicata and the Dipartimento di Fisica, Università di Genova, as well as the auspices given by the Società Italiana di Fotobiologia.

Furthermore, only due to the help of the Istituto Italiano per gli Studi Filosofici, as well as of the Società Italiana di Biofisica Pura ed Applicata and the Dipartimento di Fisica, Università di Genova, it has been possible to award grants to several deserving participants needing financial help.

I wish to thank all the scientists who agreed to lecture, who contributed to the School with their discussions and who favoured, toghether with all the participants, a charming atmosphere in a highly stimulating scientific milieu.

Finally, I thank the members of the local organizing committee: A. Cotugno, M. Izzo, A.M. Mazzarella, C. Musio, S. Piantedosi, S. Santillo, of the Istituto di Cibernetica, CNR; without their work the School could not have been realized, nor its cordial atmosphere obtained.

The beauty of the isle of Ischia and the courtesy of the staff of the Hotel Gran Paradiso at Casamicciola, where the School was held, completed the pleasantness of the environment.

Cloe Taddei-Ferretti

CONTENTS

IN MEMORY OF
A. BORSELLINO AND W.E. REICHARDT

IN MEMORY OF ANTONIO BORSELLINO

Alfonso M. Liquori

Facoltà di Scienze, Università "Tor Vergata", Roma, Italy

Antonio Borsellino was without doubt a scientist of great amiability.

Apart from his penetrating eyes, I have always been struck by his intelligence, his fantasy and above all his strict rigour (even if at time it was tinged with an excess of "physicalism" when he was confronted by biological systems of great complexity).

There is no doubt that he was a theoretical physicist of great class, who was capable of working in the field of physical chemistry and thermodynamics without difficulty and above all without presumption.

His mathematical talent was notable also, and above all the elegance and the clarity of his involvements, which I remember well, were incisive.

He came, like Fermi and Rubbia, from the Scuola Normale Superiore of Pisa, of which he retained fond memories. This initial foundation explains, at least partly, his great interest in philosophy, and not only science philosophy.

Antonio Borsellino was suddenly attracted, in his middle age, towards cybernetics, and especially towards biophysics, but it was very difficult to him to remain concentrated on any one problem in this great cognitive universe. His restleness was without boundaries.

I have had many interesting contacts with Antonio Borsellino, concerning the problems of ionic and molecular transport across biological membranes and the fundamental processes of nerve impulses. His critical comments were always stimulating for me, as were also his appreciative remarks concerning my theoretical work on the structure of polypeptides and proteins, a large sector of research with which he had never occupied himself but nevertheless considered to be at the centre of the development of molecular biology and biophysics.

Recently, at Trieste, Antonio had decided to become an "experimentalist". He had organized a laboratory of electrophysiology in an abandoned school and took to me, with great pride, to visit it.

Antonio Borsellino was a man of great human qualities and was gifted with a profound sense of scientific ethics, but unfortunately his great capacity and exceptional intelligence were not fully appreciated by the Italian scientific community. His versatility and unquenchable thirst for scientific knowledge contribute to isolate him, at least partly, from our scientific structure, dominated as it is by corporations (even though sometimes very sound). Antonio's southern Italian sensitivities certainly made his personal affirmation before mediocre individuals more difficult.

Antonio Borsellino leaves a School and many ideas to develop futher and for myself the memory of a man of great civility and intellectual charm.

ANTONIO BORSELLINO (1915-1992)

BIOGRAPHICAL SKETCH

Education

1938, Degree in Physics, Scuola Normale Superiore, Pisa, Italy.

Honors

Colombo Foudation, Milano, Award "Città di Lecco" for work in Physics.
Award "Città di Como".
Award of the Italo-Swiss Foundation "Della Riccia".
Award "City of Columbus, Ohio".

Training and Employment

1985-1992, Director of the Biophysics Sector, SISSA, Trieste, Italy.
1974-1985, Full Professor of Biophysics, University of Genova, Italy.
1950-1974, Full Professor of Theoretical Physics, University of Genova, Italy.
1939-1950, Assistant Professor, Politecnico, University of Milano, Italy.

Teaching Experiences

Courses on: Quantum Mechanics; Quantum Field Theory; Quantum Electrodynamics; Cybernetics and Information Theory; Stability Theory with Applications to Population Dynamics and Macroeconomics; Biophysics.

Official Duties

Director of the Computing Center, University of Genova, Italy.
Assistant Manager of "*Il Nuovo Cimento*".
Member of the Physics Committee, CNR (National Research Council), Italy.
President of the Biophysics Committee, CNR, Italy.
President of the National Group of Cybernetics, CNR, Italy.
Member of the Council of the International Union of Pure and Applied Biophysics (IUPAB).
Director of the Laboratory of Biophysics and Cybernetics, CNR, Camogli, Italy.
Member of the Advisory Board of the Laboratory of Cybernetics, CNR, Arco Felice, Italy.
Italian Delegate to the European Molecular Biology Conference.
President of the Italian Society of Pure and Applied Biophysics (SIBPA).
Director of the Institut of Physical Sciences, University of Genova, Italy.
Director of the International School of Biophysics, Erice, Italy.
Director of the Biophysics Section of SISSA, Trieste, Italy.

Sabbatical Leave and International Collaborations

1951, Visiting Scientist, University of Brussels, Belgium.
1952-1953, Research Associate at the Newmann Nuclear laboratories, Cornell University, Ithaca, New York, USA.

1962-1970, Several visits as Special Consultant at NIH, Bethesda, Maryland, USA.
1978-1979, Invited Professor at the Medical School, University of Geneva, Switzerland.
1982, Ivited Guest of the Sinnica Academy of Sciences, Peking, China.
1985, Fogarty Scholar at NIH, Bethesda, Maryland, USA.

Main Invited Lectures and Symposia

Int. School of Physics on "Mathematical Development in Field Theory", Varenna, Italy, 1958.
American Biophysical Soc. on "Computers in Biology", St. Louis, Missouri, USA, 1960.
Gordon Conference on "Ionic Movements", 1964.
Cold Spring Harbor Symposium on Quantitative Biology (two contributions), 1965.
Int. School of Physics on "Mathematical Development in Field Theory", Varenna, Italy, 1968.
"Information Theory in Biology" (series of lectures), NIH, Bethesda, Maryland, USA, 1970.
Symposium on "Discrete Mathematics and its Applications", Aberdeen, UK, 1975.
Int. School of Biophysics, Erice, Italy, every year from 1971 to 1982.
Chinese Academy of Sciences (series of seminars), Peking, China, 1982.

Books Edited

Development in Biological Research, New York: Plenum Press, 1980.
Photoreceptors, New York: Plenum Press, 1984.
Sensory Transduction, New York: Plenum Press, 1990.

Editorial Boards

Giornale di Fisica (Journal of the Italian Physical Society).
Il Nuovo Cimento.
Bioteorica.
Cell Physiology.

Membership

International Institute of Communications.
Italian Physical Society.
Italian Philosophycal Society.
American Physics Society.
Biophysical Society (USA).
Woods Hole MBL Corporation.

Publications

Physics: about 30 major papers.
Biophysics : about 50 major papers.

MOLECULAR RECONSTITUTION OF PHOTOSENSITIVE SYSTEMS

Isidoro Mario Pepe

Institute of Biophysics, Faculty of Medicine, University of Genoa, 16132 Genoa, Italy

The comprehension of photobiological processes involves the understanding of the membrane operating at a molecular level using functional units such as whole chloroplasts or visual cells. The complementary approach consists in the attempt to reconstitute an artificial photosensitive membrane with the final goal being to reproduce the complex biological processes occurring at this level. The rationale of the reconstitution approach is the reductionistic dissection of the complex mechanisms occurring at the membrane level into their components and the resulting attempt to reassemble a functional membrane by adding each component under controlled experimental conditions. However, the ultimate complexity of the photosensitive system can hardly be reconstituted with this method. This was a concept always present in the mind of Professor Antonio Borsellino to whom this School of Biophysics of Photoreception is dedicated. On the other hand, he was aware, that despite the rare success of the reconstitution of the functional units, the intermediate steps often give origin to a rich harvest of technical applications. Thin films of light-sensitive proteins for instance, are an expanding field with exciting applications such as optical information processing and molecular electronic devices. Mono or multi molecular layers of animal or bacterial rhodopsin can be assembled closer together than they are in native membranes, so as to build up non traditional photosensitive material, possibly capable of carrying processes of light-dependent transfers of electrons and protons with high quantum efficiency.

Photosynthesis

Tien and his collaborators (1-4) made the first attempts at reconstituting photosensitive membranes by incorporating chloroplast into black lipid membranes (BLM). They observed a photoelectric effect of a few mV, which was dramatically increased by the introduction of modifiers to one or both sides of the bathing solutions in such a way as to create asymmetrical conditions across the membrane. A photo-emf greater than 100 mV was obtained when the electron acceptor Fe^{3+} was added to one side of the bathing solution and an electron donor, such as ascorbic acid, to the other side. Moreover, the photoelectric action spectrum obtained by exciting the chloroplast-BLM with light of different wavelengths was well correlated with the absorbance spectrum of chlorophyll.

These observations, together with those of several authors (5-7), suggest that the photoeffects measured across the BLM dividing two redox solutions are due to a separation of electronic charges produced by photons absorbed by the pigments embedded in the lipid matrix.

Chen and Berg (8) observed that in the experiments with membranes containing chloroplast pigments, the crucial problem was that the photoelectric effect dissipates the pre-existing redox gradient, in antithesis with the photosynthetic process which stores energy as redox equivalents. In photosynthesis a redox gradient is being created and the electrons must flow in the opposite direction to that occurring in BLM. In thylakoid membranes an electron - transport chain system made up of accessory pigments, proteins and quinones, performs the electron transfer in the direction of energy storage. Some of these accessory pigments, phycocyanin and phycoerythrins, which are biliproteins from blue-green and red algae, were tested in a series of experiments with chloroplast-BLM (5,8) by separating a redox gradient of ferric chloride-ferrous chloride solution. A solution of biliproteins added to the reducing side of the membrane was able to reverse the sign of photovoltage thereby demostrating the electrons-directing ability of these proteins towards a creation of redox gradient.

An improvement in the artificial system has been made with a well-defined bilayer containing phosphatidylcholine, chlorophyll and b-carotene (9). This system is capable of producing photoeffects similar to those obtained from BLM containing chloroplast extract. An enhancement of photovoltages was obtained by adding plastocyanin to such a system (10), a copper protein extracted from spinach leaves known as an electron carrier associated with the thylakoid membrane. In contrast with phycocyanin, plastocyanin is effective when added to the oxidizing side of the membrane facilitating the electron transport from Fe^{3+}/Fe^{2+} aqueous phase to the chlorophyll a/beta carotene lipid phase (10).

Quinones have also been suggested as playing a role in the electron transport of higher plants. When synthetic quinones were incorporated in BLM containing purified chlorophyll and separating an asymmetric solution of oxidant and reductant, an increase in photoconductance was observed (11). It seems reasonable to explain the enhanced conductance by postulating that the small quinone molecules can easily move across the bilayer carrying electrons and protons and coupling the interfacial electron transfer reactions between chlorophyll and the redox agents. An electron transfer from chlorophyll to quinone was studied incorporating benzoquinone and ubiquinone in chlorophyll-containing liposomes (12). Laser flash photolysis has been used to show the formation of electron transfer products C+ and quinone radicals Q- which diffuse either within the bilayer or at the bilayer-water interface. A rapid recombination between chlorophyll and quinone radicals occurring within the bilayer and a slower recombination at the bilayer-water interface, result in an effective separation of radicals favouring an energy storage.

Reaction Centers (RC), the proteins of photosynthetic bacteria, have been used recently to build molecular photoelectric devices (13) The technique of Langmuir-Blodgett (LB) was very useful in preparing thin films of close-packed RC molecules, which were deposited on transparent electrodes. The light-induced electric response measured on this device was in the order of nA. LB films of RC have been prepared by several groups (14-16) However, the construction of more complex molecular organization was achieved by using the affinity of RC with quinone (17). The insertion of RC proteins into an LB film confers a unique stability to their secondary structure as a result of both specific intermolecular interactions and decreased hydration. The close packing of protein molecules into two dimensional, high ordered, crystal-like structures could dramatically change some of the chemical-physical properties commonly known for protein in solution. It was infact shown by circular dichroism (CD) measurements that the secondary structure of RC was not affected in LB films when heated up to 200 °C (18-20).

Bacteriorhodopsin

The purple membrane of *Halobacterium halobium* contains bacteriorhodopsin (21), a chromoprotein with a retinal attached via a Shiff-base linkage similar to rhodopsin. But whereas rhodopsin forms stable pigments with 11-cis and 9-cis retinal, bacteriorhodopsin forms stable pigments with all-trans and +13-cis retinal (22). The protonated Schiff-base absorbs maximally at 568 nm (B_{568}) and is bleached by light to the unprotonated form absorbing maximally at 412 mn (M_{412}) (23).The dark-reversion to the original 568-pigment is accompanied by a reprotonation. Each cycle is completed in about 10 milliseconds (24) and coupled by the uptake and release of protons (25). Bacteriorhodopsin is therefore a proton pump through which the bacteria acidify the medium when exposed to light under anaerobic conditions (26-28). The resulting proton gradient constitutes the energy storage, which is then utilized for ATP synthesis, in accordance with Mitchell's chemiosmotic theory of energy transduction (29).

A successful reconstitution of the purple membrane functions was first made by Racker and Stoeckenius (30) who succeeded in incorporating bacteriorhodopsin into liposomes. On illumination the reconstituted vesicles took up protons which were released in the dark. The incorporation of beef-heart mitochondrial ATPase into the same liposomes made the system capable of catalyzing light-dependent ATP formation. The light-induced proton uptake in reconstituted vesicles has a direction opposite to that observed in intact bacteria. The explanation for this is that the purple membrane has a definite sidedness and during reconstitution it had been turned inside out, as shown by freeze-fracture electron micrographs (31).

Other attempts in reconstituting purple membranes have been made by Skulachev and his colleagues by incorporating fragments of the purple membrane in liposomes (32) and in planar phospholipid membranes (33,34). They found that illumination of bacteriorhodopsin liposomes generated an electric potential difference, leaving the inside of vesicles positive and alkaline the incubating medium. In planar lipid bilayers containing bacteriorhodopsin they observed light-induced electric potential up to 50 mV at a current of about 0.5 pA. The addition of the protonophorous uncoupler CCCP (carbonyl cyanide m-chlorophenyl hydrazone) to the solutions decreased the electrical potential and changed the light-and-dark induced response into the characteristic response of a capacitance. This latter result suggests that at least part of the cyclic reaction is reversible giving rise to an a.c. proton current although small compared with the d.c. proton current (35).

Purple membrane characteristics were also studied by means of air-water interface films. A mixture of purple membrane and phospholipids were spread at an air-water interface forming an insoluble film which consists of membrane fragments separated by a lipid monolayer (36). Bacteriorhodopsin in these interface films seems to maintain its properties showing spectral characteristics in darkness and upon flash illumination, similar to those observed in aqueous suspension. In addition, an electron micrograph indicates that the membrane fragments are highly oriented with their intracellular surface in the direction of the aqueous phase. These interface films containing spectroscopically intact and oriented bacteriorhodopsin have been used as a model system in studying the functions of purple membranes (37). Illumination of these films produced small photopotentials in the order of few millivolts which were dramatically increased when lipid soluble proton carriers were added in the presence of decane. The action spectrum of photopotentials matches the absorbance spectrum of rhodopsin in its light-adapted state (BR_{568}).

Many attempts to study bacteriorhodopsin properties using planar lipid bilayers have been made (38-40). Various methods have been performed involving fusion of vesicles containing purple membranes with BLM. Photocurrents and photovoltages were obtained from these membranes in accordance with a proton transfer towards the bacteriorhodopsin-free solution and in close accordance with the absorbance spectrum of bacteriorhodopsin. Doubts have been raised about the actual incorporation into planar bilayers of bacteriorhodopsin which should be oriented and spans the whole membrane as in native conditions. The steady-state photocurrents observed in these model systems are dramatically enhanced only in the presence of proton carriers (38) or gramicidin (39), suggesting that the vesicles fusion leaves an aqueous compartment between the purple membrane sheet and the planar bilayer.

The stability of bacteriorhodopsin makes it a very good candidate for possible molecular technological applications based on changes in spectroscopic and electrical properties during the photocycle. Absorption of light by bacteriorhodopsin leads to a reversible colour change from purple to yellow. It can be switched back and forth with a blue and yellow light and this makes this protein attractive for optical information processing. In addition, genetic modification of the halobacterial strain allows the isolation of variants with optimized photochromic properties, adapted to the specific demands required by the biomaterials (40). Dry films of bacteriorhodopsin were structurally stable up to a temperature of 140 °C (42) and this striking result encourages technical application of this protein, for example, optical devices, biosensor and artificial memories. With the LB technique, films of bacteriorhodopsin were deposited onto quartz plates and the photocycle was studied by flash photolysis. The energy transfer from tryptophan to BR retinal takes place in both dry and wet LB films (41).

The photochromic shift between B_{568} and M_{412} is large enough (164 nm) to build bistable devices but longer lifetimes of M_{412} are required in order to get a photocontrolled system. Physiochemical methods include the lowering of temperature (44) (M_{412} is stabilized at 230 °K), the drying out of the films and changing the pH or by adding less protic solvents (45).

Rhodopsin

Phospholipid vesicles containing purified and photochemically active rhodopsin (46,47) were used in the past to study the possibility of a light-activated permeability change which could actually occur in disc membranes. When these vesicles containing internally $^{45}Ca^{2+}$ were illuminated with a light flash bleaching approximately 80% of the rhodopsin (48), the result was a release of almost all the trapped $^{45}Ca^{2+}$. Similar experiments were also performed by Darszon, Montal and Zarco (49) who loaded rhodopsin-phospholipid vesicles with radioactive markers. Exposure to light resulted in the release of Na^+, Cs^+, Ca^{2+}, glycerol and glucose indicating the formation of a wide photoinduced permeability to ions and non electrolites. The efflux of labeled ions induced by ionophores and gramicidin indicated that the vesicles were sealed and the ion release was not due to vesicles rupture. The wavelength dependence of the light-induced permeability of the vesicles showed that the efflux values fitted in rather well with the absorbance spectrum of rhodopsin. The results with liposomes therefore indicate that in these conditions light excitation of rhodopsin induces the formation of wide, nonselective transmembrane channels.

Montal, Darszon and Trissl (50) studied light-induced conductance changes in planar bilayer lipid which were formed by the apposition of two monolayers containing phospholipids and spectrally active rhodopsin. The membrane conductance increased in the light and clearly showed defined fluctuations which were interpreted by the authors as the opening and closing of channels. A latency period between the illumination and the beginning of the conductance changes varied from 3 to 200 s and was shorter the higher the rhodopsin concentration in the membrane. This delay is too long when compared with the delay of milliseconds of the photoresponse *in vivo*, and the authors suggested that this could be explained by the formation of a channel induced by the aggregation of rhodopsin. The conductance changes following the rhodopsin bleaching in reconstituted systems could simply be due to the formation of an aspecific ion leakage in the bilayer as a result of conformation changes of the protein. Similar results infact were also obtained by incorporating a photosensitive pigment in lecithin/oleylamine planar bilayers (51). The increase of membrane conductance and the formation of channels following illumination were reported and attributed to a conformation change of the incorporated pigment (52).

As for bacteriorhodopsin, fast photovoltages with no detectable latency have been recorded in rhodopsin-containing lipid layers apposed on one side of a thin teflon film (53,54). These photoelectric signals were suggested as having originated from capacitive charge displacements occurring in rhodopsin molecules during their conformational transitions. The same mechanism might underlie the physiological response to a short flash evoked in the retina and known as "early receptor potentials".

Korenbrot and Pramik (55) investigated the properties of rhodopsin osmotically shocked with lipid vescicles at air-water interface. They found that the rhodopsin molecules at the interface were spectroscopically intact and regenerable upon the addition of 11-cis retinal. Furthermore, flash illumination of rhodopsin in air-dried multilayers deposited on solid supports resulted in the formation of stable Metarhodopsin I intermediate. Bleaching of rhodopsin was obtained instead, by increasing the relative humidity of the multilayers. The result is that in dry conditions vertebrate rhodopsin is converted to a stable Metarhodopsin , which was first found by Wald et al. in 1950 (56), and is of interest because of the possibility of making a bistable system for technical applications as for bacteriorhodopsin.

The advantage of making thin films with rhodopsin rather than with bacteriorhodopsin is that rhodopsin can be extracted with a detergent from the membrane without loosing most of its properties, while the molecules of bacteriorhodopsin seem unstable in detergent (57). For this reason films of bacteriorhodopsin are usually made by using natural purple membranes. By utilizing purified rhodopsin instead, solubilized with an appropriate detergent which can be eliminated during the assembly of the film, mono or multi molecular layers can be obtained made up mainly of rhodopsin. Another advantage is that rhodopsin in dry films, at room temperature, has photoconvertible stable states (55,56), while bacteriorhodopsin has some limitations arising from different photochemical side reactions (58).

A successful attempt to build two dimensional films made up of mainly bovine rhodopsin using the Langmuir-Blodgett method was recently reported by Maxia et al. (59). In these films, rhodopsin seems organized as in native membranes with the cylindrical axe perpendicular to the plane of the film, but with the molecules assembled very close together, as indicated by the area of about 1100 $Å^2$ occupied per molecule, which is close to that calculated for rhodopsin without the surrounding lipids (60,61).

The most striking property of rhodopsin in L-B films was the thermal stability of its secondary structure, which was maintained up to about 150°C in the light and up to at least 200°C in the dark (59). This improvement of stability was also measured in disk membranes where the temperature of rhodopsin denaturation was about 56°C for bleached samples and 72°C for samples kept in the dark (62). This is due to the more compact conformation of rhodopsin when it binds its chromophore compared to that of bleached rhodopsin in which the linkage between opsin and retinal is broken. The characteristic absorbance maximum of rhodopsin at 500 nm was also preserved up to about 90 °C (59).

The increased termal stability of rhopsin in L-B films seems to be a general property shared by other proteins (18-20,42) and is likely due to the molecular order induced by the Langmuir-Blodgett film formation. In a CD study on anti-insulin antibody, a lyophilized sample was found more resistant to temperature than the corresponding antibody solution, but less resistant than the same antibody in L-B films, leading to the conclusion that, while dehydratation plays an important role in structure preservation, the main stabilizing factor is the molecular close packing of the L-B film.

References

1) Tien, H.Ti *Nature* 219 (1968) 272
2) Tien, H.Ti *J. Phys. Chem.* 72 (1968) 4512
3) Tien, H.Ti and Verma, S.P. *Nature* 227 (1970) 1232
4) Tien, H.Ti *Photochem. Photobiol.* 16 (1972) 271
5) Ilani, A. and Berns, D.S. *J. Membrane Biol.* 8 (1972) 333
6) Berns, D.S. *Photochem. Photobiol.* 24 (1976) 117
7) Hong, F.T. *Photochem. Photobiol.* 24 (1976) 155
8) Chen, C-H. and Berns, D.S. *Proc Natl. Acad. Sci. USA* 72 (1975) 3407
9) Mangel, M.D., Berns, D.S. and Ilani, A. *J. Membrane Biol.* 20 (1975) 171
10) Chen, C-H. and Berns, D.S. *J. Membrane Biol.* 47 (1979) 113
11) Master, B.R. and Mauzerall, D. *J. Membrane Biol.* 41 (1978) 377
12) Hurley, J.K., Castelli, F. and Tollin, G. *Photochem. Photobiol.* 34 (19810 623
13) Yasuda, Y., Sugino, H., Toyotama, H., Hirata, Y., Hara, M. and Miyake, J. *Bioelectrochem. and Bioenerg.* 34 (1994) 135-139.
14) Tiede, D.M., Mueller, P. and Dutton, P.L. *Biochim. Biophys. Acta*, 681 (1982) 191
15) Alegria, G. and Dutton, P.L. *Biochim. Biophys. Acta*, 239 (1991) 1057
16) Yasuda, Y., Hirata, Y., Sugino, H., Kumei, M., Hara, M., Miyake, J. and Fujihira, M. *Thin Solid Films* 733 (1994) 210-211.
17) Hirata, Y. and Miyake, J. *Thin Solid Films* 244 (1994) 865-868.
18) Nicolini, C., Erokhin, V., Catasti, P. and Facci, P. *Biochim. Biophys. Acta* 1158 (1993) 273-278.
19) Erokhin, V., Facci, P. and Nicolini, C. *Biosensors and Bioelectronics,* 9 (1994) 1-10.
20) Nicolini, C. *Biosensor & Bioelectronic* 10 (1995) 105-127.
21) Oesterhelt, D. and Stoeckenius, W.J. *Nature (New Biology)* 233 (1971) 149
22) Oerserhelt, D. and Schuhmann, L. *FEBS Lett.* 44 (1974) 262
23) Lewis, A., Spoonhower, J., Bogomolni, R.A., Lozier, R.H. and Stoeckenius, W.

Proc. Natl. Acad. Sci. USA 71 (1974) 4462

24) Lozier, R.H., Bogomolni, R.A. and Stoeckenius, W. *Biophys. J.* 15 (1975) 955

25) Henderson, R. *Ann. Rev. Biophys. Bioeng.* 6 (1977) 87

26) Oesterhelt, D. and Stoeckenius, W. *Proc. Natl. Acad. Sci.* USA 70 (1973) 2853

27) Oesterhelt, D *Ciba Found. Symp. Energy Transform. Biol. Syst.* 31 (1976) 147

28) Stoeckenius, W., Lozier, R.H., Bogomolni, R.A. *Biochim. Biophys. Acta* 505 (1979) 215

29) Mitchell, P. *Biol. Rev. (Cambridge)* 41 (1966) 445

30) Racker, E. and Stoeckenius, W. *J. Biol. Chem.* 249 (1974) 662

31) Stoeckenius, W. and Lozier, R.H. *J. Supramol. Struct.* 2 (1974) 769

32) Kayushin, L.P. and Skulachev, V.P. *FEBS Lett.* 39 (1974) 39

33) Drachev, L.A., Kaulen, A.D., Ostroumov, S.A., and Skulachev, V.P. *FEBS Lett.* 39 (1974) 43-50.

34) Drachev, L.A., Jasaitis, A.A., Kaulen, A.D., Kondrashin, A.A., Liberman, E.A., Nemecek, I.B., Ostroumov, S.A., Semenov, A.Yu. and Skulachev, V.P. *Nature* 249 (1974) 321

35) Hong, F.H. *J. Coll. Inter. Sci.* 58 (1977) 471

36) Hwang, S.B., Korenbrot J.I. and Stoeckenius, W. *J. Membrane Biol.* 36 (1977) 115

37) Hwang, S.B., Korenbrot J.I. and Stoeckenius, W. *J. Membrane Biol.* 36 (1977) 137

38) Herrmann, T.R. and Rayfield, G.W., *Biophys. J.* 21 (1978) 111

39) Bamberg, E., Apell, H.J., Dencher, N., Sperling, W., Stieve, H. and Lauger, P. *Biophys. Struct. Mechanism* 5 (1979) 277

40) Seta, P., Ormos, P., D'Epenoux, B. and Gavach, C. *Biochim. Biophys. Acta* 591 (1980) 37-46.

41) Hampp, N., Brauchle, C. and Oesterhelt, D. *Biophys. J.* 58 (1990) 83

42) Shen, Y. C., R. Safinya, K. S., Liang, A. F. Ruppert and K. J.Rothschild. *Nature.*366 (1993) 48-50.

43) Ikonen, M., Peltonen, J., Vnorimaa, E. and Lemmetyinen, H. *Thin Solid Films* 213 (1992) 277-284.

44) Becher, B., Tonunaga, F. and Ebrey, T.G. *Biochemistry* 17 (1978)

45) Birge, R.R. *Annu. Rev. Phys. Chem.* 41 (1990) 683

46) Hong, K. and Hubbell, W.L. *Proc. Natl. Acad. Sci. USA* 69 (1972) 2617

47) Applebury, M.L., Zuckerman, D.M., Lamoh, A.A. and Jovin, T.M. *Biochemistry* 13 (1974) 3448

48) Hubbell, W.L., Fung, B., Chen, Y. and Hong, K. in: "Vertebrate Photoreception" pp. 41-59, Barlow, H.B. and Fatt, P. eds., Academic Press (1977)

49) Darszon, A., Montal, M. and Zarco, J. *Biochem. Biophys. Res. Comm.* 76 (1977) 820

50) Montal, M., Darszon, A. and Trissl, H.W. *Nature* 267 (1977) 221

51) Gambale, F:, Gliozzi, A., Pepe, I.M., Robello, M. and Rolandi R. *Biochim. Biophys. Acta* 467 (1977) 103

52) Pepe, I.M., Schwemer, J. and Paulsen, R. *Vision Res.* 22 (1982) 775

53) Trissl, H.W., Darszon, A. and Montal, M. *Proc. Natl. Acad. Sci. USA* 74 (1977) 207

54) Trissl, H.W. *Photochem. Photobiol.* 29 (1979) 579

55) Korenbrot, J. I., and M. J. Pramik. *J. Membr. Biol.* 37 (1977) 235-262.

56) Wald, G., J. Durrel, and R. C. C St. George. *Science.* 17 (1950) 179-182.

57) Hampp, N. *Nature.* 366 (1993) 12.

58) Oesterhelt, D., C. Brauchle, and N. Hampp. *Quarterly Reviews of Biophysics.* 24 (1991) 425-478.

59) Maxia,L., Radicchi, G., Pepe I.M. and Nicolini, C. *Biophys. J.* (1995)(in press)

60) Dratz, E., and P.A. Hargrave. *Trend Biochem Sci.* 8 (1983) 128-151.

61) Salesse, C., Ducharme, D., Leblanc, R.M. and Boucher, F. *Biochemistry 29* (1990) 4567-4575.

62) Khan, M. A., Bolen, W., Hargrave, P. A., Santoro, M. M. and McDowell, J. H. *Eur.J. Biochem.* 200 (1991) 53-59.

WERNER E. REICHARDT (1924-1992): *IN MEMORIAM*

Roland Hengstenberg

Max-Planck-Institut für biologische Kybernetik, D-72076 Tübingen

Professor Werner E. Reichardt died on September 18, 1992 in Tübingen. He collapsed and lost consciousness at the end of a symposium that was organized in his honor by his coworkers and friends on the occasion of his retirement. He left his wife, Barbara Reichardt, two adult children, Andrea and Cornelius, and his grandson Johannes.

Werner Reichardt was the founder of the Max-Planck-Institut für biologische Kybernetik in Tübingen, and one of its directors since 1968. He was in many ways a fascinating person and an example for many young colleagues who accompanied him on parts of his way. In matters of sciences he fought for thoroughness and top quality. He was always ready to accept a good argument even if he had to change his opinion. In dealing with persons he was friendly and absolutely loyal. With friends he was warm-hearted and unfailingly sincere.

1. Youth and early experiences:

Reichardt was born 1924 in Berlin. Despite the depressing economical and political atmosphere of this time, he enjoyed a protected childhood. He became interested in physics already during school time, volunteering in the private laboratory of H.E. Hollmann, a pioneer of ultra shortwave communication. Right after his high school graduation in 1941 Reichardt was called to the air force and served in a technical unit developing long distance radio communication for weather forecasts. His parents were killed in January 1943 during the first bomb attack on Berlin. This allowed Reichardt to join a resistance group trying to establish a radio contact with the Western Allies. That was found out by the Gestapo at the end of 1944. Reichardt was arrested and expected his execution. In the last days of the Nazi regime he managed a narrow escape during a rebellion of prisoners and could hide himself in Berlin until the end of the war.

These early and horrifying experiences made Werner Reichardt an earnest person who was always aware of the presence of death and the preciousness of time where one is free and able to do something sensible.

2. The way into neuroscience:

After the war, Reichardt studied physics at the Technical University in Berlin, received his Diplom (master's degree) in 1950 and finished his studies with a doctoral thesis in solid state physics of semiconductors in 1952. He got a postdoctoral position at the Fritz-Haber-Institute of the Max-Planck-Society in Berlin, doing research on semiconductor physics.

Reichardt's way into science and especially into neurobiology was all but a routine carreer; it was repeatedly influenced by encounters with eminent scientists.

In 1950 he met again Bernhard Hassenstein, a former colleague from the air force radio unit. At that time Hassenstein had finished his doctoral thesis in the laboratory of Erich von Holst in Wilhelmshaven on the optomotor turning behavior of the beetle *Chlorophanus*. Reichardt realized that these experiments could be strictly formalized. He applied, to the beetle's visual behavior, the methods of systems analysis very much like an electrical engineer studying an unknown component of telecommunication equipment. The concept was, to infer from the experimental facts, the minimal logical structure that would produce the observed behavior in response to the applied stimuli. The model should provide non-trivial quantitative predictions about the outcome of conceivable experiments that had not been done before.

In 1950 this was quite an unusual way of thinking about a mental process like visual motion perception.

The first papers on this topic attracted the attention of Max Delbrück who invited W. Reichardt in 1954 to the California Institute of Technology in Pasadena to study the light-growth-reaction of the sporangiophore of the fungus *Phycomyces*. The year in the USA brought him many personal relations and friendships lasting his whole life. He decided to move into biology.

3. Establishment of the new field

Back to Germany, Reichardt became in 1955 Research Assistant in the department of K.F. Bonhoeffer at the MPI für physikalische Chemie in Göttingen. Together with B. Hassenstein and D. Varjú he worked out the properties of the insect motion detector. Sophisticated visual stimuli were presented to the beetle *Chlorophanus*, held by its back in a fixed position while climbing on the grass-like ribs of y-maze globe. The frequency of left/right choices revealed its intended turns.

This way it could be shown that its motion detector required at least two input sensors looking into slightly different areas of the visual field. These sensors are excited in sequence by pattern edges moving by. If one input signal is delayed appropriately the two signals become synchronous for the "preferred direction" of pattern motion but strongly asynchronous in the opposite direction. Multiplication of the two signals and time-averaging of the result yields a direction-specific motion signal irrespective of the polarity of the change of brightness achieved by a moving object. This process corresponds formally to an autocorrelation of the input signal. Hence the name "correlation-type motion detector". Since the beetle responds also to motion in the opposite direction, two antisymmetric detector subunits must be combined and their outputs subtracted. This step eliminates unspecific signal components of the subunits and thus increases directional selectivity of the motion detector.

In 1958, after Bonhoeffer had died, a Research Group for Cybernetics was established at the MPI for Biology in Tübingen. It was led by W. Reichardt, B. Hassenstein and H. Wenking who supplied the group with specially designed sophisticated electronic equipment. Young colleagues joined the group and worked on different problems of sensory transduction and visual perception in as diverse organisms as *Phycomyces*, *Limulus*, *Drosophila* and the housefly *Musca*.

In 1960, Hassenstein left the Cybernetics group to become full professor and head of the Zoology department at the University of Freiburg. Reichardt received offers from the California Institute of Technology in Pasadena, from the Massachusetts Institute of Technology in Cambridge, from the Bell Laboratories in Murray Hills and from the Max-Planck-Society which he accepted.

Reichardt convinced the Max-Planck-Society that his project needed more space. A separate building for his Department at the Max-Planck-Institute for Biology was opened in 1965. On this occasion he explained his view on how to study central nervous systems on different levels of organization, naming the kinds of phenomena to be observed and the means of analysis:

Table 1

LEVEL	PHENOMENA	DESCRIPTION
1) whole animal	behavior	system performance
2) nerve nets	signal flow	algorithms
3) single cell	membrane potential	current flow
4) membrane patch	charge transfer	molecular action

Reichardt believed that the study of nervous systems on each of these levels presents specific problems that require appropriate concepts and methods. He was convinced that results gained on lower levels are not adequate to explain specific performance on higher levels. But he was confident that insights gained on higher levels help to ask the right questions and to do experiments in the right way on lower levels. And above all he was sure that it would be necessary to study nervous systems on all levels simultaneously and perferably in the same organism. Consequently he planned and succeeded to expand the working base of his department.

In 1968, his department was transformed into the Max-Planck-Institute for biological Cybernetics. The existing building was enlarged and three co-directors were appointed for different though closely associated fields: Valentino Braitenberg from the Cybernetics Institute in Naples was intended to unravel the neuroanatomy of the fly's visual nervous system. Kuno Kirschfeld from Tübingen and his coworkers studied physiological optics, visual pigments and phototransduction. Karl G. Götz, also from Tübingen used neurological mutants of *Drosophila* to reveal how genetic instructions are translated into functional structures of the fly's central nervous system.

Until now many young colleagues have joined the four groups, have contributed to the solutions of many questions in insect vision and have spread over German and foreign universities to pursue Reichardt's concepts in their own responsibility and personal handwriting.

4. New research projects

The new collegium of directors relieved W. Reichardt from administrative duties. He embarked on new research adventures: The flight torque of a fly, flying stationarily in the center of an illuminated arena, can be measured and can be used to control the speed of the optical panorama in a negative feedback configuration. This allows the fly to chose where to fly and enables the investigator to interfere in defined ways with the structure of the panorama and the coupling between the fly's actions and their consequences. From the results of these experiments, W. Reichardt and T. Poggio developed a phenomenological theory of object fixation and tracking by the fly. The simulation of the fly's control system predicted, with surprising quantitative and qualitative accuracy, the observed free flight trajectory of a fly chasing its quarry.

In 1977 Reichardt discovered that flies use motion discontinuities in the visual surround to distinguish objects from the background. Flies turn towards a single, randomly textured stripe in front of an equally textured background, but only as long as it moves relative to the ground. Reichardt and Poggio developed a model of nonlinear lateral inhibition between motion detectors to account for these findings. This class of models is characterized by a higher nonlinear order than the elementary motion detector.

At this time, K. Hausen and myself had developed the techniques for intracellular recording and staining of motion-sensitive wide-field neurons. Some years later M. Egelhaaf discovered figure-detection neurons that are inhibited by wide-field motion of the visual background. Thus we could study the actual neural circuitry of self-motion perception and figure/ground-discrimination. The close interplay between behavioral analysis, theory and neurophysiology guided this work very efficiently and led to a profound understanding of these perceptual processes, very much as Reichardt had suggested way back in 1965.

5. Duties and Honors:

Besides his engagement in the laboratory and the Institute, W. Reichardt served as scientific advisor in committees all over the world.

In 1984 he was elected into the Senate of the Max-Planck-Gesellschaft, the top ruling board of this society.

In 1961 Reichardt had founded the "Journal of Biological Cybernetics" for which he served as Editor-in-Chief until his untimely death.

Since 1970 he was Honorary Professor at the University of Tübingen which helped to attract qualified students of Neurosciences to Tübingen.

Over the years, Reichardt was honored by election into several Academies: 1970, Akademie der Wissenschaften und Literatur in Mainz; 1971, Deutsche Akademie der Naturforscher Leopoldina in Halle; 1972, American Academy of Arts and Sciences in Cambridge MA; 1977, Koninklijke Nederlandse Akademie van Wetenschaapen in Amsterdam; 1988, National Academy of Sciences in Washington; and 1989, American Philosophical Society in Philadelphia and Academia Europea in Cambridge, England. In 1989, he received the degree of an Honorary Doctor of Engineering from the Technische Hochschule Aachen.

In 1980 W. Reichardt was elected as a member of the prestigious Order Pour le Mérite. That was originally founded in 1740 by Friedrich der Grosse of Prussia; the Class of Science and Arts was founded in 1842 by Friedrich Wilhelm IV of Prussia on advice of A. von Humboldt. In 1985 B. Julesz and W. Reichardt were awarded with the H.P. Heineken-Prize of the Royal Dutch Academy.

W. Reichardt enjoyed these honors. He felt that the scientific community had acknowledged his approach to neuroscience which was not at all guaranteed in the beginning. He was proud of the new insights produced by his and his collaborators' work and he was confident that this kind of computational neuroscience would stand the challenges of future demands.

Werner Reichardt's sudden and much too early death has deprived the scientific community of an eminent pioneer in this new field of research.

For many of us he was not only the distinguished senior of our Institute but also a warm-hearted friend over many years whose opinion and advice were seeked and respected. We will not forget him.

ANALYSIS OF VISION AND GAZE CONTROL IN INSECTS

Roland Hengstenberg

Max-Planck-Institut für biologische Kybernetik, D-72076 Tübingen

Visual perception is greatly facilitated when the eyes are properly aligned with the surroundings and gaze is held stable on the target of interest. This demand is in conflict with the desire for high mobility and maneuverability. Hence vertebrates, molluses, crustacea and some insects have evolved means to stabilize their eyes temporarily on the visual scene while moving through their habitats.

Flies, for example, have a preferred body posture like most other animals: when walking on even ground their body axis is roughly horizontal and the back is directed upwards. In flight, their attitude is similar, only the body axis may be more elevated, especially at low flight speeds. Very probably the fly's eyes, other sense organs and the sensory nervous system have been adapted during evolution to work best in this preferred orientation.

Flying insects have, in principle, all six degrees of freedom to move in space: they can translate along their body axes (thrust, lift, side-slip) and rotate about these axes (roll, yaw, pitch). After any voluntary or involuntary change of flight attitude the necessity to keep balance requires that the insect returns to its normal, upright posture. During perturbations and such corrective flight maneuvers, however, vision may become disturbed by misdirection of the gaze axis, by misalignment of the eyes and by motion blurring. Hence a means to stabilize the eyes would seem advantageous.

Flies and most other insects have two kinds of eyes: three small ocelli, each with a single underfocused wide-angle lens and several hundred photoreceptors: they are best suited to sense the mean brightness of large parts of the surroundings. In flies, the visual fields of the ocelli are directed upwards.

The two large compound eyes consist of several thousand so called ommatidia, each with a narrow-angle lens, focused to infinity and 8 photoreceptors falling in two main classes: six peripheral ones (R1-R6) that are the main inputs to the motion perception system and two central ones (R7, R8) arranged in tandem that serve other purposes. The lenses of both types of eyes are part of the head capsule and therefore rigidly coupled. Gaze can only be shifted by either turning the head or the whole body.

The blowfly *Calliphora* can turn its head horizontally (yaw ±20°) and vertically (pitch ±20°) and can rotate it around the body axis (roll ±90°). In this study only roll turns are considered which play an important part in flight control.

Head movements of insects can hardly be observed in free flight but flies can be studied during tethered flight in a wind-tunnel. Fixing the fly coaxially to the shaft of a servomotor allows to turn the animal. The walls of the tunnel are lined with varying patterns, illuminated from the outside and can be turned by a second motor.

This way either the fly and/or its optical surround can be turned arbitrarily. Head movements of the fly, relative to its trunk, are observed with a small video telescope through the entrance nozzle of the windtunnel.

When a fly is held stationary during flight and the surroundings are not moved, it nevertheless turns its head continuously in different directions, sometimes saccade-like. At the same time it varies its wingbeat amplitudes and the posture of its hind-legs, apparently in an attempt to perform one of the coordinated flight maneuvers that characterize the tortuous and unpredictable flight paths of blowflies.

When a fly is turned unexpectedly around its body axis it generates simultaneously a corrective flight maneuver and a compensatory head turn which alone may correct for about 2/3 of the imposed perturbation.

Which sense organs and sensory cues tell the fly the direction and speed of its motions? By turning the panorama, instead of the fly, it can be shown that corrective head turns are elicited visually. Conversely, turning the fly in visually featureless surroundings elicits also a corrective head turn, apparently due to a mechanosensory perception of body rotation. By systematic variation of the stimuli and surgical manipulations of different sense organs, four visual and four mechanosensory response components could be identified: (a) a response to coherent motions of extended patterns, (b) a response due to the orientation of visual contours in the frontal visual field, (c) a large tonic dorsal light response, all mediated by the compound eyes, and (d) a small phasic dorsal light response mediated by the lateral ocelli. (e) The body motion response is mediated by the halteres which are gyroscopic sense organs for rotations. They are homologous with the hindwings of other insects but have been massively transformed to serve the new purpose. (f) wing-load differences are perceived through strain receptors in the wing base, (g) the direction of gravity is sensed by the weight distribution among the legs through leg proprioceptors but only during walking and finally (h) neck-proprioceptors also influence head posture *i.e.* the alignment of head and trunk.

The different sensory components have different dynamic characteristics and complement each other: mechanosensory components have short latencies and elicit larger responses at high motion speeds than visual responses that tend to have longer latencies and operate best in the lower speed range. All components operate autonomously, which is a necessary condition, if the different components are to complement one another.

During walking and flight different combinations of sensory and motor components are used for head stabilization and locomotor control. This requires appropriate, status-dependent routing and weighting of the respective neural signals. It can be shown that the major sensory components, mentioned above, influence head posture and locomotion in parallel. At present, the major efforts are directed to determine the specific characteristics of the component responses and to identify the neural circuits subserving them.

Stabilization of the fly's eyes tends to support visual perception on several levels of processing: (a) Reduction of retinal image velocity improves contrast transfer especially for high spatial frequencies. (b) Correct localization of a target in the visual field allows specific processing in retinotopic networks. (c) Correct orientation of targets facilitates their assessment, for example in object recognition. (d) Rotational stabilization facilitates more sophisticated evaluations of retinal image flow, e.g. distance perception by utilization motion parallax. For insects, which have too small a triangulation base for stereopsis over reasonable distances, this improvement is probably the most important one.

Literature:

Collett TS, Nalbach HO, Wagner H: (1993) Visual stabilization in arthropods. In: FA Miles, J Wallman (eds). *Visual Motion in the Stabilization of Gaze. Reviews of Oculomotor Research* 5: 239-263. Elsevier Amsterdam.

Hengstenberg R: (1993) Multisensory control in insect oculomotor systems. In: FA Miles and J Wallman (eds). *Visual Motion in the Stabilization of Gaze. Reviews of Oculomotor Research* 5: 285-298. Elsevier, Amsterdam.

Hengstenberg R: (1991) Gaze control in the blowfly *Calliphora*: A multisensory two-stage integration process. *The Neurosciences* 3: 19-29.

PHOTORECEPTION AND PHOTOTRANSDUCTION IN MICROORGANISMS

MOLECULAR BASIS OF PHOTORECEPTION IN PHOTOMOTILE MICROORGANISMS

Francesco Lenci, Antonella Sgarbossa, Nicola Angelini
CNR Istituto Biofisica, via San Lorenzo 26, 56127 Pisa (Italy)

INTRODUCTION

Any physiological phenomenon which is driven, triggered or modulated by sunlight in living organisms (from vision to photosynthesis, from photoreactivation to vitamin D biosynthesis, from phototropism to seed germination) is the final result of a series of physical, chemical, biophysical and biochemical processes which originate from the same "elementary" event: the absorption of a photon by the photoreceptor molecule.

In order to efficiently capture light over a large part of the solar spectrum, biological photoreceptors have often high molar extinction coefficients and broad absorption bands. The spectral sensitivity of the whole photoreceptive system can be widened if different chromophores, absorbing in contiguous spectral ranges, are spatially assembled in molecular frameworks in such a way to maximize the probability of collective interaction among them, as suggested, f. i., for *Euglena* photosensing mechanism (see below).

Like other sensory responses of microorganisms to different environmental stimuli (such as temperature, pressure, chemical substances), this capability of sensing and reacting to illumination conditions can provide photosynthetic algae and colored ciliates with essential strategies for their survival (see, e.g., [1, 2]).

In photosensory phenomena, in general, the modification of a single receptor pigment molecule induced by the absorption of one photon can be sufficient to trigger the transduction chain. The signal associated with the absorbed photon is turned into an intracellular biochemical/biophysical signal, which is processed by the cell to eventually yield the physiological response [3, 4].

Individual chromophores of photosensing systems are usually loosely bound to flexible molecular frameworks and isolated from each other. The spatial separation of the chromophores does not allow energy transfer among them and the conformational mobility of the complex gives rise, upon absorption of even a single photon by the pigment, to local rearrangements of the molecular domain around the chromophore. An interesting exception has been suggested for *Euglena* by several Authors, and will discussed below. In any case, the local perturbation in the photosensing unit initiates the chain of molecular events which are at the basis of the sensory transduction process, ultimately ending in the physiological response.

In vivo, the various photodetecting units are usually not randomly distributed, but rather assembled in ordered arrays, this molecular architecture being functional to provide living cells with efficient photoreceptor-phototransducing systems. This can be carried through, for instance, if photoreceptor molecules are embedded into membranous structures (see, e.g., rhodopsins patches in *Halobacterium*) or packed in quasi-crystalline subcellular organelle (see, e.g., the ParaFlagellar Body in *Euglena*).

PHOTORECEPTORS FOR MICROORGANISMS PHOTOMOVEMENTS

Phototaxis (oriented movement toward or away from the light source) and photophobic responses (transient alterations of swimming pattern following sudden variations in fluence rate) of freely motile microorganisms are, apparently, mediated by a wide variety of photoreceptor pigments. Photosensing pigments can absorb from the near UV (flavins and pterins) to the blue (flavins, carotenoids), to the green and from the yellow-orange (rhodopsins) to the red (stentorins and blepharismins) (Fig. 1) [5-8].

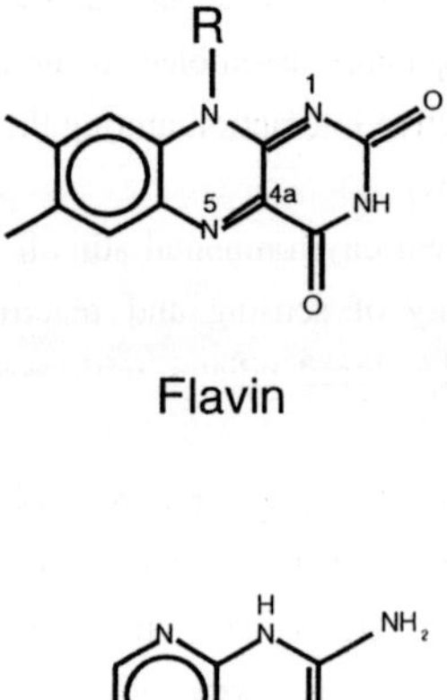
Flavin

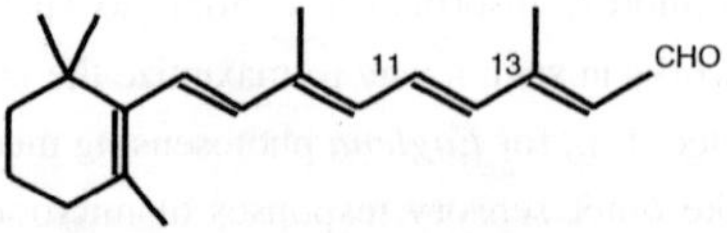
All-trans retinal

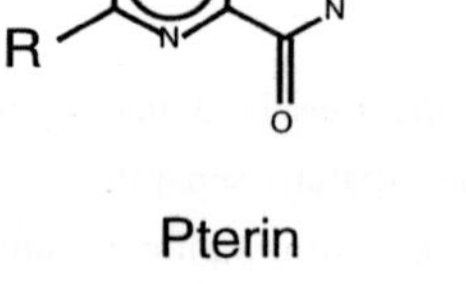
Pterin

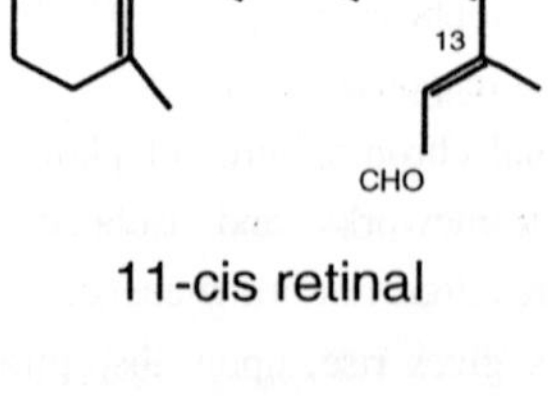
11-cis retinal

Fig. 1: Structure formulas of some photoreceptor chromophores.

In some phototactic photosynthetic bacteria, chlorophyll and some accessory pigments play the double role of driving photosynthesis and perceiving light stimuli. In what follows only "dedicated", non-photosynthetic, photoreceptor pigments will be considered.

At present, rhodopsins in halobacteria and some flagellated algae, flavins and pterins in other green algae and hypericin-type pigments in colored ciliated protozoa have been suggested, on the basis of a number of experimental evidences, as photopigments responsible for light-induced motile reactions in these microorganisms.

For some microorganisms, not only the structure of the photoreceptor pigment(s), but also the primary molecular events which follow the absorption of the photon, seem to be unambiguously established. In other unicells, even the nature of the molecule acting as light detector is still doubtful.

It is worth underlining that, in most cases, considerable progress has been achieved facing the different aspects of the problem through specific approaches, such as selection and isolation of behavioral and pigment-lacking mutants, quantitative analysis of motile responses, steady-state and time-resolved spectroscopic and microspectroscopic studies of the candidate photosensing structures.

Among photoresponsive microorganisms, whose photoreceptor pigments have been extensively studied, *Halobacterium halobium*, *Euglena gracilis* and *Chlamydomonas reinhardtii* will be considered.

Halobacterium is a prokaryotic organism in which light is used both as an energy source and as an information carrier throughout three different specialized and dedicated chromoproteins, all of them rhodopsin-like: bacteriorhodopsin and halorhodopsin for metabolic and energetic requirements and sensory rhodopsins for perceiving and reacting to environmental light conditions [9 and references therein].

Among photosynthetic unicellular algae, *Euglena* and *Chlamydomonas* are case examples of special interest, on the one hand because of the amount of knowledge achieved and, on the other hand, because of the basic questions which are still unanswered.

The case of *Halobacterium*

In the bipolarly flagellated halophylic *Halobacterium*, two sensory photopigments have been identified (Sensory Rhodopsins I and II, SRI and SRII) and their photocycles studied in detail [10-13, 14 and references therein].

Both sensory rhodopsins are all-*trans*-retinal-binding, seven helix transmembrane proteins, which undergo slow dark-reversible photocycles (approximately 100 times slower than bacteriorhodopsin and halorhodopsin photocycles). SRI exists in two spectrally different forms: a thermally stable one, with absorption maximum at 587 nm, SRI_{587}, and a long-living intermediate, with absorption maximum at 373 nm, SRI_{373}, which thermally decays back to SRI_{587} in about 750 ms. SRII, with absorption maximum at 487 nm, $SRII_{487}$, photoconverts into a species absorbing at about 520 nm, which thermally decays back to $SRII_{487}$ in about 400 ms (Fig. 2) [15, 16 and references therein].

SRI mediates both attractant and repellent responses: red-orange light (absorbed by

SRI_{587}) lowers the frequency of swimming reversal (attractant response), whereas blue light (absorbed by SRI_{373}) causes an increase of the swimming reversal frequency (repellent response). SRII (also called Phoborhodopsin) is the blue-light receptor responsible only for repellent responses [15, 16 and references therein].

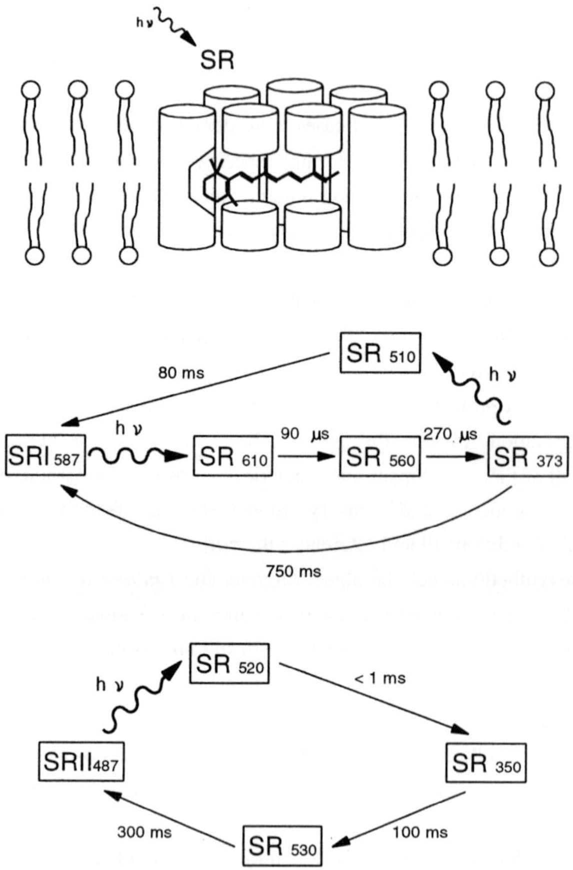

Fig. 2: Schematic drawings of Sensory Rhodopsins and their photocycles (redrawn from [11, 14]).

To clarify the primary steps of signal transduction, a retinal analog in which the all-*trans*/13-*cis* isomerization is locked was incorporated into the apoproteins of sensory rhodopsins of *Halobacterium* Flx5R and Flx3T mutants (blocked in retinal synthesis). It was shown that SRI_{373} as well as the intermediates in the photocycle of SRII are not formed from

SRI_{587} and $SRII_{487}$, respectively, and that all light-mediated motile responses of the cells are suppressed (Fig. 3A).

These findings are clearly consistent with the hypothesis that a specific all-*trans*/13-*cis* isomerization of retinal is involved in SR's photocycles and is required for photobehavior in *Halobacterium* [17].

Selective modifications of the retinal beta-ionone ring seem not to alter the physiological function of SRI [18]. The presence of the methyl group in C-13, on the contrary, plays a key role in the interaction between retinal and SRI-apoprotein: replacement of $-CH_3$ with -H, in fact, allows all-*trans*/13-*cis* photoisomerization but inhibits both the attractant and repellent response, thus suggesting that a steric trigger is required for signaling [12, 19] (Fig. 3B).

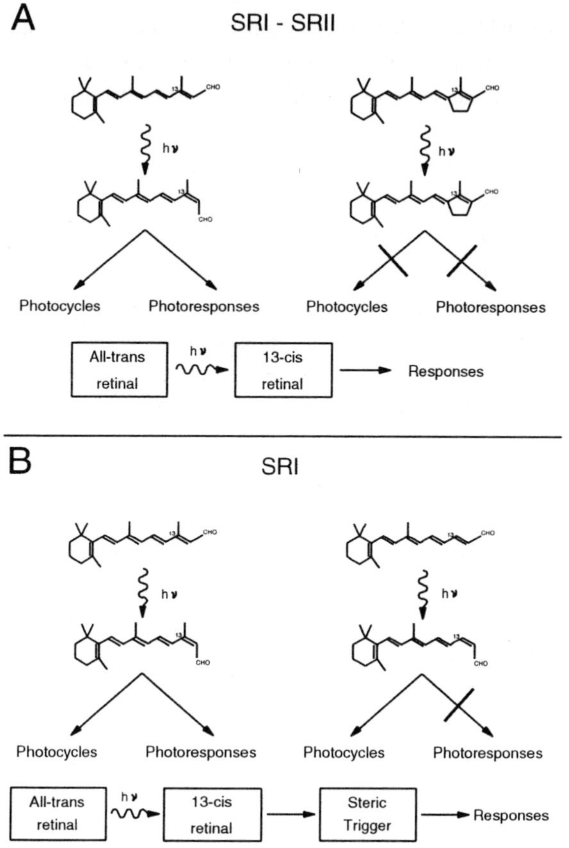

Fig. 3: Schematic representation of the effects of all-*trans*/13-*cis* photoisomerization blockage (A) and C-13 demethylation (B) on SR-I and SR-II photocycles and on cell photoresponses.

The case of *Euglena*

It seems established that the photoreceptor pigment of the photosynthetic flagellated alga *Euglena gracilis* is located in the ParaFlagellar Body (PFB), a membrane-enclosed paracrystalline structure at the basis of the emergent flagellum.

Action spectra for photoaccumulation and photodispersal [20, 21] and for step-down photophobic responses [22], as well as fluorescence emission and excitation spectra of PFB *in vivo* [23, 24], suggested flavin types chromophores as photoreceptor pigments, most likely embedded in a rigid and/or hydrophobic molecular matrix. In this physiological environment, according to the low value of fluorescence quantum yield (0.005, as compared with 0.25 of free riboflavin solutions), the first excited singlet of the pigment could undergo de-excitation pathways other than radiative decay. This low value of fluorescence quantum yield of PFB flavins is in agreement with their photoreceptor function, as no photoreceptor pigment is likely to decay from its first excited singlet state mainly through a radiative transition, but rather through reactions which can serve to trigger the first molecular steps of the photosensory transduction chain [24].

Even though no unambiguous evidence for the molecular events following light absorption by flavins was provided, the possibility of a functional connection between the photoreceptor system in the PFB and the ParaFlagellar Rod, PFR (an apparently ordered structure running along the flagellum), and the role of PFR as effector were suggested [25, 26].

Lately, the PFB of *Euglena* has been isolated, its morphology investigated by electron microscopy [27] and its optical properties measured by means of a computer-assisted microspectrophotometer. The structure of the optical absorption spectrum of an isolated PFB (absorption maximum around 500 nm; optical density around 1.4) indicates that rhodopsin-like molecules, rather than flavins, are present in this organelle [28]. The PFB fluorescence emission around 530 nm [23] and its fluorescence excitation spectrum [24] have been suggested to be due to metarhodopsin [29].

An absorption band around 500 nm, whose spectral distribution is compatible with that of a rhodopsin-like pigment, has been recently measured in the eyespot region of *Euglena* also by James *et al.* [30] by means of microspectroscopic technique. The intensity of this band was found to depend on the relative orientation of the plane of polarization of light and the long axis of the cell. This anisotropy was suggested to be due to the fact that rhodopsins are embedded in the quasi-crystalline structure of PFB.

Barsanti *et al.* [31, 32] have also reported that nicotine, which inhibits retinal biosynthesis, and hydroxylamine, which reacts with both free and opsin-bound retinal, hinder the formation of PFB in *Euglena* and suppress cells photoresponsiveness, further pointing to the rhodopsin nature of the photoreceptor pigment for photomotile responses in *Euglena*.

As all-*trans*-retinal has been detected in whole *Euglena* cells homogenates [33] and no photobleaching of *Euglena* rhodopsin has been observed [30], this retinal isomer has been suggested to be the opsin-bound chromphore and not to be released from the apoprotein upon light stimulation [32].

The output signal from the PFB photoreceptor could be transduced to get the final sensory response through the ParaFlagellar rod by means of either charge fluxes or allosteric interactions [29].

The old hypothesis of flavins as possible photoreceptor pigments in *Euglena* has been revived suggesting, on the basis of spectroscopic, biochemical and photobehavioral data, that in the PFB they could be combined with pterins [34-36].

In flagellar preparations containing the PFB, fluorescence emission bands at about 520 nm and 460 nm have been detected and attributed to flavins (absorption maxima at about 360 and 450 nm) and pterins (absorption maximum at about 350 nm), respectively [35, 36].

The presence of pterins in the PFB is suggested also by fluorescence micrographs *in vivo*: the PFB spot, in fact, fluorescing around 530 nm (green) when excited at 450 nm, appears fluorescing around 450 nm (blue) when excited in the near UV (350 nm) [35].

Convincing evidence for the contribution of both pterins and flavins to the photosensitivity of *Euglena* can be provided by the structure of the phototaxis action spectrum, in which the near UV band, where both pterins and flavins absorb, is significantly higher than the blue one, where only flavins absorb [37].

Following UV-B irradiation, pterins fluorescence from PFB-extracted proteins strongly increases and *Euglena* cells photoresponsiveness is effectively impaired [38 and references therein]. On the basis of these observations, the two chromophores have been suggested to act as photoreceptor pigments through an energy transfer mechanism. Pterins would play the role of light antenna pigments for flavins in undamaged PFB protein complexes, but would radiatively dissipate when the spoiled molecular framework does not allow efficient energy transfer to the acceptor (flavins) [38].

The role of pterins as antenna pigments for flavins, however, seems not to be fully in agreement with the fact that excitation of PFB, *in vivo* in intact cells, at about 350 nm (in the pterins absorption band) induces fluorescence in the pterin emission band (blue) rather than in the flavin emission band (green) [35].

An even more elaborate sensing mechanism has been recently proposed for *Euglena* by Sineshchekov V.A. *et al.* [39]: a system of interacting receptor pigments would in fact funnel light energy from short-wavelengths (pterins) and intermediate-wavelengths (flavins) absorbing centers to a "long wavelength absorbing pigment". According to the Authors, it is not possible, at present, to conclude if this last component of the light detecting complex is the reaction center triggering the sensory response or a further energy donor to an yet unidentified reactive molecule, possibly also a rhodopsin.

Even though not explicitly affirmed by the Authors, a direct energy transfer from pterins to the long-wavelength group has to be excluded, as the emission band of the donor and the absorption band of the acceptor do not overlap at all.

Energy transfer data, combined with fluorescence polarization results, provide some indications also about the spatial arrangements of the different pigment systems. The fact that fluorescence polarization degree is quite low after energy transfer but high within a same group of pigments (pterins or flavins or X), is compatible with an ordered array of molecules of the same species, as could be expected from the quasi-crystalline structure of the ParaFlagellar Body (Fig. 4).

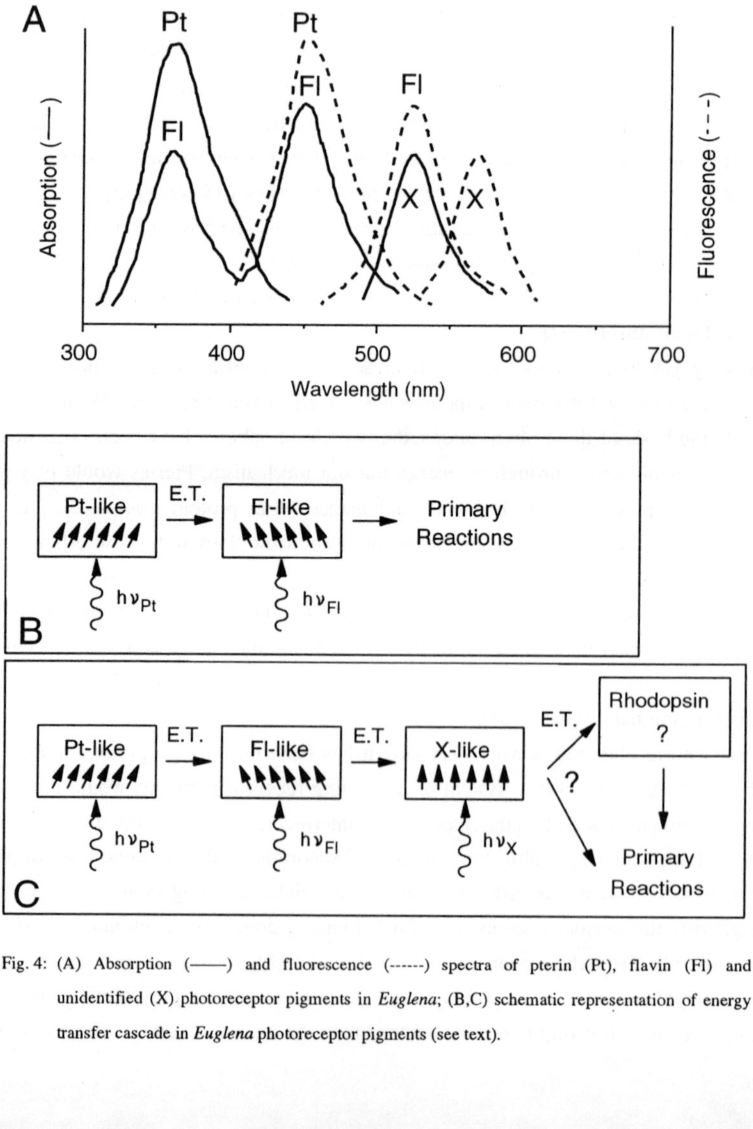

Fig. 4: (A) Absorption (——) and fluorescence (------) spectra of pterin (Pt), flavin (Fl) and unidentified (X) photoreceptor pigments in *Euglena*; (B,C) schematic representation of energy transfer cascade in *Euglena* photoreceptor pigments (see text).

Moreover, flavin-binding as well as pterin-binding proteins have been separated and characterized by chromatographic techniques from *Euglena* flagella isolated with their PFB [35, 40]. Riboflavin binding sites have been localized both in the PFB region and along the whole flagellum of *Euglena* and no clear-cut explanation for this spreading of binding sites is at present available [40-42].

Attempts to reconcile the "rhodopsin" and the "flavin" hypotheses either suggesting rhodopsin as the final acceptor of a multicomponent light-harvesting and transducing system [39] or assigning to rhodopsin the role of photoreceptor proper and to flavins bound to flagellar proteins the function of transducing the light stimulus along the flagellum [42] seem not satisfactorily supported by experimental findings.

The case of *Chlamydomonas*

The rhodopsin nature of the photoreceptor pigment for photomotile responses in the photosynthetic flagellated alga *Chlamydomonas* was initially hypothesized by Foster and Smyth [43] relating the ultrastructural and optical properties of the stigma with the structure of phototaxis action spectra previously determined [44]. That a rhodopsin-like pigment is the photoreceptor for *Chlamydomonas*, has been subsequently confirmed by a series of behavioral studies and action spectra determination.

Which is the opsin-bound chromophore and which primary molecular modifications it undergoes following light absorption are, on the contrary, still open questions. Like for *Halobacterium*, the problem has been primarily faced incorporating retinal analogs and isomers, different for their stereochemistry around their double and single bonds, into blind, retinal-lacking mutants and relating the maximum of action spectra for the photomotile responses with the absorption maximum of the added chromophore.

Foster and coworkers added several retinal analogs and short-chained compounds (including hexenal and hexanal) to FN68 mutants (almost, but not completely, blind if grown in the dark) and measured photobehavioral responses by means of a population method. As phototaxis action spectra showed that a number of retinal analogs and even the short-chained compounds restore photoresponsiveness, it was suggested that photoisomerization of the chromophore is not required in *Chlamydomonas* rhodopsin for eliciting the motile response. A charge redistribution in the excited state of the chromophore was proposed to be the primary step in rhodopsin activation. Among retinal isomers, 11-*cis* was found to be the most effective in restoring phototaxis in FN68 mutants (approximately three times more effective than all-*trans* retinal) and, consequently, suggested to be the natural chromophore of *Chlamydomonas* rhodopsin [45-49].

Quite different results were obtained by a number of Authors who used also a different, completely blind, mutant (CC-2359) and measured photomotile responses by means of cell-tracking motion analysis systems as well as population methods. No photobehavioral responses were observed after incubation with short-chained compounds and retinal analogs in which photoisomerization around C_{13}-C_{14} was hindered and maximum responsiveness was found adding all-*trans* retinal. All these findings clearly indicate that all-*trans* retinal is the functionally active chromophore and that the all-*trans*/13-*cis* photoisomerization is the primary molecular reaction responsible for initiating the transduction process in *Chlamydomonas* [50-54]. Further support to his conclusion has recently been provided also by measurements of photoinduced electric currents in *Chlamydomonas* blind mutants incubated with retinal and its analogs [55].

CONCLUDING REMARKS

Whereas the rhodopsin-like nature of *Halobacterium* and *Chlamydomonas* photoreceptor pigment seems to be ascertained, for *Euglena* both the "Flavin" and the "Rhodopsin" hypothesis sounds, *per se*, selfconsistent and based on reliable experimental findings.

The rhodopsin hypothesis can have a firm support from evolutionary considerations: rhodopsin-like pigments, in fact, are used to sense the light environment in prokaryotes, like *Halobacterium*, in other flagellated unicellular algae, like *Chlamydomonas*, and in higher organisms. If the nature of the photoreceptor pigment is highly conserved, the rhodopsin-like nature of *Euglena* photoreceptor pigment is undoubdtely to be expected [56].

The fact that rhodopsin is a seven helices transmembrane protein, like many other sensory receptors, seems not to be a so stringent argument in favor of the rhodopsin hypothesis for *Euglena*, because other photosensing receptors, like phytochrome, do not have such a conformation.

On the other hand, flavins are commonly accepted to mediate a number of sensory responses in fungi and algae induced by near-UV and blue light [42, 57]. Also in the case of photoreceptor molecules for photomotile responses of microorganisms, therefore, evolution could have advanced not only, reiterating "the good solutions", but also ramifying and compromising, ending with solutions which are not necessarily unique (or even perfect) [58].

REFERENCES

1. D. Presti and M. Delbrück (1978), Photoreceptors for biosynthesis, energy storage and vision. *Plant Cell and Environment*, 1, 81-100.
2. W. Rüdiger and F. López-Figueroa (1992), Photoreceptors in algae. *Photochem. Photobiol.*, 55, 949-954.
3. F. Lenci, F. Ghetti, G. Colombetti, D.P. Häder, P.S. Song (eds.) (1991), *Biophysics of Photoreceptors and Photomovements in Microorganisms*, New York, Plenum.
4. E.D. Lipson and B.A. Horwitz (1991), Photosensory reception and transduction in *Sensory Receptors and Signal Transduction* (J.L. Spudich and B.H. Satir, eds.), New

York, Wiley, pp. 1-64.

5. F Lenci and F. Ghetti (1989), Photoreceptor pigments for photomovement of microorganisms: some spectroscopic and related studies. *J. Photochem. Photobiol. B: Biol.*, 3, 1-16.
6. W. Haupt and D.P. Häder (1994), Photomovement in *Photomorphogenesis in Plants* - 2nd Edition, (R.E. Kendrick and G.H.M. Kronenberg, eds.), Netherlands, Kluwer Academic Publishers, pp. 707- 732.
7. G. Kreimer (1994), Cell biology of phototaxis in flagellate algae. *Int. Rev. Cytol.*, 148, 229-310.
8. F. Lenci (1994), Photoreception and photomovements in microorganisms in *Handbook of Organic Photochemistry and Photobiology*, (P.S. Song, ed.), Boca Raton, CRC, pp. 1429-1433.
9. J.L. Spudich and R. Bogomolni (1983), Spectroscopic discrimination of the three rhodopsinlike pigments in *Halobacterium halobium* membranes. *Biophys. J.*, 43, 243-246.
10. R.A. Bogomolni and J.L. Spudich (1982), Identification of a third rhodopsin-like pigment in phototactic *Halobacterium halobium. Proc. Natl. Acad. Sci. USA*, 79, 6256-6254.
11. J.L. Spudich and R. Bogomolni (1984), The mechanism of colour discrimination by a bacterial sensory rhodopsin. *Nature*, 312, 509-513.
12. J.L. Spudich and R. Bogomolni (1992), Sensory rhodopsin I: receptor activation and signal relay. *J. Bioenerg. Biomembr.*, 24, 193-199.
13. T. Takahashi, H. Tomioka, N. Kamo, Y. Kobatake (1985), A photosystem other than PS370 also mediates the negative phototaxis of *Halobacterium halobium. FEMS Microbiol. Lett.*, 28, 161-164.
14. D. Petracchi, S. Lucia, G. Cercignani (1994), Photobehaviour of *Halobacterium halobium*: proposed models for transduction and motor switching. *J. Photochem. Photobiol. B: Biol.*, 24, 75-99.
15. J.L. Spudich (1991), Color discriminating pigments in *Halobacterium halobium* in *Biophysics of Photoreceptors and Photomovements in Microorganisms* (F. Lenci, F. Ghetti, G. Colombetti, D.P. Häder, P.S. Song eds.), New York, Plenum, pp. 243-248.
16. T. Takahashi (1991), Absorption and action spetroscopy of phoborhodopsin (sensory rhodopsin II) in *Biophysics of Photoreceptors and Photomovements in Microorganisms* (F. Lenci, F. Ghetti, G. Colombetti, D.P. Häder, P.S. Song eds.), New York, Plenum, pp. 249-256.
17. B. Yan, T. Takahashi, R. Johnson, F. Derguini, K. Nakanishi, J.L. Spudich (1990), All-*trans*/13-*cis* isomerization of retinal is required for phototaxis signaling by sensory rhodopsin in *Halobacterium halobium. Biophys. J.*, 57, 807-814.
18. B. Yan, T. Takahashi, D.A. McCain, V.J. Rao, K. Nakanishi, J.L. Spudich (1990), Effects of modification of the retinal b-ionone ring on archaebacterial sensory rhodopsin I. *Biophys. J.*, 57, 477-483.
19. B. Yan., K. Nakanishi, J.L. Spudich (1991), Mechanism of activation of sensory rhodopsin I: evidence for a steric trigger. *Proc Natl. Acad. Sci. USA*, 88, 9412-9416.
20. B. Diehn (1969), Action spectra of the phototactic responses in *Euglena. Biochim. Biophys. Acta*, 177, 136-143.
21. A. Checcucci, G. Colombetti, R. Ferrara, F. Lenci (1976), Further analysis of the mass photoresponses of *Euglena gracilis* klebs (flagellata euglenoidina). *Monitore Zool. Ital. (N.S.)*, 10, 271-277.
22. C. Barghigiani, G. Colombetti, B. Franchini, F. Lenci (1979), Photobehavior of *Euglena gracilis*: action spectrum for the step-down photophobic response of individual cells. *Photochem. Photobiol.*, 29, 1015-1019.
23. P.A. Benedetti and F. Lenci (1977), *In vivo* microspectrofluorometry of photoreceptor pigments in *Euglena gracilis. Photochem. Photobiol.*, 26, 315-318.

24. F. Ghetti, G. Colombetti, F. Lenci, E. Campani, E. Polacco, M. Quaglia (1985), Fluorescence of *Euglena gracilis* photoreceptor pigment: an *in vivo* microspectrofluorometric study. *Photochem. Photobiol.*, 42, 29-33.
25. A. Checcucci, G. Colombetti, R. Ferrara, F. Lenci (1974), Modelli molecolari per la fototassi in *Euglena gracilis* in *Atti del 3° Congresso Nazionale di Cibernetica e Biofisica*, (Storace F.and Taccini P., eds.), Pisa, Lito Felici, pp. 275-282.
26. A. Checcucci, G. Colombetti, R. Ferrara, F. Lenci (1974), Meccanismo fototattico in *Euglena gracilis* in *Atti del 3° Congresso Nazionale di Cibernetica e Biofisica*, (F. Storace and P. Taccini, eds.), Pisa, Lito Felici, pp. 283-290.
27. P. Gualtieri, L. Barsanti, V. Passarelli, F. Verni (1988), Morphological investigations of *Euglena gracilis* isolated paraflagellar body, *Micron and Microsc. Acta*, 19, 241-246.
28. P. Gualtieri, L. Barsanti, V. Passarelli (1989), Absorption spectrum of a single isolated paraflagellar swelling of *Euglena gracilis*. *Biochim. Biophys. Acta*, 993, 293-296.
29. P. Gualtieri (1993), *Euglena gracilis*: is the photoreception enigma solved? *J. Photochem. Photobiol. B: Biol.*, 19, 3-14.
30. T.W. James, F. Crescitelli, E.R. Loew, W.N. McFarland (1992), The eyespot of *Euglena gracilis*: a microspectrophotometric study. *Vision Res.*, 32, 1583-1591.
31. L. Barsanti, V. Passarelli, P. Lenzi, P. Gualtieri (1992), Elimination of photoreceptor (paraflagellar swelling) and photoreception in *Euglena gracilis* by means of the carotenoid biosynthesis inhibitor nicotine. *J. Photochem. Photobiol. B: Biol.*, 13, 135-144.
32. L. Barsanti, V. Passarelli, P. Lenzi, P.L. Walne, J.R. Dunlap, P. Gualtieri (1993), Effects of hydroxylamine, digitonin and triton x-100 on photoreceptor (paraflagellar swelling) and photoreception of *Euglena gracilis*. *Vision Res.*, 33, 2043-2050.
33. P. Gualtieri, P. Pelosi, V. Passarelli, L. Barsanti (1992), Identification of a rhodopsin photoreceptor in *Euglena gracilis*. *Biochim. Biophys. Acta*, 1117, 55-59.
34. W. Schmidt, P. Galland, H. Senger, M. Furuya (1990), Microspectrophotometry of *Euglena gracilis*. Pterin- and flavin-like fluorescence in the paraflagellar body. *Planta*, 182, 375-381.
35. B. Brodhun and D.P. Häder (1990), Photoreceptor proteins and pigments in the paraflagellar body of the flagellate *Euglena gracilis*. *Photochem. Photobiol.*, 52, 865-871.
36. P. Galland, P. Keiner, D. Dörnemann, H. Senger, B. Brodhun, D.P. Häder (1990), Pterin- and flavin-like fluorescence associated with isolated flagella of *Euglena gracilis*. *Photochem. Photobiol.*, 51, 675-680.
37. D.P. Häder and E. Reinecke (1991), Phototactic and polarotactic responses of the photosynthetic flagellate *Euglena gracilis*. *Acta Protozool.*, 30, 13-18.
38. D.P. Häder and B. Brodhun (1991), Effect of ultraviolet radiation on the photoreceptor proteins and pigments in the paraflagellar body of the flagellate, *Euglena gracilis*. *J. Plant Physiol.*, 137, 641- 646.
39. V.A. Sineshchekov, D. Geiß, O.A. Sineshchekov, P. Galland, H. Senger (1994), Fluorometric characterization of pigments associated with isolated flagella of *Euglena gracilis*: evidence of energy migration. *J. Photochem. Photobiol. B: Biol.*, 23, 225-237.
40. B. Brodhun and D.P. Häder (1994), A novel procedure for the isolation of chromoproteins in the paraflagellar body of the flagellate, *Euglena gracilis*, using the SMART system. *J. Photochem. Photobiol. B: Biol.*, in press.
41. A. Nebenführ, A Schäfer., P. Galland, H Senger., R. Hertel (1991), Riboflavin-binding sites associated with flagella of *Euglena*: a candidate for blue-light photoreceptor? *Planta*, 185, 65-71.
42. R. Neumann and R. Hertel (1994), Purification and characterization of a riboflavin-binding protein from flagella of *Euglena gracilis*. *Photochem. Photobiol.*, 60, 76-83.
43. K.W. Foster and R.D. Smyth (1980), Light antennas in phototactic algae. *Microbiol. Rev.*, 44, 572-630.

44. W. Nultsch, G. Throm, I. Rimscha (1971), Phototaktische untersuchungen an *Chlamydomonas reinhardtii* dangeard in homo-kontinuierlicher kultur. *Arch. Microbiol.*, 80, 351.
45. K.W. Foster, J. Saranak, N. Patel, G. Zarrilli, M. Okabe, T. Kline, K. Nakanishi (1984), A rhodopsin is the functional photoreceptor for phototaxis in the unicellular eukaryote *Chlamydomonas*. *Nature*, 311, 756-759.
46. K.W. Foster, J. Saranak, G. Zarrilli (1988), Autoregulation of rhodopsin synthesis in *Chlamydomonas reinhardtii*. *Proc. Natl. Acad. Sci. USA*, 85, 6379-6383.
47. K.W. Foster, J. Saranak, F. Derguini, G.R. Zarrilli, R. Johnson, M. Okabe, K. Nakanishi (1989), Activation of *Chlamydomonas* rhodopsin *in vivo* does not require isomerization of retinal. *Biochem.*, 28, 819-824.
48. K.W. Foster, J. Saranak, P.A. Dowben (1991), Spectral sensitivity, structure and activation of eukaryotic rhodopsins: activation spectroscopy of rhodopsin analogs in *Chlamydomonas*. *J. Photochem. Photobiol. B: Biol.*, 8, 385-408,.
49. K.W. Foster and J. Saranak (1988), Rhodopsin activation: a novel view suggested by *in vivo Chlamydomonas* experiments. *J. Am. Chem. Soc.*, 110, 6588 6589.
50. P. Hegemann, W. Gärter, R. Uhl (1991), All-*trans* retinal constitutes the functional chromophore in *Chlamydomonas* rhodopsin. *Biophys. J.*, 60, 1477-1489.
51. M.A. Lawson, D.N. Zacks, F. Derguini, K. Nakanishi, J.L. Spudich (1991), Retinal analog restoration of photophobic responses in a blind *Chlamydomonas reinhardtii* mutant. Evidence for an archaebacterial like chromophore in a eukaryotic rhodopsin. *Biophys. J.*, 60, 1490-1498.
52. T. Takahashi, K. Yoshihara, M. Watanabe, M. Kubota, R. Johnson, F. Derguini, K. Nakanishi (1991), Photoisomerization of retinal at 13-Ene is important for phototaxis of *Chlamydomonas reinhardtii*: simultaneous measurements of phototactic and photophobic responses. *Biochem. Biophys., Res. Comm.*, 178, 1273-1279.
53. D.N. Zacks, F. Derguini, K. Nakanishi, J.L. Spudich (1993), Comparative study of phototactic and photophobic receptor chromophore properties in *Chlamydomonas reinhardtii*. *Biophys. J.*, 65, 508-518.
54. P. Kröger and P. Hegemann (1994), Photophobic responses and phototaxis in *Chlamydomonas* are triggered by a single rhodopsin photoreceptor. *FEBS Letters*, 341, 5-9.
55. O. A. Sineshchekov, E. G. Govorunova, A. Dér, L. Keszthelyi, W. Nultsch (1994), Photoinduced electric currents in carotenoid-deficient *Chlamydomonas* mutants reconstituted with retinal and its analogs. *Biophys. J.*, 66, 2073-2084.
56. P.L. Walne and P. Gualtieri (1994), Alga visual proteins: an evolutionary point of view. *Critical Rev. Planta Sc.*, 13, 185-197.
57. P. Galland (1992), Forty years of blue-light research and no anniversary. *Photochem. Photobiol.*, 56, 847-853.
58. S.M. Block (1992), Biophysical principles of sensory transduction in *Sensory Transduction*, New York, The Rockefeller University Press, pp. 1-17.

ACTION SPECTROSCOPY OF MICROORGANISM PHOTOMOVEMENTS

Francesco Ghetti and Giovanni Checcucci
CNR Istituto di Biofisica, via San Lorenzo 26, 56127 Pisa, Italy

Introduction

Action spectroscopy is an important non-destructive technique for studying *in vivo* the absorption properties and, in some cases, the primary reactions of photoreceptors involved in eliciting biological photoresponses. The use of action spectroscopy can be extended to any light-dependent phenomenon for which a standard response can be defined. It consists in the quantitative analysis of the response of the system as a function of the wavelength of the light, the outcome of this analysis, the action spectrum, being a measure of the relative efficiency of light of different wavelengths in causing a defined effect on the examined biological system.

Historically the measurement of action spectra has provided much valuable information and several reviews dealing with action spectroscopy have been published in the past [1 - 8].

The main goal of action spectroscopy is the identification of the pigment responsible for the observed photoresponse. In fact, provided some necessary conditions are met (such as the independence of the quantum yield from the wavelength, the absence of screening or scattering effects, a low influence of the molecular environment on the pigment absorption spectrum, etc.) the structure of the action spectrum will be similar to that of the pigment absorption spectrum.

As emphasised by some authors [2, 9], however, in the study of photomovements of microorganisms action spectroscopy does not allow in general a precise identification of the pigment responsible for the photoresponse. This can be due to various reasons: for example, the wavelength resolution of most of the published action spectra is poor, many pigments have similar absorption spectra in the visible range and the absorption properties of a pigment *in vivo* can be strongly influenced by its molecular environment, making difficult the comparison between action and absorption spectra.

Nevertheless, when the isolation and the biochemical characterisation of the photoreceptor are not possible, several techniques are available to clarify the nature of the photoreceptor pigment (e.g.: microspectrophotometry, microspectrofluorometry or use of specific quenchers of the excited states of the photoreceptor), but only action spectroscopy allows to establish a direct link between the functional absorption of photons by the presumed photoreceptor and the observed response.

Action spectroscopy also allows the determination of the "biological weighting function", i.e., the spectral efficiency on the observed photoresponse when the system is exposed to artificial or natural broad-band irradiation. This is obtained by multiplying the action spectrum of the observed photoresponse (determined with monochromatic irradiation) by the spectral irradiance of the light source, the effects of which one wants to test. In the field of environmental photobiology, for instance, the biological weighting function allows to estimate the effects on terrestrial plants and phytoplankton organisms of increasing levels of UV-B radiation due to the predicted depletion of the atmospheric ozone layer [7, 8, 10].

Action spectroscopy

A prerequisite for the determination of an action spectrum is the definition of a physical or numerical unit for quantifying the photoresponse. In the literature it can vary from the length or the weight of a plant to the concentration of a gas or of an enzyme, from the angle of bending of a sporangiophore to the number of cells accumulating in a light trap, etc..

An action spectrum is the plot of the effectiveness of the light in inducing the observed photoresponse as a function of wavelength.

Therefore, the stimulating light should be, in general, as much monochromatic as possible. Selecting the different wavelengths by means of interference filters, band-widths between 5 and 10 nm can normally be achieved. On the other hand, the use of gratings allows narrower bands but, in conventional monochromators, they strongly reduce the size of the incident beam. A remarkable exception is the Okazaki Large Spectrograph [11, 12], that can be considered the ideal instrument for the determination of action spectra. Tuneable dye-lasers, pumped with a continuous light source, such as, for example, an Argon laser, can provide highly monochromatic light with power of the order of 0.1 W and could represent a good alternative to conventional light sources.

Absorption spectra of biological photoreceptors molecules at room temperature show, in general, several nanometers wide bands and therefore excitation bandwidths narrower than about 5 nm could be unnecessary [3]. On the other hand, this biosensors are often rigidly embedded in spatially ordered receptive units and their absorption spectra can show *in vivo* a relatively resolved vibrational structure, thus justifying the use of narrower excitation bandwidths.

The use of highly monochromatic light, however, is only a necessary condition to solve the above-mentioned problem of the low wavelength resolution of action spectra. In order to reveal shoulders or narrow peaks, the interval between the used wavelengths should be as small as possible, compatibly with the bandwidths of the stimulating light (the intervals should be not lower than half of the excitation bandwidth [3]).

The effectiveness of the light in inducing the photoresponse can be quantified, for example, by means of the amount of light that is required by the system in order to produce a fixed level of response. Due to the stoichiometric relationship between the number of photons absorbed by the photoreceptor and the number of photoreceptor molecules producing the primary reaction initiating the phototransduction chain, the "strength" of the stimulus should be expressed as a fluence rate (number of photons or moles of photons (einstein) per square meter per second).

The basic assumption of action spectroscopy is that at each wavelength the photoresponse depends on the stimulating light only through the photoreceptor primary reaction triggering the transduction chain or, in other words, equal rate constants of primary reaction at different wavelengths imply equal levels of response [4, 5]. If the dependence of the response on the strength of the stimulus is monotonous, this relationship becomes bijective:

$$k_{\lambda_1} = k_{\lambda_2} \Leftrightarrow R_{\lambda_1} = R_{\lambda_2} \qquad (1)$$

where k_{λ_n} and R_{λ_n} are the rate constant and the response at wavelength λ_n, respectively. Being $k_\lambda = F_\lambda \cdot \varepsilon_\lambda \cdot \phi$, $_{pr}$(where F_λ is the fluence rate, expressed in einstein$\cdot$m$^{-2}\cdot$s^{-1}, ε_λ is the molar extinction coefficient of the photoreceptor and ϕ is the quantum efficiency of the primary reaction), it is possible to deduce the fundamental relationship of action spectroscopy:

$$R_{\lambda_1} = R_{\lambda_2} \Rightarrow \varepsilon_{\lambda_1} : \varepsilon_{\lambda_2} = 1/F_{\lambda_1} : 1/F_{\lambda_2} \qquad (2)$$

that states the proportionality between the molar extinction coefficients and the reciprocal of fluence rates at different wavelengths, when equal responses are measured. This means that in a hypothetical experiment where at each examined wavelength the same level of response is obtained, the plot of the reciprocal of the fluence rates used versus the wavelength should have the same shape of the absorption spectrum of the photoreceptor in its physiological environment.

In relationship (1) the response R is connected to the rate of the primary reaction. An equivalent relationship could be written linking the response to the total quantity of product and relationship (2) can be extended to fluences N_λ ($N_\lambda = F_\lambda \cdot \Delta t$) if the exposure time Δt is constant at each wavelength:

$$R_{\lambda_1} = R_{\lambda_2} \Rightarrow \varepsilon_{\lambda_1} : \varepsilon_{\lambda_2} = 1/F_{\lambda_1} : 1/F_{\lambda_2} = 1/F_{\lambda_1} \cdot \Delta t : 1/F_{\lambda_2} \cdot \Delta t = 1/N_{\lambda_1} : 1/N_{\lambda_2}$$

If the examined system obeys to the law of reciprocity [1], i.e. the response depends on the total dose, it is possible to use different irradiation times at different wavelengths. This could be necessary, for example, because the energy emitted by the light sources is not constant with wavelengths [5].

In deducing relationship (2), the quantum efficiency of the primary reaction, ϕ, has been assumed independent of wavelength. If this condition is not met, the action spectrum will result distorted relative to the absorption spectrum [4]. On the other hand, if in the photoreceptor molecule the primary reaction originates from the first singlet excited state or from the first triplet state and the rates of the various radiationless transitions have the usual values of a normal organic molecule, the assumption of the independence of ϕ from wavelength can be considered true [5].

Other sources of wavelength-dependent distortion of the action spectrum can be found in the environment where the pigment is embedded *in vivo*. In fact, if other pigments are present in the cell, they can cause screening or reflection effects, in such a way that the effectiveness of a particular spectral range on the photoreceptor could be reduced or enhanced, respectively. Moreover, in a cellular environment with a high number of scattering centers, the efficiency in eliciting the photoresponse of the shorter wavelengths can be apparently decreased, because of the dependence of the diffusion on the inverse of the wavelength. The sample should also not absorb too much light at the tested wavelengths, so that, at least ideally, every photoreceptor involved in the photoprocess could be exposed approximately to the same amount of photons and, consequently, have the same probability of responding [7]. Finally, as mentioned above, the molecular environment can drastically change the absorption characteristics of the photoreceptor pigment, thus making difficult its identification.

From relationship (2) it follows that a simple procedure to determine an action spectrum is to choose a standard level of response and to measure at each wavelength the fluence rate (or the dose) inducing that response. However, since a necessary prerequisite is that the magnitude of the response is proportional to the intensity of the stimulus, a succession of graded responses should be measured, for each wavelength, in order to construct a fluence rate-response curve and to verify that the chosen standard level of response fall on the linear part of the curve.

An important criterion to determine if the fluence rate-response curves at the different tested wavelengths are indicative of the same mechanism of action is to check if they can be reasonably superimposed, by multiplying each of them by an appropriate fluence factor, thus showing that they have the same 'shape' [7]. In other words, given any pair of fluence rate-response curves, the ratio of two fluence rates corresponding to the same level of response should be approximately a constant for any level of response and, consequently, the difference of their logarithms should be also a constant. This implies that the curves plotted on a logarithmic abscissa should appear parallel.

If a fluence rate-response curve has been determined with a sufficient number of data points, it can be fitted with an appropriate mathematical function. This, besides allowing the calculation of the fluence rates for any standard level of response, permits to check more precisely if the curves have the same shape.

The kinetic properties of the photoreceptor system determine the shape of the fluence rate-response curves, i.e. the mathematical function that best fits the data points could, in principle, give information on the kinetic mechanism of the primary step of the phototransduction process. Two general kinetic model have been discussed illustrating two important types of photoreactions, in which a single, not-photochromic photoreceptor is involved: photoproduct formation and photocatalysis [3, 5]. Biological systems and relative kinetic models based on photochromic photoreceptors, i.e. on photoreceptors existing in two ground state forms reverting to each other by effect of light (e.g. phytochrome), are discussed in details in references 3 and 4.

In the kinetic model for photoproduct formation, the photoreceptor P from its excited state P* is converted, through a one-step or a multi-step dark reaction, into a product A; the photoresponse R is proportional to the concentration of A. It can be demonstrated [3] that the relative photoresponse R_{rel} is an exponential function of the product of the fluence rate by the exposure time t with $\varepsilon_\lambda \cdot \phi$ as parameter:

$$R_{rel} = 1 - e^{-F_\lambda t \varepsilon_\lambda \phi}$$

or, as a function of the fluence ($N_\lambda = F_\lambda \cdot t$):

$$R_{rel} = 1 - e^{-N_\lambda \varepsilon_\lambda \phi}$$

As R_{rel} depends on the fluence, the reciprocity law should apply.

Examples of photobiological process following this kinetic mechanism are the light-dependent synthesis of chlorophyll in higher plants, the photoinactivation by UV light of enzymes and viruses and the UV photodamage of bacteria and protozoa [3].

In the kinetic model for photocatalysis the light-produced active state P_A of the photoreceptor P function as a catalyst regulating the photoresponse R. P_A reverts in the dark to P with a rate constant k_A. It can be demonstrated [3] that the photoresponse R is a linear function of the exposure time t multiplied by an hyperbolic function of the fluence rate with the ratio $k_A/\varepsilon_\lambda \cdot \phi$ as parameter:

$$R = (\text{const.} \times t) \times \left(\frac{F_\lambda}{F_\lambda + k_A/\varepsilon_\lambda \phi} \right)$$

As R depends linearly on the exposure time and hyperbolically on the fluence rate, the reciprocity law is not valid. If the exposure time is kept constant for all the experiments the relative photoresponse R_{rel} becomes:

$$R_{rel} = \left(\frac{F_\lambda}{F_\lambda + k_A/\varepsilon_\lambda \phi} \right)$$

Examples of photobiological process following this kinetic mechanism are the photosynthesis and the photoactivation of enzymes, such as the cytochrome oxidase or the nitrate reductase [3]. Hyperbolic functions have been used also to fit the fluence-response curves for various photoprocesses occurring in *Phycomyces*, such as photomorphogenic responses and photogravitropic equilibrium of sporangiophores [6].

Action spectroscopy of microorganism photomovement

Action spectroscopy has been widely used in the study of photomovements of microorganisms, mainly with the aim of identifying or, at least, classifying in a particular biochemical class (carotenoids, flavins, etc.) the photoreceptor molecules involved in the process. In the following, an overview is presented of experimental works in which action spectroscopy was applied with this purpose to the study of eukaryotic unicellular organisms, with some emphasis on more recent publications. For a more complete information on the photoreceptor pigments for these photoresponses, that take into account results obtained also with other techniques than action spectroscopy, the reader can refer to some general review on this topic [2, 13-20]. As discussed above, in fact, action spectra can provide only a first hint about the nature of the photoreceptor involved and other experimental approaches should be adopted in order to enrich the information obtained from action spectroscopy.

For phototactic algae, where the interpretation of photomovement action spectra is complicated by the presence of screening pigments or reflecting organelles (stigma or eyespot), Foster and Smyth [21] proposed, in order to identify the photoreceptor, the determination of an action spectrum obtained by plotting against wavelength the reciprocal of the fluence rate at which the maximum response occurs, showing how this value can be obtained from the fluence rate-response curves. Under some conditions [21], this spectrum should approximate the photoreceptor absorption spectrum, without any distortion due to screening or reflecting pigments.

Foster and Smyth, reinterpreting on the basis of this model previous data by Nultsch *et al.* [22], produced an action spectrum suggesting a rhodopsin-like photoreceptor for the phototactic response of the flagellated alga *Chlamydomonas reinhardtii* [21]. Later, this hypothesis was confirmed using a blind mutant, which exhibits photoresponse only at very high fluence rates: incorporating analogues of the retinal chromophore into the mutant cells, it was shown, in fact, that the photoresponse was restored also at lower fluence rates and that the maxima of the action spectra were shifted consistently with the absorption maxima of the different incorporated retinal analogues [23].

By means of action spectroscopy measurements, performed at the Okazaki Large Spectrograph [11, 12], the photophobic and phototactic responses of *Dunaliella salina* were showed to be triggered by two different photoreceptors, that in both cases were suggested to be a rhodopsin or a carotenoprotein [24].

In the case of the flagellated alga *Euglena gracilis* action spectra for the phototactic and the photophobic response, based either on population or on single cell measurements, indicated that flavin molecules may be responsible for photobehaviour [25-27]. More recently, phototaxis action spectra based on computer-assisted tracking of individual cells showed, both for green and dark bleached *E. gracilis*, a good agreement with the hypothesis that, besides flavins, also pterin molecules act as photoreceptors [28]. On the basis of action spectroscopy, a flavin-type photoreceptor was proposed also for the photophobic response of the euglenoid flagellate *Astasia fritschii* [29].

In brown algae, for the phototaxis of the gametes of *Ectocarpus siliculosus* and of the zoospores of *Pseudochorda gracilis*, action spectra in the range 400-500 nm, with two bands at about 420 and 460 nm, were determined [30, 31]. A combined action of a free and a protein-bound flavin (8-hydroxy-5-deazaflavin) was suggested to explain the 420 nm and the 460 nm bands, respectively [31].

For the photoorientation of the dinoflagellate *Peridinium gatunense* an action spectrum was determined showing a marked effect in the spectral range 550-720 nm; the involvement of photosynthetic or accessory pigments was excluded on the basis of experiments with inhibitors of photosynthetic electron transport chain and of the lack of action bands in the blue-green spectral range and no hypothesis on the nature of this photoreceptor was put forward [32].

Also in the case of the ciliate protozoan *Loxodes* (*L. striatus* and *L. magnus*) a flavin was proposed as photoreceptor, on the basis of the action spectrum for the escape of the cells from the oxycline [33], whereas action spectra of the photoresponses of various species of the genus *Paramecium* could be interpreted both in favour of a flavin- and a rhodopsin-like pigment (see reference 17 for an extensive discussion on this subject and further experimental references). For the marine ciliate *Fabrea salina* a more puzzling action spectrum has been determined and a rhodopsin, possibly with the cooperative effect of a second pigment absorbing around 420 nm, was proposed as photoreceptor [34].

Pigments similar to hypericin were, on the contrary, found to be involved in triggering the photoresponses of the coloured ciliates *Stentor coeruleus* and *Blepharisma japonicum.* For the former, a good agreement was shown between the absorption spectrum of the endogenous pigment stentorin and the action spectra for the step-up photophobic response [35-38] and for phototaxis [39]. Recently, a markedly different action spectrum for photokinesis was determined and the presence of a second photoreceptor system was suggested [40].

For the step-up photophobic response in the ciliate *Blepharisma japonicum*, a first action spectrum, in which the light dependent time-lag between the application of the stimulus and the appearance of the reaction was used for quantifying the response, indicated that the photoresponse was triggered by a blue absorbing pigment [41]. It was tentatively identified with an yellow pigment, extracted from the cell, that IR spectroscopy seemed to indicate as a different form of the endogenous pigment blepharismin [41]. Later, two other action spectra, based on the time-lag and on the percentage of responding cells, showed a relevant action of the wavelengths in the range 550-600 nm and blepharismin in its native red form was proposed as photoreceptor [42]. This pigment, that is mainly localised in subpellicular granules spread all over the cellular body, undergoes an *in vivo* photooxidation, converting it in a grey-blue form and altering its absorption spectrum [43]. In recent years, action spectra, based again on the percentage of responding cells and on the latency of the response, not only confirmed red blepharismin as photoreceptor for the red cells [44-46], but also suggested that the photooxydation process does not alter the photosensing capability of the pigment. In fact, the action spectrum for the photophobic response of the blue cells of *B. japonicum* shows, in comparison to that of the red ones, a shift of the maxima reflecting the shift of the peaks observed in the absorption spectrum [44, 45, 47]. This means that the molecular transformation of blepharismin, following light-induced oxidation, does not modify its reactivity as far as its photoperception and photosignalling function is concerned.

Finally, both in *Stentor coeruleus* and *Blepharisma japonicum*, recently determined action spectra for the light-induced receptor potential confirmed the role of stentorin and blepharismin as the photoreceptors triggering the ciliary reversal responsible for the step-up photophobic response [39, 46].

References

1. W. Shropshire, Jr., (1972) Action Spectroscopy, in: *Phytochrome* (K. Mitrakos and W. Shropshire, Jr., eds.), London, Academic Press, pp. 161-181.
2. G. Colombetti and F. Lenci (1980) Identification and Spectroscopic Characterization of Photoreceptor Pigments, in: *Photoreception and Sensory Transduction in Aneural Organisms* (F. Lenci and G. Colombetti, eds.), New York, Plenum Press, pp. 173-188.
3. K. M. Hartmann (1983) Action Spectroscopy, in: *Biophysics* (W. Hoppe, W. Lohmann, H. Markl and H. Ziegler, eds.), Berlin, Springer-Verlag, pp. 115-144.
4. E. Schäfer, L. Fukshansky and W. Shropshire, Jr., (1983) Action Spectroscopy of Photoreversible Pigment System, in: *Encyclopaedia of Plant Physiology, New Series, Volume 16. Photomorphogenesis* (W. Shropshire, Jr., and H. Mohr, eds.), Berlin, Springer-Verlag, pp. 39-68.
5. P. A. Galland (1987) Action Spectroscopy, in: *Blue Light Responses: Phenomena and Occurrence in Plants and Microorganisms* (H. Senger, ed.), Boca Raton, CRC Press, pp. 37-52.
6. E. D. Lipson (1991) Action Spectroscopy, in: *Biophysics of photoreceptors and photomovements of microorganisms* (F. Lenci, F. Ghetti, G. Colombetti, D.-P. Häder and P.-S. Song, eds.), NATO ASI Series A211, New York, Plenum Press, pp. 293-309.
7. T. P. Coohill (1991) Action Spectra Again?, *Photochem. Photobiol.* 54, 859-870.
8. T. P. Coohill (1992) Action Spectra Revisited, *J. Photochem. Photobiol., B: Biol.* 13, 95-98.
9. P. Gualtieri (1993) *Euglena gracilis*: is the photoreception enigma solved?, *J. Photochem. Photobiol., B: Biol.* 19, 3-14.
10. T. P. Coohill (1989) Ultraviolet action spectra (280 to 380 nm) and solar effectiveness spectra for higher plants, *Photochem. Photobiol.* 50, 451-457.
11. M. Watanabe, M. Furuya, Y. Miyoshi, Y. Inoue, I. Iwahashi and K. Matsumoto (1982) Design and performance of the Okazaki Large Spectrograph for photobiological research, *Photochem. Photobiol.* 36, 491-498.
12. M. Watanabe (1991) High-fluence rate monochromatic light sources, computerized analysis of cell movements, and microbeam irradiation of a moving cell: current

experimental methodology at the Okazaki Large Spectrograph, in: *Biophysics of photoreceptors and photomovements of microorganisms* (F. Lenci, F. Ghetti, G. Colombetti, D.-P. Häder and P.-S. Song, eds.), NATO ASI Series A211, New York , Plenum Press, pp. 327-337.

13. W. Nultsch and D.-P. Häder (1989) Photomovement in motile microorganisms - II, *Photochem. Photobiol.* 47, 837-869.
14. G. Colombetti and D. Petracchi (1989) Photoresponse mechanisms in flagellated algae, *Critical Reviews in Plant Sciences* 8, 309-355.
15. F. Lenci and F. Ghetti (1989) Photoreceptor pigments for photomovements of microorganisms. Some spectroscopic and related studies, *J. Photochem. Photobiol., B: Biol.* 3, 1-16.
16. P.-S. Song and K. L. Poff (1989) Photomovement, in: The science of photobiology (2nd ed.) (K. C. Smith, ed.), New York , Plenum Publishing Corp., pp. 305-346.
17. G. Colombetti (1990) Photomotile responses in ciliated protozoa, *J. Photochem. Photobiol., B: Biol.* 4, 243-259.
18. L. Kuznicki, E. Mikolajczyk and P. L. Walne (1990) Photobehavior of euglenoid flagellates: theoretical and evolutionary perspectives, *Critical Reviews in Plant Sciences* 9, 343-369.
19. F. Lenci, F. Ghetti, G. Colombetti, D.-P. Häder and P.-S. Song (eds.) (1991) *Biophysics of photoreceptors and photomovements of microorganisms*, New York, Plenum Press.
20. G. Kreimer (1994) Cell biology of phototaxis in flagellated algae, *Int. Rev. Cytol.* 148, 229-310.
21. K. W. Foster and R. D. Smyth (1980) Light antennas in phototactic algae, *Microbiol. Rev.* 44, 572-630.
22. W. Nultsch, G. Throm and I. von Rimscha (1971) Phototaktische untersuchungen an *Chlamydomonas reinhardtii* dangeard in homokontinuierlicher kultur, *Arch. Mikrobiol.* 80, 351-369.
23. K. W. Foster, J. Saranak, N. Patel, G. Zarilli, M. Okabe, T. Kline and K. Nakanishi (1984) A rhodopsin is the functional photoreceptor for phototaxis in the unicellular eukaryote *Chlamydomonas*, *Nature* 311, 756-759.
24. R. Wayne, A. Kadota, M. Watanabe and M. Furuya (1991) Photomovement in *Dunaliella salina*: fluence rate-response curves and action spectra, *Planta* 184, 515-524.
25. B. Diehn (1969) Action spectra of the phototactic responses in *Euglena*, *Biochim. Biophys. Acta.* 177, 136-143.
26. A. Checcucci, G. Colombetti, R. Ferrara and F. Lenci (1976) Action spectra for photoaccumulation of green and colorless *Euglena*: evidence for identification of receptor pigments, *Photochem. Photobiol.* 23, 51-54.
27. C. Barghigiani, G. Colombetti, B. Franchini and F. Lenci (1979) Photobehavior of *Euglena gracilis*: action spectrum for the step-down photophobic response of individual cells, *Photochem. Photobiol.* 29, 1015-1019.
28. D.-P. Häder and E. Reinecke (1991) Phototactic and polarotactic responses of the photosynthetic flagellate *Euglena gracilis*, *Acta Protozool.* 30, 13-18.
29. E. Mikolajczyk and P. L. Walne (1990) Photomotile responses and ultrastructure of the euglenoid flagellate *Astasia fritschii*, *J. Photochem. Photobiol., B: Biol.* 6, 275-282.
30. H. Kawai, D. G. Muller, E. Folster and D.-P. Häder (1990) Phototactic responses in the gametes of the brown alga *Ectocarpus silicosus*, *Planta* 182, 292-297.
31. H. Kawai, M. Kubota, T. Kondo and M. Watanabe (1991) Action spectra in zoospores of the brown alga *Pseudochorda gracilis*, *Protoplasma* 161, 17-22.

32. S. M. Liu, D.-P. Häder and W. Ullrich (1990) Photoorientation in the freshwater dinoflagellate *Peridinium gatunense* Nygaard, *FEMS Microbiol. Ecol.* 73, 91-102.
33. B. J. Finlay and T. Fenchel (1986) Photosensitivity in the ciliated protozoon *Loxodes*: pigment granules, absorption and action spectra, blue light perception and ecological significance, *J. Protozool.* 33, 534-542.
34. R. Marangoni, S. Puntoni, L. Favati and G. Colombetti (1994) Phototaxis in *Fabrea salina* I. Action spectrum determination, *J. Photochem. Photobiol., B: Biol.* 23, 149-154.
35. D. C. Wood (1976) Action spectrum and electrophysiological responses correlated with the photophobic response of *Stentor coeruleus*, *Photochem. Photobiol.* 24, 261-266.
36. P.-S. Song, D.-P. Häder and K. L. Poff (1980) Step-up photophobic response in the ciliate *Stentor coeruleus*, *Arch. Microbiol.* 126, 181-186.
37. C. B. Hong, R. K. Prusti and P.-S. Song (1987) Light adaptation in the photophobic response by *Stentor coeruleus*, *Arch. Microbiol.* 147, 117-120.
38. P.-S. Song, D.-P. Häder and K. L. Poff (1980) Phototactic orientation by the ciliate *Stentor coeruleus*, *Photochem. Photobiol.* 32, 781-786.
39. S. Fabczak, H. Fabczak, N. Tao and P.-S. Song (1993) Photosensory transduction in ciliates. I. An analysis of light-induced electrical and motile responses in *Stentor coeruleus*, *Photochem. Photobiol.* 57, 696-701.
40. K. Iwatsuki (1992) *Stentor coeruleus* shows positive photokinesis, *Photochem. Photobiol.* 55, 469-471.
41. M. Kraml and W. Marwan (1983) Photomovement responses of the heterotrichous ciliate *Blepharisma japonicum*, *Photochem. Photobiol.* 37, 313-319.
42. P. Scevoli, F. Bisi, G. Colombetti, F. Ghetti, F. Lenci and V. Passarelli (1987) Photomotile responses of *Blepharisma japonicum* I: Action spectra determination and time-resolved fluorescence of photoreceptor pigments, *J. Photochem. Photobiol., B: Biol.* 1, 75-84.
43. F. Ghetti (1991) Photoreception and photomovements in *Blepharisma japonicum*, in: *Biophysics of photoreceptors and photomovements of microorganisms* (F. Lenci, F. Ghetti, G. Colombetti, D.-P. Häder and P.-S. Song, eds.), NATO ASI Series A211, New York, Plenum Press, pp. 257-265.
44. T. Matsuoka, S. Matsuoka, Y. Yamaoka, T. Kuriu, Y. Watanabe, M. Takayanagi, Y. Kato and K. Taneda (1992) Action spectra for step-up photophobic response in *Blepharisma*, *J. Protozool.* 39, 498-502.
45. G. Checcucci, G. Damato, F. Ghetti and F. Lenci (1993) Action spectra of the photophobic response of blue and red forms of *Blepharisma japonicum*, *Photochem. Photobiol.* 57, 686-689.
46. S. Fabczak, H. Fabczak and P.-S. Song (1993) Photosensory transduction in ciliates. III. The temporal relation between membrane potentials and photomotile responses in *Blepharisma japonicum*, *Photochem. Photobiol.* 57, 686-689.
47. Y. Murakami, Y. Kato and T. Matsuoka (1993) Blue form of *Blepharisma* photoreceptor caused by light irradiation, *Cytobios* 73, 39-48.

LIGHT SIGNAL TRANSDUCTION IN CILIATE *STENTOR* AND *BLEPHARISMA*. 1. STRUCTURE AND FUNCTION OF THE PHOTORECEPTORS

Pill-Soon Song

Department of Chemistry and Institute for Cellular and Molecular Photobiology, University of Nebraska-Lincoln, Lincoln, NE 68588, USA

Abstract

Stentor coeruleus and *Blepharisma japonicum* are photoresponsive ciliates, exhibiting photophobic responses. The former also exhibits a negative phototaxis. The photoreceptors for the light-avoiding motile responses are stentorin and blepharismin, respectively. We recently determined the structure of stentorin to be octahydroxydiisopropylnaphathodianthrone. Blepharismin, although not fully elucidated, is also a hypericin analogue. The photochemistry of this class of novel photoreceptors is unknown; however, proton and/or electron transfer may play a significant role in the primary photoprocess.

INTRODUCTION

The colored ciliates *Stentor coeruleus* and *Blepharisma japonicum* accumulate in shady or dark areas (Jennings, 1904; Mast, 1906: Giese 1973, 1981). They swim freely under normal conditions. However, *Stentor* stops and swims backwards momentarily when suddenly exposed to a higher intensity light. *Blepharisma* rotates briefly around an axis perpendicular to its swimming direction and then swims forward in a new direction (Figure 1). These light responses have been characterized as a step-up photophobic response (Song *et al.*, 1980a,b; Scevoli *et al.*, 1987; Matsuoka *et al.*, 1992a; Checcucci *et al.*, 1993; Gioffre *et al.*, 1993). Additionally, *Stentor coeruleus* shows a negative photokinesis (Iwatsuki, 1991), and *Blepharisma japonicum* exhibits a step-down photoresponse (Matsuoka and Taneda, 1992).

As in any other stimulus-response system, light stimulus in these ciliates is perceived by their intrinsic receptors; namely photoreceptors which absorb the light of a particular wavelength. Based on the action spectra for the photophobic response and for the light-induced bioelectrical responses, stentorin and blepharismin have been identified as the photoreceptor molecules that are responsible for the light avoiding behavior in *Stentor* and *Blepharisma, respectively* (Wood, 1976; Song *et al.*, 1980a,b, 1981, 1983; Scevoli *et al.*, 1987; Matsuoka *et al.*, 1992a; Checcucci *et al.*, 1993). These photosensor molecules are imbedded just beneath the pellicle in the pigment granules which are arranged longitudinally along the cell body parallel to the ciliary rows (more details later).

Unlike *Stentor coeruleus, Blepharisma japonicum* changes its color from red to blue upon light exposure. The visual color change results from the red fluorescence of the dark-grown red *Blepharisma* cell that is quenched in the blue cell upon light exposure. The blue-turned cell also shows photophobic response with oxyblepharismin as its photosensor molecule. The photoreceptor molecules have been identified as hypericin derivatives based on the similarities in spectroscopic characteristics. These hypericin compounds represent a novel class of photoreceptor molecules with a significantly different chromophore structure than the known photoreceptors; such as rhodopsin, bacteriorhodopsin, chlorophylls, phytochrome, and flavins.

Investigations of the photosensory transduction in these ciliates would shed light on our understanding of the mechanistic nature of stimulus-response systems. *Stentor coeruleus* and *Blepharisma japonicum* are particularly suitable for this purpose since they are unicellular and the problems associated with complex interactions that occur between mutually interdependent cells in higher organisms can be avoided. In addition, *Stentor coeruleus* is one of the largest ciliates. Its typical size is 700 μm in length. *Blepharisma japonicum* is not quite as large and has a typical size of 400 μm in length. The large size makes it amenable for analyzing individual cell motility and microelectrode measurements.

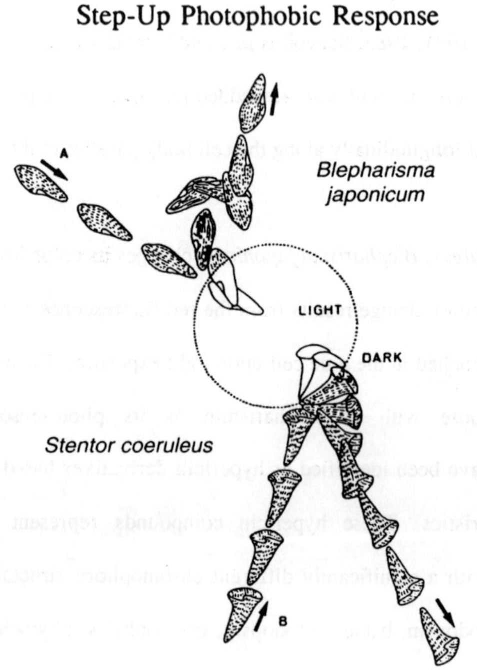

Figure 1 - Schematic drawing of the photophobic response of (A) *Stentor coeruleus* and (B) *Blepharisma japonicum*. The size of the cells is 250-350 μm.

Chemical Structures of the Ciliate Photoreceptors

The pigment of *Stentor coeruleus* was named stentorin (Lankester, 1873). Stentorins belong to the meso-naphthodianthrone class of compounds (Møller, 1962). Walker *et al.* (1979) proposed that the native chromophore was linked to a 65 kDa protein and that the released chromophore upon acid hydrolysis was hypericin. However, he failed to notice that even the absorption maximum of acid treated stentorin was not entirely identical to that of hypericin at low pH's. In fact, stentorin, which was HPLC-purified from a crude extract of *Stentor coeruleus* cells, has a λ_{max} at 595 nm (in methanol), while hypericin has one at 588 nm (Tao *et al.*, 1993). When acidified, the λ_{max} for stentorin and hypericin shifted to 577 nm and 578 nm, with somewhat different molar extinction coefficients at the absorption maxima, respectively.

In our negative ion fast atom bombardment (FAB) mass spectrometry (Tao *et al.*, 1993), stentorin showed an $(M-1)^-$ ion at 591.1304, consistent with the $(M-1)^-$ of $C_{34}H_{24}O_{10}$ [$(M-1)^-$ calculated to be 591.1291]. Results from acetylation of stentorin, in combination with FAB-MS, indicated that there are eight hydroxyl groups in stentorin. NMR spectrum of stentorin revealed the presence of isopropyl groups in stentorin. Thus, we proposed that the structure of stentorin either **1** or **2** (Figure 2; Tao *et al.*, 1993). Recently, these two structures have been synthesized (Cameron and Riches, 1995). A spectroscopic comparison with our native stentorin chromophore revealed that structure **1** is the actual chromophore for stentorin.

Arcichovskij (1905) studied the pigments from *Blepharisma laterilium* and named it zoopurpurin. Emerson (1930) and Giese (1953) investigated the solubility and absorption maxima of the pigments from an American species of *Blepharisma undulans* in various solvents. Møller showed that the pigment of an American species of *Blepharisma undulans* strongly resembled hypericin in absorption maxima. Sevenants (1965) carefully analyzed the pigment of *Blepharisma undulans* and reported that the pigment possessed the same mesodianthrone structure. Giese (1973, 1981) carried out a series of studies on the pigments from *Blepharisma americanum*,

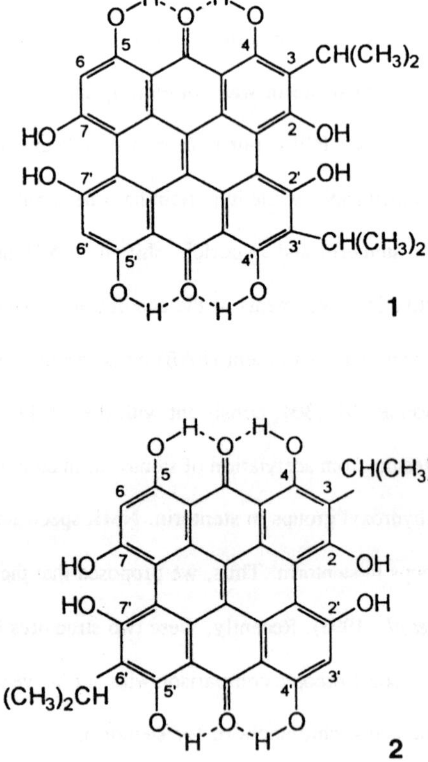

1

2

Figure 2 - The possible chemical structures of stentorin (Tao *et al.*, 1993). Structure **1** is the actual structure of stentorin (Cameron and Riches, 1995).

Blepharisma intermedium and *Blepharisma japonicum* and other species. He found that the pigments existed in red and blue form in all of the species that were tested. They were also essentially alike in spectrophotometric and solubility properties as well as in their sensitivity to light. Due to the very limited amount of material, the pigment of *Blepharisma* has not been purified and its structure has not been elucidated chemically. However, we recently accumulated enough *Blepharisma japonicum* cells and the preliminary study of the HPLC-purified pigments from the cell extract showed that the red to blue transformation of blepharismin occurred via a loss of two mass units. Detailed results will be reported in a separate paper.

Photoreceptor Proteins

Two groups of pigments, stentorin-1 and stentorin-2, exist in *Stentor coeruleus*. Stentorin 1 can be readily extracted in ethanol from the cell suspension. Stentorin-2 was obtained by sonication of a cell suspension in phosphate buffer, followed by ammonium sulfate fractionation and sucrose density gradient centrifugation (Walker *et al.*, 1979). Subsequently, two pigment fractions were isolated and purified from the *Stentor coeruleus* buffer extract by column chromatography (Kim *et al.*, 1990). It should be noted that a neutral detergent, reduced Triton X-100, increased the solubility of membrane-bound stentorin-2. Partial purification of stentorins was performed by using a hydroxyapatite (HA) column and a Bio-Gel A-1.5m gel filtration column (Kim *et al*, 1990). Stentorin-1 (λ_{max} at 608 nm, strongly fluorescent) appeared to be a relatively small complex composed of at least two heterologous subunits corresponding to apparent molecular weights of 46,000 (fluorescent, Coomassie blue negative) and 52,000 (fluorescent, Coomassie blue positive) on 13% SDS-PAGE gels. Stentorin-2 (λ_{max} at 620 nm, weakly or non-fluorescent) is a large protein complex which is eluted out of the Bio-Gel A-1.5m column near the void volume.

Since stentorin-2 exhibits an ultrafast photoprocess, and stentorin-1 does not, it was proposed that stentorin-2 may serve as the functional photoreceptor (Song *et al.*, 1990). Recently, stentorin-2B was isolated from stentorin-2 by hydrophobic interaction chromatography (Dai, 1994). It contains the chromophore covalently bound to an approximately 50 kDa protein, as determined by SDS-urea-PAGE (Dai, 1994). Further characterization of stentorin 2B is underway in this lab. Stentorin-1 and stentorin-2, in their native states within the *Stentor* cell, are spectrophotometrically indistinguishable and the relation between them in the pigment granules remains elusive (Kim *et al.*, 1990). However, a photoacoustic spectroscopic study clearly demonstrated the existence of the two forms of the pigment in the cell (reported later in this paper).

As mentioned earlier, *Blepharisma japonicum*, when exposed to light of moderate intensity, turns from red to blue (Giese, 1981). Both red- and blue-pigmented cells show step-up photophobic responses (Matsuoka *et al.*, 1992a; Checcucci *et al.*, 1993). The action spectrum of the response in the red cells closely resembles the absorption spectrum of the red pigment, while the action spectrum of the response in the blue cells closely resembles the absorption spectrum of the blue pigment (Matsuoka *et al.*, 1992a; Checcucci *et al.*, 1993). This finding strongly suggests that blepharismin is involved in the step-up photophobic response in *Blepharisma japonicum*, and rules out a rhodopsin as the possible photoreceptor.

In an attempt to isolate the native photoreceptor pigment from *Blepharisma japonicum*, Gioffré and coworkers in Lenci's group, and Dai in Song's lab, isolated a presumed photoreceptor protein of apparent molecular mass of 38 kDa (Gioffré *et al.*, 1993). However, Matsuoka *et al.* (1993) reported that a 200 kDa protein was the native photoreceptor. Further work will be necessary in order to verify these contradicting results. It should be mentioned that different detergents were used by these two groups [CHAPS by Gioffré *et al.* (1993) and sodium cholate by Matsuoka *et al.* (1993)]. Caution should be exercised in determining the molecular weight of chromoproteins (especially those containing hypericin-like chromophores) by SDS-

PAGE or gel-filtration chromatography, since stentorin chromophore (molecular weight 592) dissolved in a detergent-containing buffer may appear to be bigger than 100 kDa (Tao and Song, unpublished results).

Photoreceptor Localization

The pigment granules in *Stentor coeruleus* are 0.3-0.7 μm in diameter and are distributed longitudinally along the cell body between ciliary rows (Huang and Pitelka, 1973). They are localized near the ciliary basal bodies fused into the pellicle, or plasma membranes (Kennedy, 1965; Newman, 1974). A recent scanning electron micrograph study showed that pigment granules were closely packed in a random array within each of their rows. A transmission electron micrograph study revealed that the pigment granules were striated with alternating layers having high and low electron density; nonetheless, amorphous pigment granules were also evident. It is uncertain whether the amorphous appearance of the pigment granule resulted from artificial loss of the striated fine structure. No mitochondrial fine structure was observed in the pigment granules (Newman, 1974; Small and Marszalek, 1969), although the pigment granules were thought to be mitochondria-like (Weisz, 1949, 1950). Meza-Keuthen (1992) observed a distinctly membrane stack-like structure (Fig. 3), consistent with an earlier observation (Randall and Jackson, 1958).

Stentorin-1 and stentorin-2, in their native states within the *Stentor* cell, are spectrophotometrically indistinguishable and the relation between them in the pigment granules has not been established (Kim *et al.*, 1990). To test the possible existence of two chromoproteins in a native *Stentor* cell, we used photoacoustic spectroscopy. Photoacoustic spectroscopy is applicable to non-transparent and highly scattering media. The photoacoustic signal involves the absorption of modulated light by the sample. Radiationless deactivation processes produce heat, which is distributed to the sample surface. This results in a modulated pressure change, detected

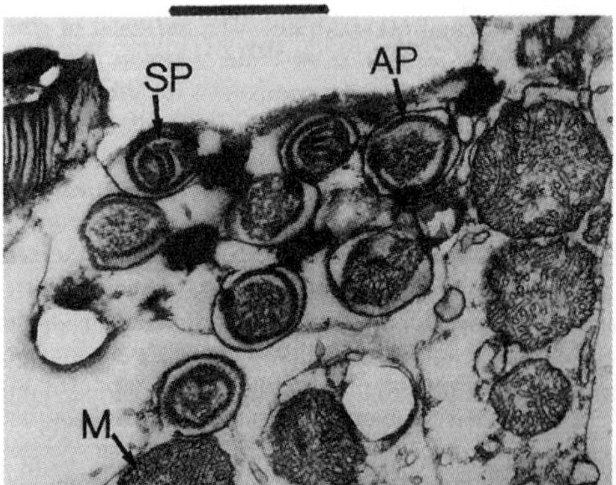

Figure 3 - Transmission electron micrograph showing both striated (SP) and amorphus (AP) pigment granules. Mithocondria (M) are also seen on the lower part and upper right-hand side of the photo. The upper horizontal bar represents ca. 1.1 μm in length.

by a microphone (Rosencwaig, 1980). The applicability of the photoacoustic spectroscopy for determining *in vivo* spectra of biological species is well known (for example, see O'Hara *et al.*, 1983; Romanowski *et al.*, 1987; Tao *et al.*, 1994). This spectroscopic technique can also be used for resolving electronic transition bands that are hidden in absorption spectra (Pottier *et al.*, 1988). Photoacoustic spectroscopy was useful in studying the excitation energy path in biological systems (Frackowiak *et al.*, 1985; Romanowski *et al.*, 1987), and it is a well suited tool for the study of pigmented ciliate cells.

Photoacoustic spectra can be measured using a single beam photoacoustic spectrometer (PAS). A light source for a PAS is a 1000 W xenon arc lamp or alternative source including tunable laser. Light from the monochromator is mechanically chopped and focused into the photoacoustic cell, which is a gas-microphone system based on a design described by Ducharme *et al.* (1979). In the PAS system, constructed in the author's laboratory, the frequency of light modulation is 26 Hz. The incident monochromatic light flux is 3 mW/cm^2 at 550 nm. The photoacoustic signal is collected by a ceramic microphone, equipped with a preamplifier. An amplified and filtered signal is converted by an analog-to-digital converter, and processed by a personal computer. The photoacoustic spectra are corrected for the spectral output of the excitation system. Contribution to the photoacoustic signal from the filtration paper, used as a support for the sample, was flat over the wavelength range covered.

We developed a computer-based method of photoacoustic signal analysis. We incorporated some of the concepts for photoacoustic data analysis, as presented by Anjo and Moore (1984). The depth into the sample probed by the photoacoustic technique may be approximated as a thermal diffusion length μ_T (Rosencwaig, 1980):

$$\mu_T = (2\alpha/\omega)^{1/2}$$

where α is thermal diffusivity of the sample, and $\omega = 2\pi \times f_M$ where f_M is the frequency of light modulation. Thermal diffusivity of most biological samples is ca. 0.1 mm^2/s (Touloukian *et al.*, 1973). At the frequency $f_M = 26$ Hz being used, the depth of penetration is approximated as $\mu_T = 35$ μm, which is an appropriate value for a morphological study of the *Stentor* cell, whose diameter does not exceed 100 μm. This kind of analysis shows individual contributions to the absorption from different depths and is specially useful in analyzing heterogenous systems (O'Hara *et al.*, 1983).

The photoacoustic spectra and their analysis are described in detail elsewhere (Tao *et al.*, 1994). Briefly, the photoacoustic spectroscopy has demonstrated that there are indeed two light absorbing species *in vivo*. One is presumably closer to the outer cell membrane and its photoacoustic spectrum is red-shifted. Stentorin-2 is localized within the pigment granules that are distributed between the ciliary rows. The other, buried into the cell volume, is contributing to energy dissipation; therefore, it is assumed to be less efficient photochemically. Their spectral characteristics are similar to those of isolated pigment species, stentorin-1 and stentorin-2 (Kim *et al.*, 1990). This may indicate that the red-shifted component, detected in the native cell by photoacoustic spectroscopy, corresponds to stentorin-2 and the blue-shifted component belongs to stentorin-1. According to the presented interpretation of photoacoustic data, stentorin 2 is preferentially located at the outer membrane and involved in a more efficient process of the utilization of absorbed energy of light.

As is the case with photoacoustic spectroscopy, microspectrophotometry (on intact samples) has the advantage of preserving the integrity of biological structures, and allowing reliable *in vivo* spectroscopic investigation of pigment distribution and composition (Schmidt *et al.*, 1990). We recorded absorption spectra with a visible- and infrared-light microspectrophotometer (Schmidt *et al.*, 1990). We resolved two microspectrophotometric maxima at 609 ± 2 and 619 ± 2 nm (Tao *et al.*, 1994). The peak positions correspond to the reported maxima for stentorin-1 and stentorin-2, respectively (Kim *et al.*, 1990).

The pigment granule in *Blepharisma* has been described previously (Inaba *et al.*, 1958;

Kennedy, 1965). The pigment granules are 0.5 μm in diameter and lying just below the pellicle (Giese, 1973). Two types of granules (one dense, homogeneous and spherical; the other less dense, granular and irregular in outline) were observed by Inaba *et al.* (1958), Kennedy (1965) and Giese (1973). A recent study by Matsuoka *et al.* (1993) showed that the granules surrounded by membrane were connected to a plasma membrane. The pigment granules contain a honeycomb-like structure and a stacked lamella structure. No photoacoustic or microspectrophotometric data are available to describe the localization and the spectroscopy of blepharismin.

Photochemical Mechanisms

The photochemical mechanisms of stentorin and blepharismin are unknown at the present time. Preliminary steady state photolysis of stentorin and blepharismin failed to yield spectrophotometrically observable photochemical cycles (Dai, 1994, and unpublished results). In contrast to hypericin, stentorin is a poor singlet oxygen generator under physiological conditions (Dai *et al.*, 1992). Thus the reactive molecular oxygen might play an insignificant role in the light signal transduction in *Stentor coeruleus*.

Stentorin contains several hydroxyl groups. Their pK_a values are likely to decrease upon light excitation (Walker *et al.*, 1979). 2- or 2'-Hydroxyl group is the preferred site of deprotonation of hypericin and the pK_a value decreased by 2 units upon light excitation (Falk *et al.*, 1992). Similar deprotonation of stentorin may serve as the initial photosensory transduction step, as suggested earlier (Song, 1981). The proton dissociation may be facilitated via an acid-base network of amino acid residues of the apoprotein, analogous to the charge relay network of serine proteases (Blow and Steitz, 1970). In fact, the photoresponses of *Stentor coeruleus* were shown to be dependent on pH (Walker *et al.*, 1979; Fabczak *et al.*, 1993a) and a light-induced pH decrease was observed in *Stentor coeruleus* (Walker *et al.*, 1981). Additionally, intracellular pH-modulating reagents [including ammonium chloride, protonophores carbonylcyanide *m*-chlorophenyl-hydrazone (CCCP) and carbonylcyanide *p*-(trifluoromethoxy)-phenyl-hydrazone (FCCP)] were found to reduce the photoresponsiveness (Fabczak *et al.*, 1993b).

A recent pump-probe experiment showed that stentorin-2 undergoes a rapid absorbance decrease (bleaching) in less than 3 ps; whereas free stentorin chromophore did not exhibit the ultrafast photoprocess (Savikhin *et al.*, 1993). This process is faster than the fluorescence decay occurring with a lifetime of about 10 ps (Song *et al.*, 1990). The ultrafast bleaching process, monitored at 565-630 nm, may be an intermolecular proton transfer to an appropriately situated amino acid residue in the photoreceptor molecule. Alternatively, the ultrafast process may involve electron transfer (see below).

Electron transfer may also be involved in the primary photoprocess of stentorin. It was shown that hypericin photoreduced ferric ions to ferrous ions in DMSO (Song *et al.*, 1991). On the contrary, hypericin and stentorin were found to be unreactive in the presence of electron donor, such as dithionite (Kim *et al.*, 1990). Recently the formal potentials for hypericin were measured using cyclic voltammetry (Redepenning and Tao, 1993). Two reversible one-electron reductions were observed at -0.87 V and -1.18V. Oxidation is observed at +0.90 V. These formal potentials are consistent with the absorption maximum at 599 nm (2.1 eV). Thus, hypericin is not only a good electron acceptor, but also a strong electron donor. As an analog to hypericin, stentorin is expected to be electrochemically similar. However, whether electron transfer plays a primary role in the sensory transduction is still unknown. The 3-ps absorption transient observed may result from electron transfer.

Since the photoreceptor molecule in *Blepharisma japonicum* has not been purified, and its structure has not been fully elucidated the photochemistry of blepharismin is even more difficult to describe. Since blepharismin is structurally similar to hypericin and stentorin, we expect that similar photochemistry occurs in blepharismin. An added proof is that the photophobic response in *Blepharisma japonicum* can be modulated by ammonium chloride, CCCP, and FCCP, as in *Stentor coeruleus* (Fabczak *et al.*, 1993b). Also, an intracellular pH decrease, upon light irradiation, was observed in *Blepharisma japonicum* (Matsuoka *et al.*, 1992b).

ACKNOWLEDGEMENTS

This work was supported in part by the U.S Army Research Office (28748-LS-SM and 29597-LS-EPS) and NIH (R01-NS15426). This lecture is based on the author's recent review article, and I wish to acknowledge the co-authors' contributions to the work described here (Tao *et al.*, 1994).

REFERENCES

Anjo, D.M., and Moore, T. A. (1984) A photoacoustic depth profile of β-carotene in skin. *Photochem. Photobiol. 39:* 635-640.

Arcichovskij, V. (1905) Uber das Zoopurpurin ein neues pigment der protozoa (*Blepharisma lateritium*, Erbhb.). *Arch. Protistenk.* **6**: 227-229.

Blow, D.M., and Steitz, T.A. (1970) X-ray diffraction studies of enzymes. *Annu. Rev. Biochem.* **39**: 63-100.

Cameron, D.W., and Riches, A.G. (1995) Synthesis of stentorin. *Tetrahedron Lett.*, in press.

Checcucci, G., Damato, G., Ghetti, F., and Lenci, F. (1993) Action spectra of the photophobic response of the blue and red forms of *Blepharisma japonicum*. *Photochem. Photobiol.* **57**: 686-689.

Dai, R., Song, P.-S., Anderson, J.L., Selke, M., and Foote, C.S. (1992) Hypericin and silkworm chlorophyll metabolites as anti-retroviral and anti-tumor photosensitizers, IUBMB Conference on Biochemistry and Molecular Biology of Diseases, Nagoya, Japan.

Dai, R. (1994) Further investigation of the initial photoprocess in stentorin and further characterization of the ciliate photoreceptor. *Ph.D. Dissertation*, Univ. of Nebraska, Lincoln, NE.

Ducharme D., Tessier A., Leblanc R. M. (1979) Design and characteristics of a cell for photoacoustic spectroscopy of condensed matter. *Rev. Sci. Instrum.* **50**, 1461-1462.

Emerson R. E. (1930) Some properties of the pigment of *Blepharisma*. *J. Gen. Physiol.* **13**: 159-161.

Fabczak, S., Fabczak, H., Tao, N., and Song, P.-S. (1993a) Photosensory transduction in ciliates. I. An analysis of light-induced electrical and motile responses in *Stentor coeruleus*. *Photochem. Photobiol.* *57:* 696-701.

Fabczak, H., Fabczak, S., Song, P.-S., Checcucci, G., Ghetti, F., and Lenci, F. (1993b) Photosensory transduction in ciliates. Role of intracellular pH and comparison between *Stentor coeruleus* and *Blepharisma japonicum*. *J. Photochem. Photobiol. Part B.* **21**: 47-52.

Falk, H., Meyer, J., and Oberreiter, M. (1992) Deprotonation and protonation of hydroxyphenanthroperylenes. *Monatsh. Chem.* **123**: 277-284.

Frackowiak D., Hotchandani S., Leblanc R. M. (1985) Photoacoustic spectra of phycobiliproteins and chlorophyllin in isotropic and anisotropic polyvinyl alcohol films. *Photochem. Photobiol.* **42**,559-556.

Giese, A.C. (1953) Some properties of a photodynamic pigment from *Blepharisma*. *J. Gen. Physiol.* **37**: 259-269.

Giese, A.C. (1973) *Blepharisma: The Biology of a Light-Sensitive Protozoan*, Stanford University Press, Stanford, CA; pp. 366.

Giese A.C. (1981) The photobiology of *Blepharisma*. *Photochem. Photobiol. Rev.* *6:* 139-180.

Gioffre, D., Ghetti, F., Lenci, F., Paradiso, C., Dai, R., and Song, P.-S. (1993) Isolation and characterization of the presumed photoreceptor protein of *Blepharisma japonicum*. *Photochem. Photobiol. 58:* 275-279.

Huang, B., and Pitelka, D. R. (1973) The contractile process in the ciliate, *Stentor coeruleus*. I. The role of microtubules and filaments. *J. Cell Biol. 57:* 704-728.

Inaba, F., Nakamura, R., and Yamaguchi, S. (1958) An electron-microscopic study on the pigment granules of *Blepharisma*. *Cytologia* (Tokyo) *23*: 72-79.

Jennings, H.S. (1904) Contributions to the study of the behavior of lower organisms. *Publ. Carneg. Instn. Washington* No. 16, 1-256.

Kennedy, J.R. (1965) The morphology of *Blepharisma undulans* Stein. *J. Protozool. 12:* 542-561.

Lankester, E.R. (1873) Blue stentorin, the coloring matter of *Stentor coeruleus*. *Quart. J. Microsc. Sci. 13:* 139-142.

Mast, S. O. (1906) Light reactions in lower organisms. I. *Stentor coeruleus*. *J. Expl. Zool. 3:* 359-399.

Matsuoka, T., and Taneda, K. (1992) Step-up and step-down photoresponses in *Blepharisma*. *Zool. Sci. 9:* 529-532.

Matsuoka, T., Matsuoka, S., Mamaoka Y., Kuriu, T., Watanabe, Y., Takayanagi, M., Kato, Y., and Taneda, K. (1992a) Action spectra for step-up photophobic response in *Blepharisma*. *J. Protozool. 39:* 498-503.

Matsuoka, T., Murakami, Y., Furukohri, T., Ishida, M., and Taneda, K. (1992b) Photoreceptor pigment in *Blepharisma*: H^+ release from red pigment. *Photochem. Photobiol. 56:* 399-402.

Matsuoka, T., Murakami, Y., and Kato, Y. (1993) Isolation of blepharismin-binding 200kDa protein responsible for behavior in *Blepharisma*. *Photochem. Photobiol. 57:* 1042-1047.

Meza-Keuthen, M. S. (1992) Pigment granule distribution; histological staining and possible implication for a phototactic mechanism in *Stentor coeruleus*. *M.S. Thesis*, University of Nebraska-Lincoln, NE.

Møller, K.M.(1962) On the nature of stentorin. *Compt. Rend. Trav. Lab. Carlsberg. 32:* 471-497.

Newman, E. (1974) Scanning electron microscopy of the cortex of the ciliate *Stentor coeruleus*. A view from the inside. *J. Protozool. 21:* 729-737..

O'Hara, E. P., Tom, R. D., and Moore, T.A. (1983) Determination of the *in vivo* absorption and photosynthetic properties of the lichen *Acarospora schleicheri* using photoacoustic spectroscopy. *Photochem. Photobiol. 38:* 709-715.

Pottier R., Lachaine A., Pierre M., Kennedy, J. C. (1988) A new electronic absorbance band in concentrated aqueous solutions of hematoporphyrin IX detected by photoacoustic spectroscopy. *Photochem. Photobiol.* **47**, 669-674.

Randall, J.T., and Jackson, S. (1958) Fine structure in *Stentor polymorphus*. *J. Biophys. Biochem. Cytol. 4:* 807-830.

Redepenning, J., and Tao, N. (1993) Measurement of formal potentials for hypericin in dimethylsulfoxide. *Photochem. Photobiol. 58:* 532-535.

Romanowski M., Erokhina L. G., Frackowiak D., Leblanc, R. M. (1987) Polarized spectra of allophycocyanin and phycocyanin-allophycocyanin mixture in polymer matrix. *Photosynthetica* **21**, 71-75.

Rosencwaig, A. (1980) Photoacoustics and Photoacoustics Spectroscopy. *Chem. Anal. 57:* 207-244.

Savikhin, S., Tao, N., Song, P.-S., and Struve W.S. (1993) Ultrafast pump-probe spectroscopy of the photoreceptor stentorins from the ciliate *Stentor coeruleus*. *J. Phys. Chem. 97:* 12379-12386.

Scevoli, P., Bisi, F., Colombetti, G., Lenci, F., and Passarelli, V. (1987) Photomotile responses of *Blepharisma japonicum*. I: Action spectra determination and time-resolved fluorescence of photoreceptor pigments. *J. Photochem. Photobiol. B. Biol. 1:* 75-84.

Sevenants, M. R. (1965) Pigments of *Blepharisma undulans* compared with hypericin. *J. Protozool. 12:* 240-245.

Small, E.B., and Marszalek, D.S. (1969) Scanning electron microscopy of fixed, frozen, and dried protozoa. *Science 163:* 1064-1065.

Song, P.-S., Häder, D.-P., and Poff, K.L.(1980a) Step-up photophobic response in the ciliate, *Stentor coeruleus. Arch. Microbiol. 126:* 181-186.

Song, P.-S., Häder, D.-P., and Poff K.L. (1980b) Phototactic orientation by the ciliate, *Stentor coeruleus. Photochem. Photobiol. 32:* 781-786.

Song, P.-S.(1981) Photosensory transduction in *Stentor coeruleus* and related organisms. *Biochim. Biophys. Acta 689:* 1-29.

Song, P.-S. (1983) Protozoan and related photoreceptors: Molecular aspects. *Annu. Rev. Biophys. Bioengin. 12:* 35-68.

Song, P.-S., Kim, I.-H., Florell, S., Tamai, N., Yamazaki, T., and Yamazaki, I. (1990) Structure and function of the photoreceptor stentorins in *Stentor coeruleus*. II. Primary photoprocess and picosecond time-resolved fluorescence. *Biochim. Biophys. Acta 1040:* 58-65.

Song, P.-S., Kim, I.-H., Rhee, J.S., Huh, J.W., Florell, S., Faure, B., Lee, K.W., Kahsai, T., Tamai, N., Yamazaki, T., and Yamazaki, I. (1991) Photoreception and photomovement in *Stentor coeruleus*. In: *Biophysics of Photoreceptors and Photomovements in Microorganisms*: NATO ASI Series A: 211. (F. Lenci, F. Ghetti, G. Colombetti, D.-P. Häder and P.-S. Song, Eds.). Plenum Press, New York. pp. 267-279.

Tao, N., Orlando, M., Hyon, J.-S., Gross, M., and Song, P.-S. (1993) A new photoreceptor molecule from *Stentor coeruleus. J. Am. Chem. Soc. 115:* 2526-2528.

Touloukian Y. S., Powell R. W., Ho C. Y., Nicolau M. C. (eds.) (1973) *Thermophysical Properties of Matter*, IFI/Plenum, New York.

Walker, E.B., Lee, T.Y., and Song, P.-S. (1979) Spectroscopic characterization of the *Stentor* photoreceptor. *Biochim. Biophys. Acta 586:* 129-144.

Walker, E.B., Yoon, M., and Song, P.-S. (1981) The pH dependence of photosensory response in *Stentor coeruleus* and model system. *Biochim. Biophys. Acta 634:* 289-308.

Weisz, P.B. (1949) A cytochemical and cytological study of differentiation in normal and reorganizational stages of *Stentor coeruleus*. *J. Morphol. 84:* 335-363.

Weisz, P.B. (1950) On the mitochondrial nature of the pigmented granules in *Stentor* and *Blepharisma*. *J. Morphol 86:* 177-184.

Wood, D.C. (1976) Action spectrum and electrophysiological responses correlated with the photophobic response of *Stentor coeruleus*. *Photochem. Photobiol. 24:* 261-266.

LIGHT SIGNAL TRANSDUCTION IN CILIATE *STENTOR* AND *BLEPHARISMA*. 2. TRANSDUCTION MECHANISM

Pill-Soon Song

Department of Chemistry and Institute for Cellular and Molecular Photobiology, University of Nebraska, Lincoln, NE 68588, USA

Abstract

Stentor coeruleus and *Blepharisma japonicum* are capable of detecting light (with respect to intensity and wavelength) with high efficiency. The nature of the primary photoprocess involved in the photosensory transduction is still unknown. External free Ca^{2+} ions, Ca^{2+} channel-blockers, and -ionophores modulate the light-dependent motile responses in both *Stentor coeruleus* and *Blepharisma japonicum*. Either decrease in the concentration of external free Ca^{2+} ions or addition of Ca^{2+} channel blockers, pimozide or diltiazem, significantly inhibit the photosensitivity in *Stentor* and *Blepharisma* (Fabczak *et al.*, 1994). It appears that Ca^{2+} ions play a crucial role in the light transduction chain in *Stentor* and *Blepharisma*. We propose that light irradiation of the cells generates a pH gradient which induces the Ca^{2+} ion influx and subsequently leads to the reversal of ciliary stroke. Preliminary studies suggest that the photo-signal transduction in both organisms utilizes G-protein(s) as an initial transducer and a cGMP-phosphodiesterase as the effector system, analogous to the visual system of higher animals.

Introduction

Since the photoreceptors of *Stentor coeruleus* and *Blepharisma japonicum* are both hypericin derived, it is reasonable to assume that the photochemical and signal transduction mechanisms involved in their photosensory responses are mechanistically similar. In this lecture we will describe the photo-signal transduction in *Stentor coeruleus* preferentially. Results reported in the literature suggest that both organisms share similar or common signal transduction pathways.

The Possible Signal Transduction Cascade

Initial event: Signal generation. Figure 1 shows a microelectrode recording of the membrane potentials in *Stentor coeruleus* when stimulated by light. The depolarizing action potential, followed by a prolonged "after-depolarization potential", suggests an ionic process triggered by light stimulus. The prolonged after-depolarization potential may indicate occurrence of a slow photocycle in the cell. The light-induced intracellular pH change, as a cellular signal associated with the depolarizing receptor potential (Fig. 2), will be important in interpreting the structure-function relationship of stentorin, particularly with respect to the role of a transient proton release by the photoreceptor complex and/or pigment granules. Recent studies have further suggested that a light-induced transient drop in intracellular pH leads to a photoreceptor potential, which precedes the influx of Ca^{2+} ions associated with the action potential (Fabczak *et al.*, 1993a). These changes apparently leads to the reversal of ciliary stroke, as schematically shown in Fig. 3 (Song and Poff, 1989).

We still do not understand how photoexcitation of stentorin, imbedded in the pigment granules, induces a drop in intracellular pH. The stentorin-imbedded liposome preparation generates pH gradients across model membrane as monitored by the quenching of 9-amino acridine fluorescence (Yoon, 1981; Walker *et al.*, 1981). A slow drop in intracellular pH has also been observed in *Stentor coeruleus*, although the evidence is inconclusive (Yoon, 1981). It is unknown if the stentorin-liposome reconstitution system can be studied in a manner similar to the seminal bacteriorhodopsin-liposome system that exhibits light-driven pH changes (Racker and Stockenius, 1974).

The stoichiometry of proton released to photon absorbed remains to be determined. Major

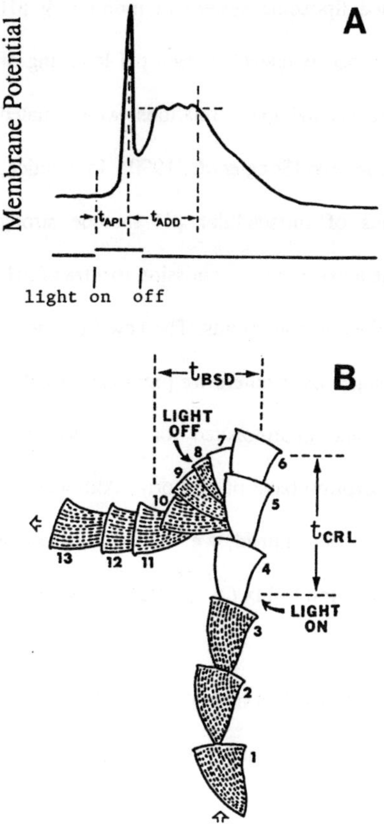

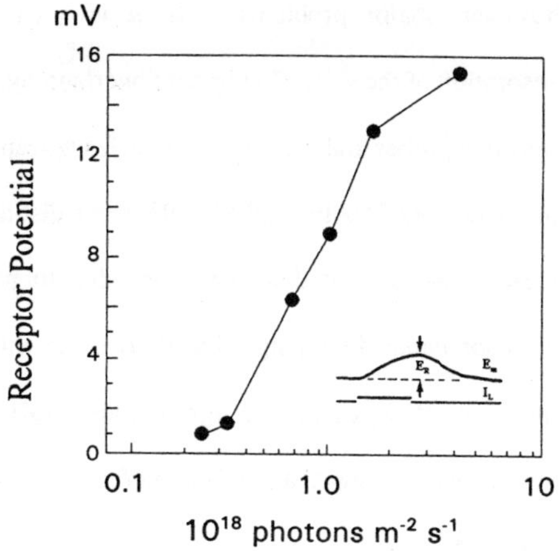

Figure 1 (left) - (A) A typical microelectrode recording of the membrane potentials in *Stentor coerulens* elicited by light stimulus, as indicated. After a delay time (t_{APL}), a spike potential (action potential) appears, which is followed by a prolonged after-depolarization potential (during t_{ADD}). (B) Time course of the photomotile response in *Stentor coerulens*. The delay time (latency) corresponds to t_{APL} in (A), whereas t_{BSD} corresponds to t_{ADD} in (A), See Fabczak et al. (1993b) for details.

Figure 2 (right) - Amplitudes of photoreceptor potentials (E_R) as a function of photon fluence rate (I_L). See Fabczak et al. (1993b) for details.

difficulties are poor reproducibility of the reconstituted liposome system in monitoring pH changes 9-amino acridine (the 9-amino acridine fluorescence is quenched by a pH lowering of inside the liposome; Yoon, 1981). With several assumptions and approximations, we estimated a theoretical maximum for the light-induced ΔpH of 0.5 *in vivo* (Song *et al.*, 1990). In practice, however, major problems with *in vivo* measurements of intracellular pH include strong absorption of the stentorin pigment overlapping with both absorption and emission spectra of pH-sensitive probes and the unknown buffering capacity of the cytosolic fluids. The new fluorescent pH indicators SNAFL and SNARF offer distinct advantages over others we previously used as these reagents emit dual emissions due to protonated and deprotonated forms. Also, their emission bands do not significantly overlap with the absorption band of stentorin. Although the quantum efficiency for the light-induced ΔpH is yet to be determined, we were able to obtain an approximate signal amplification factor in terms of Ca^{2+} ion influx ($\geq$133 Ca^{2+} ions/photon absorbed).

The possible involvement of electron transfer from the excited state of stentorin is difficult to ascertain as virtually nothing is known about the biochemical characteristics of the pigment granule and its membrane and thylakoid-like stacks within the granule (Dai *et al.*, 1995). Previous electron microscopic studies (Randall and Jackson, 1958; Newman, 1974; Tartar, 1961) described the pigment granules as true membrane vesicles of amorphous molecular contents. Although evidence was lacking, these pigment granules were also thought to be mitochondrion-like organelles, as they can be stained with Janus Green and contain labile cytochrome oxidase (Small and Marszalek, 1969). In the absence of more definitive information about the possible electron transport system, it will be difficult to establish the functional linkage between the photoreceptor molecule and the electron transport system in the pigment granule, if it exists. One possible avenue of study on this question would be to prepare the pigment granule suspension in polyethylene glycol and investigate its redox reactions induced by light of the wavelength absorbed by stentorin (action spectroscopy) and monitored by absorbance changes based on the cytochrome oxidase assay (Small and Marszalek, 1969).

We suggest the following possibilities for the initial transduction event: (1) Proton dissociation from the stentorin chromophore in the excited state, facilitated by the proton transfer network of stentorin and the as-yet unidentified associated proteins/pigment granular membrane (Song, 1981, 1983; Song *et al.*, 1990; Falk *et al.*, 1992), (2) electron transfer in the excited state, coupled with proton transfer, as suggested recently (Song *et al.*, 1991), and (3) electron transfer in the excited state, coupled with the electron transport system of pigment granule and subsequent proton release to the cytoplasm. The possible role of proton transfer/release pathways (1) and (2) in the photo-signal transduction initiated by the photoexcitation of stentorin must be assessed on the basis of the photoreactivity of the photoreceptor molecule. There is now published precedence for proton transfer (uptake) to be the possible primary photoprocess of a photoreceptor (Baca *et al.*, 1994): Proton release and uptake are known to occur during the thermal relaxation steps subsequent to the primary photoprocesses in rhodopsin, bacteriorhodopsin, and phytochrome (Mathies and Kakitani, 1992; Tokutomi *et al.*, 1982)· However, proton release from the excited state of stentorin-2 is mechanistically attractive. For example, naphthylammonium and phenylanthrylammonium ions enclosed within a crown ether exhibit a one-way proton release from their excited states with very slow reassociation of the protons (Shizuka *et al.*, 1985, 1986). We suggest that a unique topographic juxtaposition of the acid/base groups within the stentorin complex is conducive to efficient proton release during the excited state of stentorin-2.

We propose that a primary cellular signal resulting from the photo-excitation of the photoreceptors in *Stentor* and also in *Blepharisma* is a transient intracellular pH change (Song *et al.*, 1980, 1981;, Walker *et al.*, 1981; Fabczak *et al.*, 1993a). Matsuoka *et al.* (1992) proposed a similar role of proton transfer in blepharismin. A transient increase in intracellular free H^+ may activate an electrogenic pump for H^+/Ca^{2+} exchange (Song, 1981; Wood, 1991). The net Ca^{2+} inward fluxes then cause a depolarizing receptor potential (Fabczak *et al.*, 1993b,c). The photoreceptor potential eventually triggers an action potential triggering the subsequent photophobic response (Fabczak *et al.*, 1993b,c). It would be useful to perform electrophysiological experiments with *Stentor* and *Blepharisma* using voltage clamping in combination with ion substitution in extracellular media to ascertain the functional role of Ca^{2+} ions in the photomotile responses of the ciliates, in analogy to the mechanosensory transduction in *Paramecium* (Eckert and Naitoh, 1972; Naitoh, 1973) and *Stentor* (Wood, 1982).

Signal amplification. Consistent with the working model proposed above, correlation between the light-induced membrane potentials and the photomotile responses suggests that light-triggered calcium ion influx (and the reversal of ciliary beat stroke) is induced by a transient proton gradient across the membrane, or by a specific H^+-dependent opening of the calcium channels in the membrane of *S. coeruleus* (Chang *et al.*, 1981; Eckert and Naitoh, 1972; Fabczak *et al.*, 1993a). We are exploring two possible hypotheses: (1) The photoexcitation of stentorin resulting in a transient drop in intracellular pH (Song *et al.*, 1990; Fabczak *et al.*, 1993a) may then be coupled to the activation of a transducin-like G-protein. (2) A meta-stable stentorin activates the G-protein, followed by a transient drop in intracellular pH for the generation of the photoreceptor potential (Fig. 2). How either one of these, or an entirely unexplored mechanism, is coupled to the generation of an action potential resulting from the influx of Ca^{2+} ions is also unknown. Also, cGMP-dependent phosphodiesterase is a possible signal amplifier in the photosensory signal cascade in *Stentor coeruleus* and *Blepharisma japonicum* (Fabczak *et al.*, 1993d).

Low $[Ca^{2+}]_o$ and Ca^{2+} channel blockers, such as diltiazem and pimozide, inhibit the photosensitivity of *Stentor* and *Blepharisma.* A23187 equilibrates concentrations of Ca^{2+} ions at both sides of the cell membrane (Pfeifer *et al.*, 1978; Pohl *et al.*, 1980; Lee and Vidaver,

1984). Normally, cytoplasmic or axonemal Ca^{2+} ion concentration in ciliated protozoa is maintained at nM level (Naitoh, 1973). The effects of Ca^{2+} and A23187 on the photomotile responses of *Stentor* and *Blepharisma* are consistent with the intracellular or intraciliary free Ca^{2+} ion concentration of around 10^{-7} M. In analogy to the ionic mechanism of mechanosensory transduction in *Paramecium*, the receptor potential elicits voltage-dependent Ca^{2+} fluxes (action potential) and leads to the ciliary beat reversal (Machemer, 1977).

Figure 3 presents a schematic representation of the photosensory transduction chain in *S. coeruleus* (Song and Poff, 1989). To study the molecular mechanism of light signal transduction mediated by photoexcitation of stentorin we carried out a series of experiments demonstrating that a G-protein-coupled phosphodiesterase is likely to be a signal amplifier in *Stentor coeruleus* (Fabczak *et al.*, 1993d). Chemical activation of G-protein(s) (for example by fluoroaluminate) specifically enhanced the photosensitivity of the cell, whereas inactivation of G-protein(s) *in vivo* suppressed the cell's photophobic response. It is emphasized here that the enhancement of the photosensitivity is over and above the native level of the photophobic response. One can reasonably argue that complete activation of the G-protein by fluoroaluminate would, in fact, be expected to block the light response, rather than enhance it and that the enhanced photoresponse by fluoroaluminate could be evidence for a separate G-protein on a regulatory pathway. We cannot rule out these possibilities, especially in view of the apparent contradiction with the inhibitory effect of cholera toxin on the photoresponse. Tentatively, our interpretation of the stimulatory effect of G-protein activation by fluoroaluminate is as follows: Unlike the efficient photoreceptor cells such as rod cells, not every photon leads to signal transduction in *Stentor*, as the photosensitivity of the ciliate is approximately $1 \sim 2$ orders of magnitude less sensitive to light than vertebrate rod cells. Thus, when the photoresponse of the ciliate cell is assayed at a light intensity well below the maximum photoresponsive intensity, an increase in light intensity results in a sharply enhanced photoresponse. Thus, activation of a G-protein by fluoroaluminate

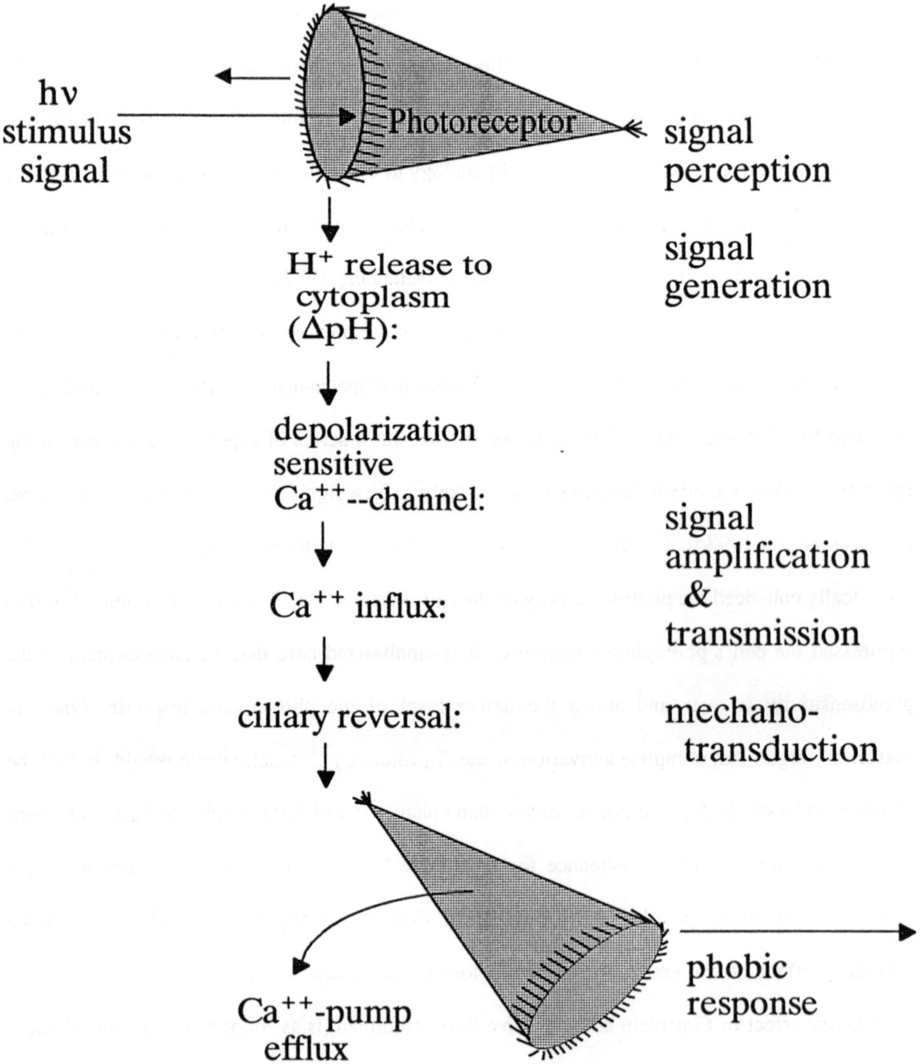

Figure 3 -. The schematic signal transduction cascade based on the membrane and ionic events in the step-up photophobic response. See Song and Poff (1989) for details.

is additive and its effect on the photoresponse is equivalent to the effect of an increase in stimulatory light intensity. It is possible that there are two types/fractions of G-proteins, one activated by light/stentorin and another by fluoroaluminate, with these two paths leading to an overall stimulatory effect of the latter. This explanation is nevertheless tenuous mechanistically. Therefore, to gain a better understanding of the phenomenon described, it will be necessary to examine the effects of external G-protein modulators on the photoresponses under varying conditions. For example, it will be desirable to study whether a higher concentration of fluoroaluminate begins to mimic the photoresponse even in the dark. Experiments with phosphodiesterase inhibitors and activators also suggest that the cytosolic cGMP level and its modulation play an important role in the photosensory transduction chain involved in both the photophobic and phototactic responses (Fabczak *et al.*, 1993d).

Future Research

A fundamental problem in photosensory biology is the unavailability of *in vitro* assay of the functional activities of photoreceptors, unlike enzymes and proteins with activities that can be readily assayed. Once the photoreceptor is separated and decoupled from the signal cascade chain, its functional activity is lost, making it difficult to assay its activity. The photoreactivity *per se* of a photoreceptor, such as the well established photocycle for rhodopsin, is not necessarily a measure of its functional activity. Thus, the discovery of transducin and the modulation of its activation by rhodopsin represented a major step in establishing a functional assay of rhodopsin *in vitro* (Stryer, 1986, 1989). It appears that the light signal cascade in the photophobic response of *Stentor coeruleus* also utilizes a G-protein (Fabczak *et al.*, 1993d). To explore the involvement of a G-protein in the photosensory transduction in the aneural ciliate cell, we are carrying out the following experiments.

(a) Cloning of 36- and 39-kDa G_α-proteins. The presence of G-proteins (36- and 39-kDa) from a membrane fraction of *Stentor* cell was confirmed by using an antiserum raised against bovine transducin's carboxyl-terminal decapeptide (KENLKDCGLF). The degree of amino acid sequence homology of G-protein α-subunits is 50% with respect to bovine transducin. They have several highly conserved motifs, particularly those regions involved with GTP-binding and hydrolysis sites (Lochrie and Simon, 1988).

One can use epitope peptides as competition controls to confirm that the immunoblot bands at 36- and 39-kDa, detected from the cellular pellicular membrane fractions, are in fact G-proteins. There is likely to be more than one G-protein. Part of the gene was isolated by using PCR based on the conserved regions of G-protein α-subunit sequences (Fabczak *et al.*, 1993d). Recently, we picked up a second G-protein by PCR (P. B. Park and P.-S. Song, unpublished results). The AS7 antibody raised against the above transducin decapeptide reacts with not only transducin-α, but also with G_i-α from rat, which contains the sequence KNNLKDCGLF. Thus, we have two G-proteins from *Stentor* membrane fractions. The second G-protein sequence is more homologous to the rat G_i (67%) than to the transducin (56%); sequence identity between stentorin G-protein-1 and -2 is 50%.

(b) Primary structure and cloning of stentorin gene. To facilitate deduction of the primary structure of stentorin, it is necessary to clone the stentorin-2B gene for structure-function studies of stentorin. We have been able to microsequence three peptide fragments of the N-blocked stentorin-2B (Dai, 1994; Dai *et al.*, 1995). After stentorin 2B was digested with trypsin, three main fractions were collected and further subjected for sequencing. The resulting sequences are listed as follows:

(1) N P F T A E L V E T A; (2) G I L A A D E S T G T I G; (3) I L E N S H G L A R

Sequences (1) and (2) were analyzed for homology searches. These sequences were found to have high homology with membrane proteins in the photoreceptor cells (Montel and Rubin, 1988; Reilander *et al.*, 1992). Sequences (1) and (2) showed 40-55% sequence identify and 50-75% sequence similarity with the photoreceptor cell proteins. The above sequences were obtainable only after proteolysis. Furthermore, the 50-kDa band contained the covalently linked stentorin chromophore. Thus, if the above homology is further confirmed, stentorin 2B would turn out to be an extremely interesting protein, both structurally and functionally.

Interestingly, both drosophila and bovine rod photoreceptor cell proteins (Montel and Rubin, 1988; Reilander *et al.*, 1992) contain unusually high contents of Asp and Glu residues; in the latter protein, 26 acidic amino acid residues cluster together near the C-terminus: DSEDEEEEDEEEEDEEEEEEEEEEEEEEENEQPLSLEWPET. Stentorin-2B is a strongly acidic protein with a pI ≤ 3.0. Perhaps, stentorin 2B contains myosin- and kinase-like domains in its primary structure. This remains to be seen.

(c) G-Protein:stentorin interactions. Using a GTP photoaffinity label, such as azido-anilido GTP, it should be possible to demonstrate a functional linkage for light/stentorin-induced GTP binding to the 36- and 39-kDa G-proteins. As a preliminary test, we are currently using bovine transducin as a model for the *Stentor* "transducin" for the study of the photoactivity of stentorin *in vitro*.

ACKNOWLEDGEMENTS

The work described in this lecture was in part supported by the U.S. Army Research Office (28748-LS-SM). The lecturer wishes to acknowledge the contributions of Drs. S. and H. Fabczak to the work reviewed in this chapter.

References

Baca, M., Borgstahl, G.E.O., Boissinot, M., Burke, P.M., Williams, D.R., Slater, K.A., and Getzoff, E.D. (1994) Complete chemical structure of photoactive yellow protein: Novel thioester-linked 4-hydroxycinnamyl chromophore and photocycle chemistry. *Biochemistry 33:* 14369-14377.

Chang, T.-H., Walker, E. B., and Song P.-S. (1981) Effects of ionophores and ruthenium red on the phototaxis of *Stentor coeruleus* as measured by simple devices. *Photochem. Photobiol. 33*: 933-936.

Dai, R. (1994) Ph.D. Dissertation, University of Nebraska, Lincoln, NE.

Dai, R., Yamazaki, T., Yamazaki, I., and Song, P.-S. (1995) Initial spectroscopic characterization of the ciliate photoreceptor stentorin. *Biochim. Biophys. Acta*, submitted.

Eckert, R., and Naitoh, Y. (1972) Bioelectric control of locomotion in the ciliates. *J. Protozool. 19*: 237-243.

Fabczak, H., Fabczak, S., Song P.-S., Checcucci, G., Ghetti, F. and Lenci, F. (1993a) Photosensory transduction in ciliates. Role of intracellular pH and comparison between *Stentor coeruleus & Blepharisma japonicum. J Photochem. Photobiol. Part B. 21*: 47-52.

Fabczak, S., Fabczak, H., Tao N., and Song P.-S. (1993b) Photosensory transduction in ciliates. I. An analysis of light-induced electrical and motile responses in *Stentor coeruleus. Photochem. Photobiol. 57*: 696-701.

Fabczak, S., Fabczak, H., and Song, P.-S. (1993c) Photosensory transduction in ciliates. III. The temporal relation between membrane potentials and photomotile responses in *Blepharisma japonicum. Photochem. Photobiol. 57*: 872-876.

Fabczak, H., Park, P.B., Fabczak, S., and Song, P.-S. (1993d) Photosensory transduction in ciliates. II. Possible role of G-protein and cGMP in *Stentor coeruleus. Photochem. Photobiol. 57:* 702-706.

Falk, H., Meyer, J., and Oberreiter, M. (1992) Deprotonation and protonation of hydroxyphenanthroperylenes. *Montsch. Chem. 123:* 277-284.

Lee, J.W., and Vidaver, G. A. (1984) Transport and control of Ca^{2+} by pigeon erythrocytes. I. Survey of some cell responses to a range of A23187 doses in the presence of Ca^{2+}. *Cell Calcium 5:* 501-524.

Lochrie, M.A., and Simon, M.I. (1988) G-protein multiplicity in eukaryotic signal transduction systems, *Biochemistry 27:* 4957-4965.

Machemer, H. (1976) Interaction of membrane potential and cations in regulation of ciliary activity in *Paramecium*. *J. Exp. Biol. 65*: 427-448.

Mathies, R., and Kakitani, T. (Eds.) (1992) Biophysical chemistry of retinal proteins. *Photochem. Photobiol. 56:* 857-1187.

Matsuoka, T., Murakami, Y., Furukohri, T., Ishida, M., and Taneda, K. (1992) Photoreceptor pigment in *Blepharisma*: H^+ release from red pigment. *Photochem. Photobiol. 56:* 399-402.

Montell, C., and Rubin, G.M. (1988) The drosophila ninaC locus encodes two photoreceptor cell specific proteins with domains homologous to protein kinases and the myosin heavy chain head. *Cell, 52*: 757-772.

Naitoh, Y. (1973) Control of ciliary activities by adenosinetriphosphate and divalent cations in Triton-extracted models of *Paramecium caudatum*. *J. Exp. Biol. 58*: 657-676.

Newman, E. (1974) Scanning electron microscopy of the cortex of the ciliate *Stentor coeruleus*. A view from the inside. *J. Protozool. 21:* 729-737.

Pfeifer, D.R., and Lardy, H.A. (1976) Ionophore A23187: The effect of H^+ concentration on complex formation with divalent and monovalent cations and the demonstration of K^+ transport in mitochondria mediated by A23187. *Biochemistry 15*: 935-943.

Pohl, W. G., Kreikenbohm, R., and Seuwen K. (1980) The specificity of ionophore A23187 in cation transport across lipid membranes. Studies with lecithin vesicles. *Z. Naturforsch. C. Biosci. 35*: 526-568.

Racker, E., and Stoeckenius, W. (1974) Reconstitution of purple membrane vesicles catalyzing light-driven proton uptake and adenosine triphosphate formation. *J. Biol. Chem. 249:* 662-663.

Randall, J.T., and Jackson, S. (1958) Fine structure and function in *Stentor polymorphus. J. Biophys. Biochem. Cytol. 4:* 807-829.

Reilander, R., Achilles, A., Friedel, U., Maul, G., Lottspeich, F., and Cook, N.J. (1992) Primary structure and functional expression of the Na/Ca,K-exchanger from bovine rod photoreceptors. *EMBO J., 11*: 1689-1695.

Shizuka, H., Kameta, K., and Shinozaki, T. (1985) Excited-state proton-transfer reactions of Naphthylammonium ion—18-crown-6 complexes. *J. Am. Chem. Soc. 107*: 3956-3963.

Shizuka, H., and Serizawa, M. (1986) Dynamic behavior in the excited state of phenanthrylammonium ions—18-crown-6 complexes: A one-way proton-transfer reaction. *J. Phys. Chem. 90:* 4573-4578.

Small, E.D., and Marszalek, D.S. (1969) Scanning electron microscopy of fixed, frozen, and dried protozoa. *Science 163:* 1064-1065.

Song, P.-S. (1981) Photosensory transduction in *Stentor coeruleus* and related organisms. *Biochim. Biophys. Acta 639:* 1-29.

Song, P.-S. (1983) Protozoan and related photoreceptors: Molecular aspects. *Annu. Rev. Biophys. Bioengin. 12*: 35-68.

Song, P.-S., Kim, I.-H., Florell, S., Tamai, N., Yamazaki, T., and Yamazaki, I. (1990) Structure and function of the photoreceptor stentorins in *Stentor coeruleus*. II. Primary photoprocess and picosecond time-resolved fluorescence. *Biochim. Biophys. Acta 1040*: 58-65.

Song, P.-S., Kim, I.-H., Rhee, J.S., Huh,, J.W., Florell, S., Faure, B., Lee, K.W., and Kahsai, M. (1991) Photoreception and photomovements in *Stentor coeruleus*, In: *Biophysics of Photoreceptors and Photomovements in Microorganisms:* NATO ASI Series A:211. (F. Lenci, F. Ghetti, G. Colombetti, D.-P. Häder and P.-S. Song, Eds.), Plenum Press, New York, pp. 267-279.

Song, P.-S., Häder, D.-P., and Poff, K.L. (1980a) Step-up photophobic response in the ciliate, *Stentor coeruleus*, *Arch. Microbiol. 126*: 81-186.

Song, P.-S., Häder, D.-P., and Poff, K.L. (1980b) Phototactic orientation by the ciliate, *Stentor coeruleus*, *Photochem. Photobiol. 32:* 781-786.

Song, P.-S., and Poff, K.L. (1989) Photomovement, In: *The Science of Photobiology* (Edited by K.C. Smith), Plenum, New York, pp. 305-346.

Stryer, L. (1986) Cyclic GMP cascade of vision. *Annu. Rev. Neurosci. 9:* 87-119.

Stryer, L. (1989) Molecular mechanism of visual transduction, In: *Membrane Proteins: Targeting and Transduction*: R. A. Welch Foundation Conf. on Chem. Research, Vol. 33, Houston, TX, pp. 291-300.

Tartar, V. (1961) *The Biology of Stentor*, Pergamon Press, New York, pp. 250-251.

Tokutomi, S., Yamamoto, K.T., Miyoshi, Y., and Furuya, F. (1982) Photoreversible changes in pH of pea phytochrome solutions. *Photochem. Photobiol. 35:* 431-433.

Walker, E.B., Yoon, M., and Song, P.-S. Song (1981) The pH dependence of photosensory response in *Stentor coeruleus*. *Biochim. Biophys. Acta 634:* 289-308.

Wood, D.C. (1982) Membrane permeabilities determining resting, action and mechanoreceptor potentials in *Stentor coeruleus*. *J. Comp. Physiol. 146:* 537-550.

Wood, D.C. (1991) Electrophysiology and photomovement of *Stentor*. In: *Biophysics of Photoreceptors and Photomovements in Microorganisms* (Edited by F. Lenci, F. Ghetti, D.-P. Hader and P.-S. Song), Plenum Press, New York, pp. 281-291.

Yoon, M. (1981) Photosensory transduction in *Stentor coeruleus*: Liposome model systems. *Ph.D. Dissertation*, Texas Tech University, Lubbock, TX.

A MODEL SYSTEM FOR PHOTOSENSORY PERCEPTION IN PROTOZOA: THE MARINE CILIATE *FABREA SALINA*

Roberto Marangoni, Spartaco Puntoni and Giuliano Colombetti
Istituto di Biofisica del C.N.R., Via S. Lorenzo 26, 56127 Pisa, Italy.

Light sensing is a biological feature common to a very wide variety of organisms; it has appeared very soon in the evolution, so that bacteria, algae, protozoa and almost all the higher organisms are able to process light signals, in order to extract information about their surrounding world. This capability has grown greatly during the evolution, leading to the selection of very complex, highly specialized visual systems.

In microorganisms, the sensory perception of the light is limited to the detection of light presence/absence, of temporal changes in light intensity and, in some cases, of light direction. In some bacteria and algae (such as *Halobacterium halobium, Euglena viridis, Chlamydomonas reihnardtii, Haematococcus pluvialis* etc...) the pigment identity is reasonably well known and there are indications on the light transduction chain[1].

In protozoa, instead, and in ciliates in particular, work is still at its initial stage also with respect to pigment identification. With respect to this, photoresponsive ciliates may be divided into two major classes:

1) pigmented ciliates, such as *Blepharisma japonicum* and *Stentor coeruleus*, where a large amount of a hypericin-like photosensitizing pigment is present in the cytoplasm;
2) colorless ciliates, such as *Paramecium bursaria*, where no cytoplasmic pigments are detectable.

To the first class, it is generally accepted that the hypericin-like pigment is the photoreceptor molecule[2,3]; in a colorless ciliate, instead, a rhodopsin-like molecule has been identified[4]. At present, the discussion on the identity of the photoreceptor molecule is still more complicated because of some preliminary evidences showing the possible presence of a rhodopsin gene also in *Blepharisma japonicum*[5].

The marine ciliate *Fabrea salina* represents a very interesting model system, because there is evidence suggesting the contemporary presence, in these cells, of a rhodopsin-like molecule, and of granules containing a hypericin-like pigment.

In the following we will briefly describe and discuss the experimental results that have led us to this conclusion.

1. Notes on *Fabrea salina.*

1.1: Cultivation

Fabrea salina were grown in artificial sea water (density 1022) at a constant temperature of 22 °C in a light-dark cycle (15 h light - 9 h dark). *F. salina* were fed the green marine alga *Dunaliella salina*, which are grown in the growth medium suggested by Johnson[6], and on the same light-dark cycle. *D. salina* used to feed the *F. salina* were centrifuged and washed three times in the ciliate growth medium.

1.2: Morphology

Cells in good conditions observed under the optical microscope (Fig. 1) show a typical pear-shaped body, the surface of which is densely covered with cilia; the adoral zone is characterized by a deep spiral shaped invagination, on the edge of which adoral membranelles are clearly visible. Normal dimensions range from 100 up to 300 μm for the major axis, and from 50 up to 200 for the minor axis. The cell cytoplasm is apparently colorless; sometimes green food vacuoles, containing the rests of ingested *Dunaliella salina* are visible (Fig. 5).

1.3: Behavior and photomotile responses

Fabrea salina is a very active swimmer that moves along a left-handed helix, the radius of which is very small; when observed at low optical resolution, the helical tracts are well-approximable with straight-line tracts. Ii is thus possible to describe the cell motion by means of a random-walk model: a series of straight-lines tracts, interrupted by sudden directional changes. We have measured, by means of a laboratory developed automatic image analysis system, several parameters of cell motion, such as speed, frequency of directional changes and frequency distribution of the angles of directional changes. We have found that speed depends on temperature (unpublished observations), but it does not seem to be affected by light stimuli. In *Fabrea salina*, in other words, no photokinesis is detectable[7].

Fabrea salina reacts to light stimulation by showing:

1) positive phototaxis: it orients its swimming course toward the light source, when exposed to unilateral illumination;

2) step-down photophobic response: when the light intensity level decreases suddenly, *Fabrea salina* performs a complex motile reaction, the pattern of which is shown in Fig. 2.

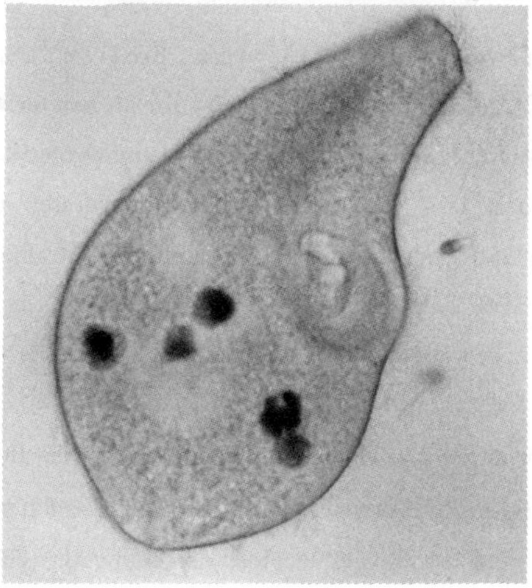

Figure 1. *Fabrea salina* observed under the otpic microscope at 200×.

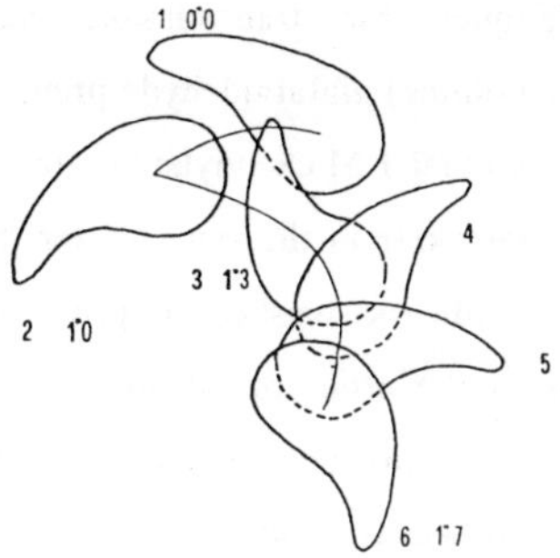

Figure 2. A step-down photophobic reaction. Immediately after the light turning off the cell stops for about one sec. (position 2), and then it starts to swim backward, performing many tumblings. This reaction takes about two sec.; when finished the cell starts again its normal swimming.

1.4: The pigment problem

Since *Fabrea salina* reacts to light, it must have some photoreceptor pigment. In the old literature[8,9] *F. salina* has been described as red-pigmented, because of the presence of a molecule, originally called *fabrein*, probably similar to *blepharismin* and *stentorin*. More recently, Møller[10] describes *F. salina* cells as filled by a dark pigment, probably protecting the cells from UV-B irradiations. Our strain, and all the other strains we are aware of presently in other laboratories, are, instead, apparently colorless. A small quantity of a pigment is however detectable by means of microscopic fluorescence techniques. We cannot actually establish whether this pigment is the photoreceptor pigment: in the following section we will show the evidence indicating that *F. salina* possesses an endogenous pigment, which is very similar (with respect to his localization and fluorescence spectra) to those of *B. japonicum* and *S. coeruleus*.

2. The "hypericin-like" hypothesis

2.1 Electronic microscopy.

We have performed ultrastructural studies of *Fabrea salina* under T.E.M.: in the following we briefly describe the techniques and the main results obtained from these studies (a full paper is in preparation).

Samples for transmission electron microscopy (T.E.M.) were fixed in 2% (volume/volume) glutaraldehyde prepared in 0.1 M cacodylate buffer, pH 7.4 for 30 min. at 4 °C, washed in 0.1 M cacodylate buffer with 7% sucrose, and post-fixed in 1% (weight/volume) osmium tetroxide in the same buffer, for 40 min at 4 °C. The cells were then dehydrated in ethanol gradients, washed in propylene oxide and embedded in Epon araldite. Ultrathin sections of the order of 40 nm were obtained with a Ultratome LKB ultramicrotome, and stained with uranyl acetate and lead citrate. The observations were made with a Philips CM-10 transmission electron microscope operated at 80 kV.

Figure 3 shows a picture at 60,000× taken in the adoral region: a pigment granule, the dimensions of which are about 0.35 μm, is clearly visible. We have found that these granules are uniformly diffused in the whole cell body, and are always located just below the cell membrane. These granules are very similar, as far as dimensions, shape, location and electron-density are concerned, to those found in *Blepharisma japonicum*[11]; the major difference is represented by the total amount: *B. japonicum* contains a large quantity of granules, whereas *F. salina* shows only a few granules (measurements are in progress to quantify granule density). This can explain why *B. japonicum* appears red-pigmented, whilst *F. salina* appears colorless.

2.2: Fluorescence microscopy and microspectrofluorometry

F. salina fixed cells, observed under the fluorescence microscope, show a red-orange fluorescence apparently similar to that showed by *B. japonicum*. We have performed microspectrofluorometry studies in order to obtain single-cell fluorescence spectra.

Microspectrofluorometric experiments were conducted by means of an Optical Multichannel Analyzer (OMA-EG&G) attached to a Leitz microscope. The excitation light was provided by a nitrogen pumped dye laser at 440 nm. The OMA displays real time spectra by using an intensified diode array to acquire the light dispersed by a grating spectrometer (resolution = 0.5 nm). In our experiments the light from the microscope was guided to the spectrometer using a fiber bundle. Samples for microspectrofluorimetry were prepared collecting the cells by suction and fixing them on glass slides with 20 μl of a glacial acetic acid, formalin and 1% chromic acid solution (1:8:16 v/v/v).

Figure 4 shows the fluorescence emission spectrum for an intact cell of *Fabrea salina*: two peaks are present, one at about 610 nm, the other, localized at about 656 nm. The fluorescence spectrum of *B. japonicum*, determined under the same experimental conditions[12], is very similar to that of *F. salina*.

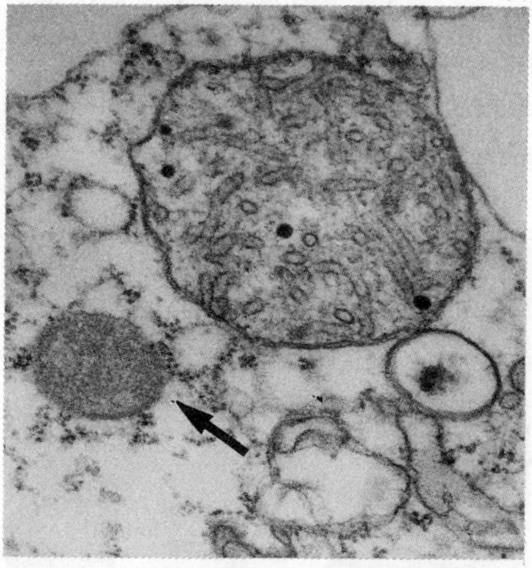

Figure 3. T.E.M. section in the adoral zone at 60,000×. The arrow indicates the pigment granule; in the upper corner is clearly visible a mitochondrion.

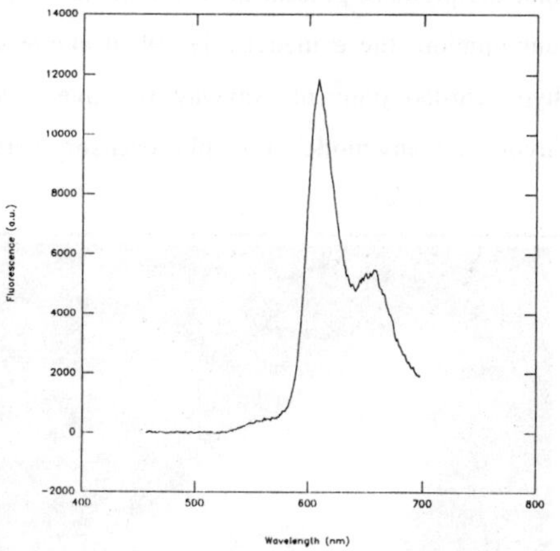

Figure 4. Single-cell emission spectrum for excitation at 440 nm. The two maxima are located at 616 and 656.

2.3: Confocal microscopy

We have also performed confocal microscopy studies in order to compare the pigment spatial distribution both in *F. salina* and in *B. japonicum*. The samples for observation are treated in the same way as for microspectrofluorimetry. The observations were made by means of a laser confocal microscopy, a Multiprobe 2001 CLSM (Molecular Dynamics), with confocal pinholes of 50 and 100 μm, equipped with an inverted microscope (Nikon) and with a graphic workstation (Silicon Graphics, Personal Iris). The image analysis has been performed by means of Imagespace, a software from Molecular Dynamics.

Figure 6 shows the CLSM image of *Fabrea salina*, taken when the scan plane crosses the cell center. It is clearly visible that the fluorescence is located at the edges of the cell body, just below the cell membrane. The same is true for *B. japonicum* observed in the same conditions.

2.4: Summary of the hypericin-like hypothesis.

As described in detail in the previous paragraphs, the autofluorescence of *F. salina* shows a spectrum very similar to that in *B. japonicum*; this fluorescence is localized in the same region, in which, in both these cells, are detectable some pigment granules. It is known that the pigment present in *B. japonicum* contains a group hypericin-like as chromophore: to our opinion, the evidences described above suggest that also the fabrein is identifiable as a hypericin-like pigment. Anyway, we have no additional evidences showing that this pigment is involved in any mode, in the photosensory perception.

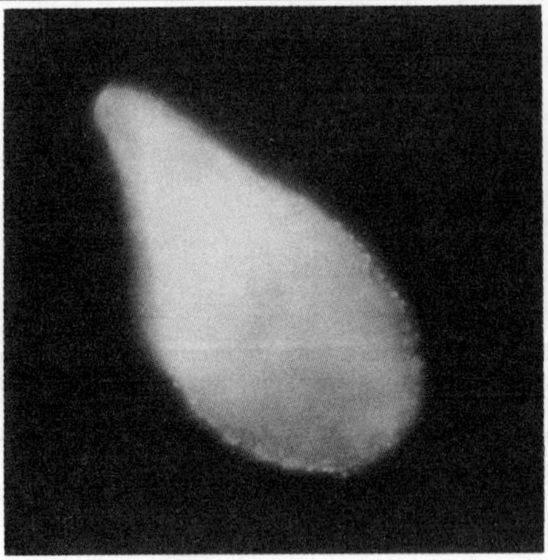

Figure 5. *Fabrea salina* observed under fluorescence optic microscope at 200×.

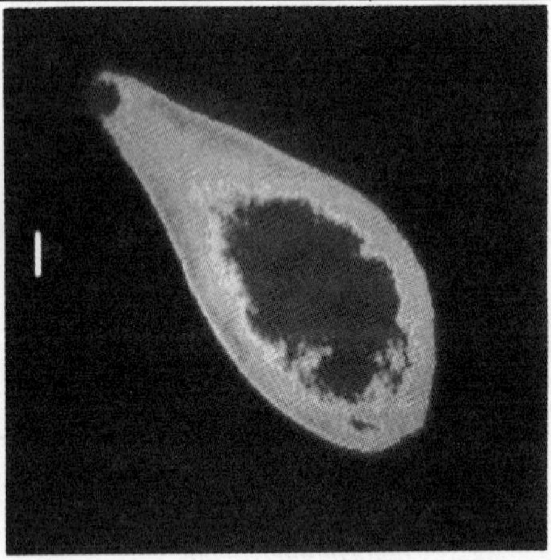

Figure 6. CLSM pseudocolor image of *Fabrea salina*. The whitest regions represent the most fluorescent ones. It is clearly visible that the fluorescence maximum is located just below the cell membrane

3. The opsin-like hypothesis

Two major evidences suggest the possible involvement of an opsin-like pigment in the photoreactions of *F. salina*: the former derives from behavioral studies, the latter from immunocytochemical assays.

3.1: The action spectrum for positive phototaxis in F. salina

In our laboratory we have developed an automatic cell tracking system which allows us to measure some motion parameters of the cells: we have employed this system to quantitatively measure the dependence of the phototactic response on the wavelength of the actinic light, i. e., to determine the action spectrum for positive phototaxis in *F. salina*.

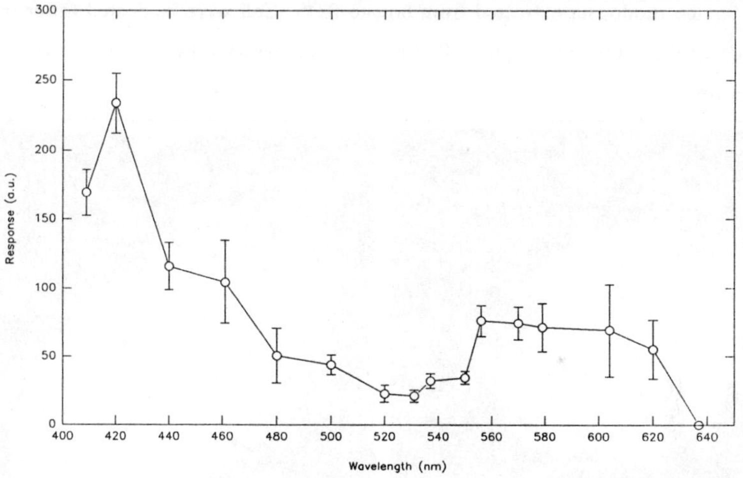

Fig. 7. The action spectrum for positive phototaxis in Fabrea salina. The abscissa represents the wavelength, and the ordinate the response, in arbitrary units. The error bar refers to the standard error of the mean.

A detailed description of the experimental apparatus can be found in [13]. Briefly, we have determined the phototactic response in the range 400 - 600 nm, at intervals of 20 nm and determined the fluence-rate curves at each wavelength, at 5 different light intensities. Every point has been repeated for at least five times; considering that each sample contained about 200 cells, we obtain that every point of the action spectrum presented has been computed on about 5,000 cells.

Figure 7 shows the resulting action spectrum for the positive phototaxis in *Fabrea salina*: this spectrum results very different from that of *Blepharisma japonicum*, and very similar, instead, to that of the photoaccumulation reaction in *Paramecium bursaria*[14]. In this ciliate, the presence of an opsin-like protein on the cell membrane has been demonstrated by immunocytochemical techniques; biochemical investigations have, moreover, revealed the presence of retinal molecules[4].

3.2: Immunocytochemical studies

Immunocytochemical studies were performed by testing the specific binding of anti-rhodopsin antibodies on the cell membrane of *Fabrea salina*. We have used a polyclonal sheep antiserum raised against bovine rhodopsin as specific marker, and non-immune sheep serum as non-specific marker[15]. Both immune and non immune serum were marked with FITC (fluoresceine isothiochianate). The rhodopsin binding capability of the specific serum has been tested on bovine rhodopsin extracted from bovine ROS. Cell were incubated (after the same fixation as described in the microspectrofluorimetric section) with the immune or with the non-immune serum.

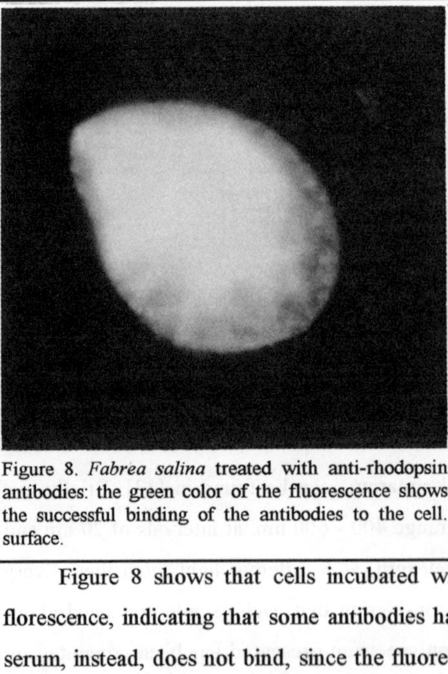

Figure 8. *Fabrea salina* treated with anti-rhodopsin antibodies: the green color of the fluorescence shows the successful binding of the antibodies to the cell. surface.

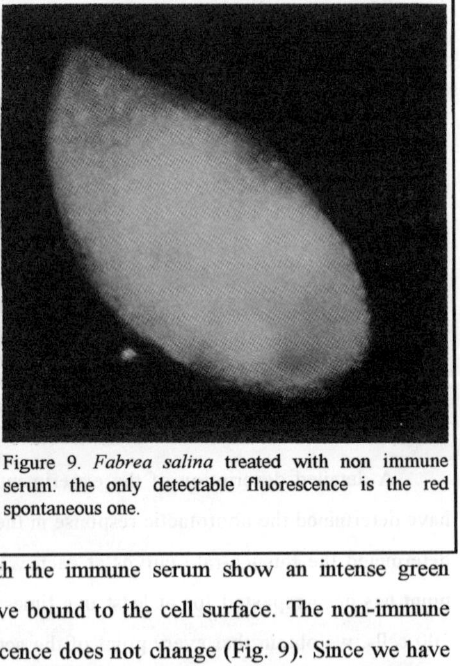

Figure 9. *Fabrea salina* treated with non immune serum: the only detectable fluorescence is the red spontaneous one.

Figure 8 shows that cells incubated with the immune serum show an intense green florescence, indicating that some antibodies have bound to the cell surface. The non-immune serum, instead, does not bind, since the fluorescence does not change (Fig. 9). Since we have operated with the serum, and not with the purified antibody, we have performed a series of tests, aimed at ruling out the possibility that the reactive antibodies contained in the immune serum were different from the anti-rhodopsin one. Moreover, since rhodopsin belongs to a gene-related protein family, including adrenergic or cholinergic receptors, the cell response to acute exposure to muscarinic and adrenergic agonist and antagonist drugs was investigated. Acetylcholine, atropine, hexamethonium, noradrenaline, or phenoxybenzamine in the culture medium (10^{-4} M) did not influence motion behavior and phototaxis of *Fabrea salina*.

These findings raise the possibility that a rhodopsin-like molecule is present at the surface of *Fabrea salina*. Nevertheless, they do not prove the identity of this surface antigen as a rhodopsin, or that this molecule is responsible for photoresponsiveness in this organism.

Conclusions

The evidence discussed above suggests that *Fabrea salina* possesses two different pigment molecules one belonging presumably to the class of hypericin-like compounds, and the other to the rhodopsin-like.

These studies do not allow us, however, to draw any definite conclusion as to the identification of the pigment responsible for the photoperception, even if the action spectrum of phototaxis is very similar to that one of *Paramecium bursaria*, where a rhodopsin is thought to be he photoreceptor pigment.

Studies are actually in progress, aimed at the identification of a rhodopsin gene in the *F. salina* genome. If this gene is present, we will try to modulate its expression and correlate it with the strength of the photomotile reactions. If this correlation could be demonstrated, it would constitute reasonable proof of the photoreceptor role of rhodopsin in *Fabrea salina*.

REFERENCES

[1] G. Kreimer (1994) Cell biology of flagellate algae. *Int. Rev Cytol.,* 148: 229-311.

[2] H. Fabczack, N. Tao, S. Fabczack, P.S. Song (1993) Photosensory transduction in ciliates. IV. Modulation of the photomovement response of *Blepharisma japonicum* by c-GMP. *Photochem. Photobiol.,* 57:889-892.

[3] F. Ghetti, G. Checcucci, F. Lenci (1992) Photosensitized reactions as primary molecular events in photomovement of microorganisms. *J. Photochem. Photobiol. B,* 15:185-198.

[4] Y. Nakaoka, R. Tokioka, T. Shinozawa, J. Fujita, J. Usukura (1991) Photoreception of *Paramecium* cilia: localization of photosensitivity and binding with anti-frog-rhodopsin IgG, *J. Cell Sci.,* 99:67-72.

[5] M. Salvini, T. Dalle Lucche, L. Barsanti, V. Passarelli, P. Gualtieri (1995) Identification of rhodopsin genes in microorganisms and higher plants. *Medecine, Biologie, Environnement* (in press).

[6] M.K. Johnson, E.J. Johnson, R.D. MacElroy, H.L. Speer, B.S. Bruff. (1968) Effects of salts on the halophilic alga *Dunaliella viridis. J. Bacteriol.,* 95:1461-1468.

[7] G. Colombetti, R. Marangoni, H. Machemer (1992) Phototaxis in *Fabrea salina. Medecine Biologie Environnement,* 20:93-100.

[8] E. Fauré-Fremiet (1911) La structure intime de *Fabrea salina. Comptes Rend. Acad. Sci.,* 71:419-420.

[9] J.M. Ellis (1937) The morphology, division and conjugation of the salt marsh ciliate, *Fabrea salina* Henneguy. *Univ. Calif. Publ. Zool.*, 46:343-388.
[10] K.M. Møller (1962) On the nature of stentorin. *Compt. Rend. Lab. Carlsberg*, 32:471-498.
[11] A.C. Giese (1973) *Blepharisma*. Stanford Univ. Press.
[12] P. Taroni, R. Ramponi, F. Lenci, F. Ghetti, R. Cubeddu (1991) Microspectrofluorimetry of *Blepharisma japonicum* red and albino strain: time-gated spectra and decay times in single cells. *Medecine Biologie Environnement*, 19:61-64.
[13] R. Marangoni, S. Puntoni, L. Favati, G. Colombetti (1994) Phototaxis *in Fabrea salina* I: action spectrum determination. *J. Photochem. Photobiol. B (Biol.)*, 23:149-154.
[14] Y. Nakaoka, R. Tokioka (1988) Photoreceptor potential causing phototaxis in *Paramecium bursaria*, *J. Exp. Biol.*, 137:477-485.
[15] A. Podestà, R. Marangoni, C. Villani, G. Colombetti (1994) A rhodopsin-like molecule on the plasma membrane of *Fabrea salina*, *J. Eukar. Microbiol.*, (september issue).

OPTICAL TRAPPING AND MANIPULATION OF SINGLE CELLS BY LASER DIODE RADIATION

C. FREDIANI

ISTITUTO DI BIOFISICA DEL CNR , VIA S. LORENZO 26, 56127 PISA

P. VERKERK, A. FIORETTI, L.GUIDONI, E. ARIMONDO

DIPARTIMENTO DI FISICA, UNIVERSITA' DI PISA, PIAZZA TORRICELLI 2, 56126 PISA

1. ABSTRACT

Optical trapping and manipulation of motile unicellular microorganisms are described, Optical trapping is obtained by focusing a solid state laser emitting at 850 nm with a power of 100 mW. The principles of the theory of the radiation-matter interaction on which the optical trapping is based are briefly illustrated.

1. INTRODUCTION

In the last years electromagnetic radiation has been used for controlling the motion of atoms and microscopic particles and for trapping them spatially. This has been achieved because of the increasing availability of more and more intense laser sources at different wavelengths. Actually neutral atoms, as alkali and alkali-earth's, can spatially confined in the minima of a three-dimensional standing wave forming an optical lattice with temperature a few microkelvins [1], or they can be trapped at the same temperature by magneto-optical trapping [2]. The use of the same interaction with laser radiation allows to realize, in a completely different range of particles dimensions, the trapping and the optical manipulation of biological macromolecules and single cells. This optical trap, suggested and realized for the first time by Ashkin [3, 4],,has been used to trap dielectric microspheres both in air and in solution [5, 6].

Here it will be explained, in a very schematic way, how radiation-matter interaction allows to built optical traps of many different particles like neutral atoms, dielectric beads, cells. Moreover we will show a simple experiment of trapping and optical manipulation of *Dunaliella salina*, an unicellular alga, and of *Halobacterium salinarum*, a bacterium. Both these microorganisms are swimming microorganisms, living in salt environments.

The experiment has been performed by using a 100 mW power laser diode emitting at 850 nm (near infrared).

2. THEORY

In the radiation-matter interaction energy and momentum are exchanged. Thus, by using intense and higly monochromatic laser sources, mechanical effects on atoms and small particles can be obtained. The fundamental properties of this interaction can be understood by a semiclassical theory in which the electromagnetic field is treated classically while the physical quantities of matter are treated quantistically.

An electromagnetic wave, interacting with atoms or molecules induces an electric dipole on them; this dipoleis subjected, in te wave field, to the Lorentz force. Let us schematize a laser beam as a monochromatic, linearly polarized wave of frequency ω; then the Electric Field (EF) can be written as:

$$\mathrm{E}\left(\vec{r},t\right) = \mathrm{E}\left(\vec{r}\right)\cos\left[\omega t + \phi\left(\vec{r}\right)\right]$$

where $E_0(\vec{r})$ e $\phi(\vec{r})$ are the waveamplitude and the wave phase respectively; these actually are dependent of the spatial coordinates. In particular:

in the case of a plane wave $E(\vec{r},t) = E_0\cos(\omega t - \vec{k}\bullet\vec{r})$, thus the amplitude E_0 is a constant indipendent of $\vec{r}$; while in the case of a standing wave $E(\vec{r},t) = E_0\cos(\vec{k}\bullet\vec{r})\cos(\omega t)$, thus the phase $\phi(\vec{r})$ is identically null.

The dipole induced on the particle is $P(\vec{r},t) = (\chi' + \chi'')E(\vec{r},t)$, where χ' e χ'' are the susceptibilies. The induced dipole oscillates at the frequency ω and normally is not in phase with the inducing electric field; the mean force acting on the dipole consists of two contributions with different features and effects on the particle:

$$\vec{F}_{tot}(\vec{r}) = \vec{F}_{press}(\vec{r}) + \vec{F}_{dip}(\vec{r}) = P_{\pi/2}\cdot E(\vec{r})\cdot\vec{\nabla}\phi(\vec{r}) + P_0\cdot\vec{\nabla}E_0(\vec{r}),$$

where $P_{\pi/2}$ e P_0 are the components of the dipole π/2 out of phase and in phase with respect to the EF. $\vec{F}_{press}$, usually called *radiation pressure*, depends on the gradient of the EF phase and on the dipole component π/2 out of phase; thus it is zero for a standing wave. $\vec{F}_{dip}$, usually indicated as *dipolar force*, depends on the gradient of the amplitude of the EF and on the in phase component of the induced dipole; it is zero for a plane wave.

The different features of the two forces can be illustrated, describing the radiation-atom interaction quantistically as photons exchange between the electromagnetic field and the atoms.

The *radiation pressure*, is related to the process of absorption and successive spontaneous emission, fig. 1.

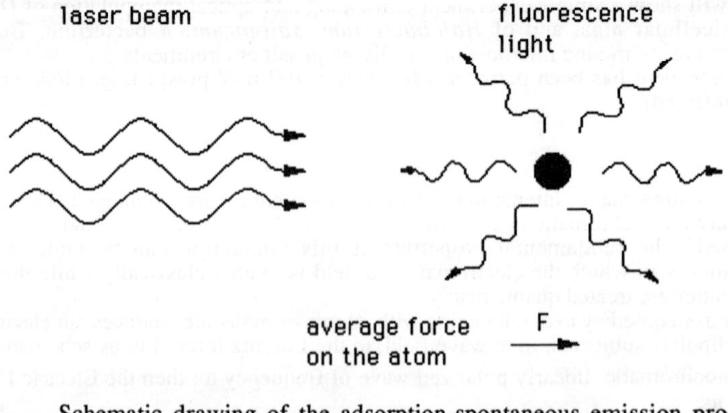

Figure 1. Schematic drawing of the adsorption-spontaneous emission process. A force along the direction of the beam is acting on the atom.

As the spontaneous emission is isotropic, the resulting exchanged momentum is not zero and the atom is affected by a force along the direction of the wave propagation, in this way the light pulls the atom. The force has a maximum in correspondence of the maxima of absorption of the atom and decrease rapidly far from these maxima.

The *dipolar force* is instead related to the absorption-stimulated emission process; if the incident light has components of different wave vector $\vec{k}$, as it occurs if the gradient of the EF amplitude is not zero, the total exchanged momentum is not zero and a force, which is directed as the gradient of the EF amplitude, is acting on the atom, fig. 2.

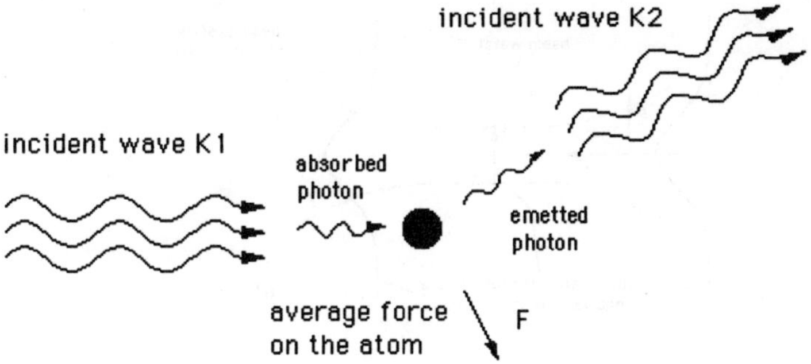

Figure 2. Schematic drawing of the absorption-stimulated emission for two incident wave components with wavevectors $\vec{k}_1$ and $\vec{k}_2$ respectively. If the photon is emitted along a direction different from the absorbed photon, an average force directed along $\Delta\vec{k} = \vec{k}_1 - \vec{k}_2$ is acting on the atom.

The *dipolar force* results from a process of coherent redistribution of photons among the different components of the lightwave.

As a consequence, in the presence of an inhomogeneous electric field, the force is directed along the maximum of the light intensity if the frequency of the laser light is lower than the absorbption frequency of the atom; the force is zero if the laser frequency is equal to the absorption frequency, the force is directed along the minimum of the light intensity if the laser frequency is higher than the absorption frequency of the atom.

In other words, as the energy of a particle in an electric field is $W = -\vec{P}\cdot\vec{E}$ and the induced dipole $\vec{P}$ is $< \pi/2$ out of phase with respect to the electric field when the laser frequency is lower than the resonance frequency, the energy has a minimum where the field intensity has a maximum, so the focus of the laser beam is a position of stable equilibrium for the particle; whereas, the induced dipole $\vec{P}$ being $> \pi/2$ out of phase with respect to the electric field when the laser frequency is higher than the resonance frequency, the energy has a maximum where the field intensity has a maximum, so the focus of the laser beam is a position of unstable equilibrium for the particle.

During the laser light-matter interaction, both forces are present and their contributions are added; with suitable choices of either the beam frequency or the beam spatial distribution one component is more evident than the other one.

In experiments on atomic beams, i.e.; the *radiation pressure* was used for the slowing down of the atoms and the *dipolar force* for the transverse focusing or the defocusing the atomic beam [7].

In the case of microscopic bodies, like dielectric spheres or single cells, the resulting force should be be obtained by integrating the contributions from all the atoms of the particle, but it seems easier to solve the problem by classical optics methods [8, 9]. The *radiation pressure* is related to the imaginary part of the refraction index and then to the light absorbtion while the *dipolar force* is related to the real part of the refraction index and then to the light dispersion. However the force exerted by a laser beam on a microsphere can be hardly estimated. In reference [9] the force was calculated to be a few pico-newtons for an homogenous dielectric sphere.

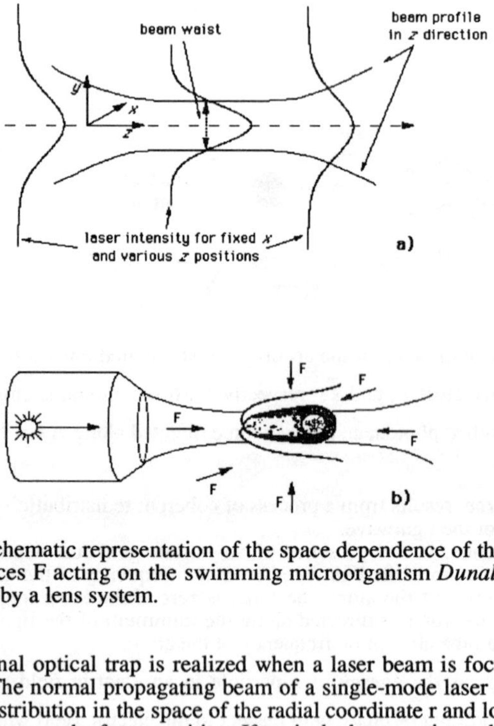

Figure 3. In a) schematic representation of the space dependence of the laser intensity at the beam waist. In b) forces F acting on the swimming microorganism *Dunaliella salina* when the laser beam is focused by a lens system.

A tridimensional optical trap is realized when a laser beam is focused to a diffraction-limited beam waist. The normal propagating beam of a single-mode laser (TEM_{00}) is described through a Gaussian distribution in the space of the radial coordinate r and longitudinal coordinate z, with z=0 the coordinate at the focus position. If w_0 is the beam waist at the focus, near z=0 and r=0 the amplitude $I(r,z)$ of the electric field is given by:

$$I(r,z) = I_0 \exp\left[-\frac{r^2}{w_0^2}\right]\frac{1}{1+z^2/b^2}$$

with I_0 the maximum intenisity and b the Rayleigh length given by $b = \pi w_0^2/\lambda$ The space dependence of the electric field is schematically represented in Fig. 3. If the laser is tightly focused, the Rayleigh length may become comparable to w_0, and the gradient forces along the radial and longitudinal directions are comparable. If the laser frequency is below the dipole transition frequency the beam waist becomes a trap and the force, present in the waist is always directed toward the beam focus, as it is shown in fig. 3b. This last fact is a little counterintuitive, reminding that the laser beam also exerts a *radiation pressure* force along the propagating direction, but in this experimental condition, laser frequency far from the absorption frequency of the particle, it results negligible.

For microspheres and cells the trapping *dipolar force* can be explained as the result of the interaction between the laser and all the dipoles (i.e. chromophores in the cell) absorbing at frequencies higher than the laser frequency or in terms of classical optics as caused by the light refraction in a medium (the particle) with a refraction index higher than that one of the surrounding medium (vacuum, air, water usually), [9].

In our experiments a laser at 850 nm was used and this wavelenght ia a good compromise between tobe not too far the absorbing peaks of the cell and to avoid the maxima of the medium (water) absorption [10, 11].

3. EXPERIMENT

The experimental apparatus for the trapping and the manipulation of cells based on laser diodes is presented in Fig. 4.

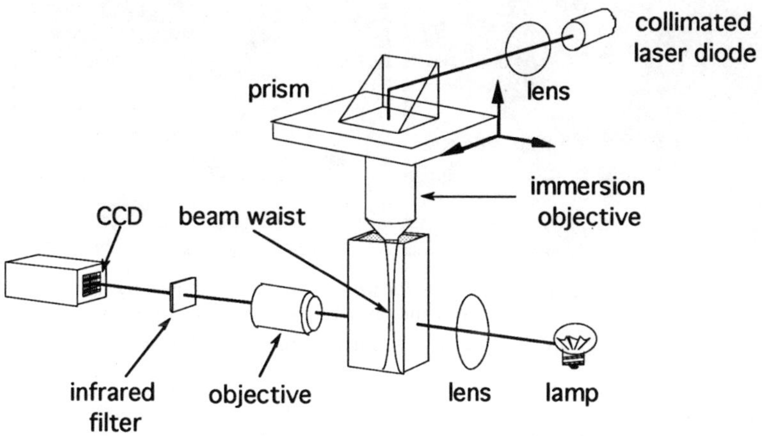

Figure 4. Scheme of the experimental apparatus for the optical trapping with laser diodes.

The radiation emitted from a diode laser (model SDL-5410-G1 Spectra Diodes Labs) was collimated through an objective (model 06-GLC-02 Melles Griot) and reshaped by a pair of anamorphic prisms. The resulting laser beam nearly circular in shape, was focused through an f= 50 cm lens and a microscope objective for water immersion with large numerical aperture (Nikon 40X, 0.65 NA) into a cuvette containing the biological specimens contained in a physiological water solution. The microscope objective was mounted on an x-y translator system, so that its image within the cuvette could be displaced laterally without modifying the focal length.

The cuvette was illuminated laterally by a 20 W quartz-iodine lamp, filtered for its long wavelength components to avoid heating of the cuvette. The image of the central part of the cuvette, near the laser focus was projected through a microscope objective of large front distance (Mitutoyo Plan Apo 20X,∞ , 0.42 NA) into a CCD video camera. In front of the telecamera a near infrared filter eliminated the large amount of laser light diffused by the optical trapped biological species. The CCD camera images were collected through a frame grabber.

The laser beam waist was determined on the basis of the propagation laws for Gaussian beams, [12], taking into account the influence of all the optical elements present along the laser beam propagation. An independent check of the laser beam diameter was performed by replacing the biological sample with water containing milk droplets. Those droplets diffusing the laser light allowed a measurement of the beam waist. It was noticed that in conditions of tight focusing even the milk droplets were optically trapped. The final estimated beam waist was 3-4 μm.

This apparatus has allowed to realize optical trapping of both *Dunaliella salina* and *Halobacterium halobium,* keeping a very low concentration of the cells in the cuvette in order to avoid the simultaneous presence of several cells in the observation region. Optical trapping has been realized with laser power around 80 mW focused on a 3-4 μm diameter region. As soon as optical trapping was realized the laser power could be reduced down to around 50 mW, maintaining the optical trap still operating.

The optical trapping was initially inhibited by the presence of a large convection motion inside the cuvette produced by the thermal heating from the quartz-iodine lamp. That convection was largely decreased using a cuvette with a reduced level of water and applying a better filtering of the infrared components in the quartz-iodine lamp.

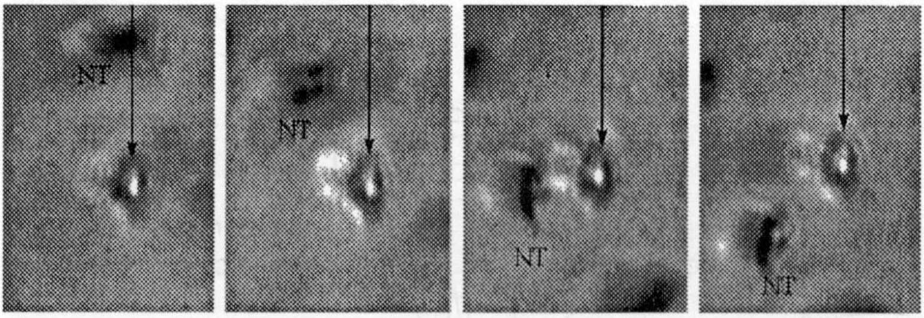

Fig.5. Frame grabber images of one *Dunaliella salina* optically trapped by the 850 nm laser, and another one noted as NT, passing near the laser focus without being stopped by the laser. The dimensions of the image are 30x50 μm^2. The images are collected at an interval of 1 sec.

Fig. 5 reports images from the frame grabber output in conditions where a sample of *Dulaniella salina* was optically trapped. The images collected at an interval of , reports the presence of several biological species inside the illuminated region. However only one biological species, present inside the beam waist, is optically trapped and scatters a large amount of laser light, while the other species move freely within the imaged region It was verified that, as soon the laser beam was blocked, the trapped species are swimming again without any sign of damaged produced by the laser trapping.

When the optical trapping was performed, the motion through micrometers of the focusing objective allowed to displace the trapped microorganisms for several tens of microns inside the observation region.

4. CONCLUSIONS

We have shown that an optical trap can be realized through the use of low power laser diode at 850 nm, without apparent damage to the trapped biological species, and also without heating of the physiological solution. A limited use of the device as optical tweezers has been also demonstrated: the main limitation to the present set-up is produced by the shift in the focus longitudinal position for a lateral movement of the optical trap.

The present experimental effort is to modify the configuration of the trapping beam and of the illumination source in order to realize a more flexible device. Thus we plan to make use of a conventional optical microscope with two laser diodes propagating from the ocular down focused the beam with the microscope objective. Moreover we plan to illuminate the sample from the bottom, by separating through a cold mirror the illumination light from the laser light. The optical relay system for the two laser diode should allow a flexible lateral movement of the diode images.

5. REFERENCES

1. P. Verkerk, B. Lounis, C. Salomon, C. Cohen-Tannoudji, J.Y. Courtois and G. Gryndberg, "Dynamics and spatial order of cold cesium atoms in a periodic optical potential", *Phys.Rev. Lett.* **68**, 3861-3863 (1992).
2. a recent review is"*Laser Manipulation of Atoms ans Ions*", E. Arimondo, W.D. Phillips and F. Strumia eds. North Holland, Amsterdam (1992).
3. A. Ashkin, "Application of laser radiation pressure", *Science*, **210**, 1081-1087 (1980)
4. A. Ashkin and J.M. Dziedzic, "Optical trapping of viruses and bacteria", *Science* **235**, 1517-1520 (1987)
5. A. Ashkin, "Acceleration and trapping by radiation pressure", *Phys. Rev. Lett.* **4**, 156-159 (1970)
6. A. Ashkin and J.M. Dziedzic, "Optical levitation by radiation pressure", Appl. Phys. Lett. **19**, 283-285 (1971).
7. T.W. Hansch and A.L. Schawlow, "Cooling of gases by laser radiation", *Opt. Commun.* **13**, 68-69 (1975)
8. H.C. Van der Hulst, "*Light scattering by small scattering*", (Wiley, New York, 1957)
9. W. H. Wright, G.J. Sonek, Y. Tadir and M. W. Berns, "Laser trapping in cell biology", *IEEE J. Quant. Electron*,. **QE26**, 2148-2157 (1990).
10. G.M. Hale and M.R. Querry, "Optical constants of water inthe 200-nm to 200-μm wavelenght region", *Appl. Opt.* **12**, 555-563 (1973)
11. J.L. Boulnois, in *Optronics Techniques in Diagnostic and Therapeutic Medicine*, pag. 29-44, R. Pratesi ed. (Plenum Press, New York, 1991).
12. H. Kogelnik and T. Li, "Laser beams and resonators", *Appl. Opt.* **5**, 1550-1567 (1966)

INTERPRETATION OF FLUENCE RESPONSE CURVES AT DIM STIMULI: THE RELEVANCE OF SHOT NOISE IN CHEMICAL PROCESSES

Donatella Petracchi*, Giovanni Cercignani®, Sabina Lucia*

Istituto di Biofisica del CNR, Via S. Lorenzo 26, 56127, Pisa (Italy)
®Dip. di Fisiologia e Biochimica, Via S. Maria 55, Università di Pisa, 56126 Pisa (Italy)

INTRODUCTION

The variability of the responses to weak stimuli is a common feature of sensory systems. It was typically observed both in the probability of perception by the human eye [1] and in the amplitude and latency of bumps in the *Limulus* [2]. The first attempt at interpreting the randomness of the responses to dim light stimuli was made in the early fourties by Hecht *et al.* [1]. Their approach was based on the consideration that at very low light intensity the stimulation itself occurs randomly, due to the quantum nature of light.

Since then, the variability intrinsic to weak stimuli has been widely used as the basic interpretative key for response variability. However, chemical signals are themselves made of particles (molecules or ions) and therefore, using the words of Borsellino and Fuortes [3], when elementary responses are considered "it is essential to take into account the stochastic fluctuations of the response in addition to the fluctuations of the stimulus". As shown in the following, we do not agree with the results of their analysis [3, 4], but we take their view of the stochastic features of transduction processes a basic point.

During the three decades following those papers, there has generally been little stress on the noise due to the chemical transmission of signals, although the presence of noise in the visual transduction pathway has been demonstrated, by a clear analysis of experimental data, by Lillywhite and Laughlin [5] and taken into account in studies on the latency of bumps in the ventral photoreceptors of *Limulus* [6]. Recently, the variability of the response to single photons in the ventral photoreceptors of *Limulus* has been measured and related to theoretical predictions in two cognate papers [7, 8]. On the other hand, people working on Vertebrate photoreceptor hold that the responses of rods to single photons are locally saturated and that the "quantal" responses in rods are surprisingly stereotyped; thus, the variability in responses to dim light flashes is thought of as basically due to the absorption of 1, 2, ... few photons [9].

Our interest in the analysis of the noise intrinsic to chemical transduction came from the available data on the dose-response curves of *Halobacterium halobium*, an archæbacterium whose swimming behaviour is influenced by light; a review on this subject has been recently issued [10]. *H. halobium* swims along an almost straight line by means of rotating flagellar bundles, making reversals in the swimming direction both spontaneously and in response to light stimuli. These bacteria are endowed with two rhodopsin-like sensory pigments, SrI and SrII, displaying different absorption spectra and signalling properties. Roughly speaking, when the stimulus is blue light, the response is an increased frequency of reversals (photophobic response), while the opposite occurs when the stimulus is red-orange light. Reversals occur through changes in the direction of rotation of flagella at the cell pole: thus, bacterial sensory rhodopsins must convey some kind of chemical signal to a switch controlling the changes in the flagellar motor rotation.

Marwan *et al.* [11] analysed the photophobic response of *H. halobium* to dim blue-light flashes. The Poisson curve for a single photon fitted well their experimental data (frequency of reversals *vs.* photon fluence); thus, they took this result as evidence that a single photon can excitate the transduction chain signalling from the sensory pigment to the flagellar motor switch.

There are, however, two technical difficulties in supporting this interpretation. The first one was pointed out also by Marwan *et al.* [11]: although the Poisson curve for a single photon has the same shape and steepness as the experimental dose-response curve, the absolute value of α (mean energy or number of photons per flash) must be shifted by a factor ranging from 10 to 20 along the x axis; *i.e.* the experimental data describe a less sensitive response than that of the theoretical curve. Marwan *et al.* [11] suggest that only 1 out of 10-20 excited photoreceptor molecules gives rise to a successful transduction event, but they do not discuss how this "decimation" occurs.

A second problem comes out from data obtained in mutants unable to synthesize retinal and was highlighted by Takahashi *et al.* [12]. In these "blind" mutants, it is possible to restore photobehaviour by introducing retinal or its analogues in the medium; it is thus possible to change the life time of the excited pigment, as tested by flash photolysis experiments. A relevant result of such experiments is that the absolute sensitivity of this sensory system depends on the life time of the excited pigment, whereas the slope of the dose-response curves does not (or varies slightly with it). Since the shape of the curve and the absolute sensitivity are closely connected in a plain Poisson model, one may wonder whether some other effects are to be considered in order to account for the experimentally found response.

An attempt in this direction has been made [12] by including an independent probabilistic process occurring in the effector process (flagellar motor switching) of the photoresponse of *H. halobium*. The details of this model have been discussed [10], but a general point at this regard is that the added probabilistic process is a specific hypothesis based on empirical evidence, not a fundamental property due to basic physical reasons.

We moved from a different start, by considering the contribution of the noise arising from chemical reactions, which is actually relevant when the number of molecules taking part in the process is low ($1 \div 10^2$). This noise is due just to the fact that molecules can be counted as discrete quantities; therefore, we call this kind of noise *chemical shot noise (CSN)* [13]. To discuss the relevance of CSN, we analyse here a simple basic scheme of enzymatic amplification. Unfortunately, very little is known about the chemical steps involved in the transduction chain(s) signalling from sensory rhodopsins to motor switches in *H. halobium*. Evidence for the identity of signalling factors has been reported only for one of them [14-17]. Nevertheless, the existence of at least one amplifying catalytic step in this bacterial system is generally accepted [16, 10], and was actually proved in higher organisms.

Here we will not extend our analysis to more complex schemes (for instance that used by Forti *et al.* [18] for the visual cascade in rods), since at this step we are interested in analysing a simple model, with only few parameters, in order to get criteria to understand the effects of CSN. What we like to discuss is how CSN is generated and what is its relevance compared to the input shot noise, in order to get a satisfactory interpretation of dose-response curves at dim stimuli.

ANALYTICAL STUDIES AND COMPUTER SIMULATIONS

A) Analytical computation

The scheme in Fig. 1 shows an accepted hypothesis on the first step in the transduction chain of archæbacterial sensory rhodopsin: any scheme proposed to account for *H. halobium* photobehaviour contains, explicitely or implicitely, a similar step catalysed by an activated pigment P^*. In this scheme, the activated pigment P^* comes back to its native state spontaneously by thermal decay, as shown by flash photolysis experiments [19, 20]. In the simulations, we will use for P^* life time the value of 80 ms which is that of SrI in its blue form. In higher organisms, the shut off of the activated pigment requires more complex mechanisms, but this is not really important for the computations carried out here. In fact, the relevant point is that the activated pigment disappears with the rate constant k_4 and its subsequent fate does not affect the results.

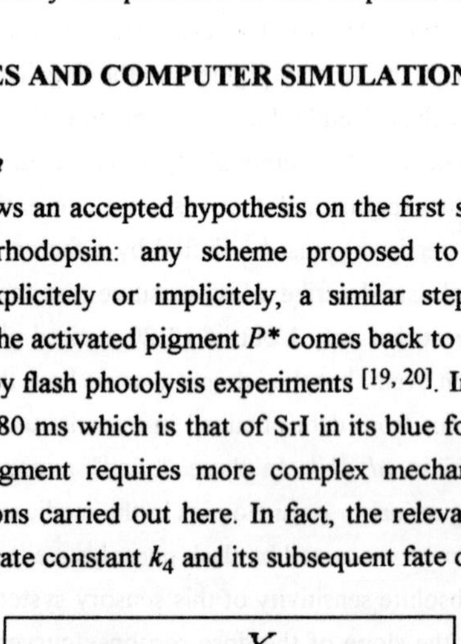

Fig. 1. *A simple kinetic scheme for phototransduction in* H. halobium. *For details see text.*

We will assume $k_2 > k_4$, so that at each time the activated pigment P^* has a greater probability to convert X into S than to decay to P. The activated pigment P^* lasts on average a time k_4^{-1}, assuming $k_3 >> k_4$; in this way, the time spent by the pigment in the catalytic cycle is negligible with respect to its decay process, and the loop acts simply as an amplifying step, without changing the decay kinetics. The last assumption is only taken for sake of simplicity, in order to reduce the number of parameters and because very often the catalytic loops are described giving their overall amplification and not their detailed rate constants (see for instance [18]). Note that k_2 is an apparent first-order rate constant, containing the concentration of *X*.

We shall now consider in detail what happens upon the absorption of a photon. It is important to realise that each chemical step is a probabilistic process regarding a single molecule. For instance, the average life time of the activated pigment is k_4^{-1}, but the actual life time of *P** is exponentially distributed according to the following equation:

$$f(t) = k_4 \cdot e^{-k_4 t} \qquad (1)$$

Thus, *f(t)dt* is the probability that *P** decays between *t* and *dt*.

In the scheme of Fig. 1, *S* is the signalling factor, or transducer, acting on some effector; *X* is the precursor of *S*. We may think that the number of *S* molecules must cross some threshold in order to produce a response; by making this assumption, we can clearly highlight the consequences of CSN on the behaviour. Initially, we will assume that *S* can accumulate indefinitely; this is by no means a realistic feature, but we shall deal later with the fate of the signal molecule. The problem is therefore to evaluate how many molecules of *S* are formed per *P** molecule. Since the activated pigment lasts a time *t*, $k_2 t$ molecules of *S* are formed on the average, k_2 being the probability per unit time that an *S* molecule is formed. The probability that an excited pigment molecule, lasting *t* seconds in the activated state, generates *n* molecules of *S* is given by:

$$P_1(n,t) = \frac{(k_2 t)^n}{n!} e^{-k_2 t} \qquad (2)$$

$P_1(n,t)$ means "probability of formation of *n* molecules within *t* after the absorption of a single photon under the condition that the activated pigment lasts a time *t*". In order to consider all the possible *t* values, we must combine eq. (2) with the probability of the actual life time *t* for the activated pigment given by eq. (1). Thus:

$$P_1(n) = \int_0^\infty \frac{(k_2 t)^n}{n!} e^{-k_2 t} k_4 e^{-k_4 t} dt = P_1(n) = q^n(1\text{-}q) \qquad (3)$$

where $q = k_2/(k_2+k_4)$. This is the probability of producing *n* molecules of the transducer *S* when a single photon is absorbed. The mean value and the standard deviation, which can be computed starting from eq. (3), are [13]:

$$\overline{n_1} = \frac{k_2}{k_4} = G \quad \text{and} \quad \sigma_1 = \sqrt{\overline{n_1}(\overline{n_1}+1)} \cong \overline{n_1} \qquad (4)$$

Note that in eq. (4) the standard deviation is greater than that expected in a Poisson process (which should be $\sqrt{n_1}$). In fact, the process which generates the signalling species *S* is a Poisson process, lasting for a time that in its turn is a random variable. This is in disagreement with the result which can be evinced from eq. (28A) of Borsellino and Fuortes [3]. In fact, from their expression a standard deviation equal to $\sqrt{n_1}$ is expected.

In the case that *m* photons are absorbed, we get the general result:

$$P_m(n) = q^n (1-q)^m \frac{(n+1)(n+2)\dots(n+m-1)}{(m-1)!} \tag{5}$$

Being the system linear and according to eq. (4), $\overline{n_m} = m \cdot \overline{n_1}$ while the standard deviation is:

$$\sigma_m = \sqrt{m} \cdot \sigma_1 \tag{6}$$

It is worth to recall that the noise we are speaking about is *completely* due to the CSN in the transduction process, since m is a fixed number of photons. A simple numerical example may help to make this point more clear: for $m = 10$ and $n_1 = 10$, $n_{10} = 100$, with standard deviation $\sigma_m = 30$. This noise is quite relevant: the variation coefficient is the same as that of the activated pigment number when 10 photons *on average* are absorbed.

The following problem was then considered: is CSN relevant when compared to the input noise? To answer this question, we carried out the computation of the standard deviation expected when α photons on average are absorbed, by combining σ_m with the Poisson law. Obviously, we have $n_\alpha = G\alpha$, while for the standard deviation we obtain:

$$\sigma = \sqrt{2G^2\alpha} \tag{7}$$

Eq. (7) tells us that, at the output of the enzymatic loop, the variation coefficient (ratio between the standard deviation and the mean) is increased by the factor $2^{1/2}$ when compared with the input, thus showing that CSN is not negligible.

We tested these conclusions by a computer simulation of scheme 1, using a finite time constant for the decay of S to make the model more realistic. The rate constants of the scheme are used as the probability densities of transition for the corresponding species (for each single molecule). The simulation is performed step by step, choosing a time interval Δt much shorter than the shortest time constant in the scheme; in this way, $k_n\Delta t \ll 1$ can be used as a probability. At each step, a random number (uniformly distributed in the range 0-1) is generated; each molecule of each chemical species can either undergo a transition or remain in its original state, according to whether the random number is in the range $k_n\Delta t$ or not.

Fig. 2 shows the raw traces of the time course of S in the case that a fixed number of photons (1, 10 or 100 photons) are absorbed by the cell: note that the noise in the output is *entirely due to chemical shot noise*. In Fig. 3 we report the average and standard deviation (dashed lines) of 100 trials obtained with the same parameters as in Fig.2 b). We may note that the standard deviation reported in Fig. 3 varies in time and also differs at two time values corresponding to identical values of the mean. At this point, the results of our simulation are at variance from those predicted by Borsellino and Fuortes [3]. Their expression (28A) implies that the standard deviation, displayed by the solid line in Fig. 3, should be the square root of the mean. We note that our result can be intuitively accepted, since the spread of different time courses *inevitably increases* with the time elapsed since their common start. Moreover, although we do not give here a general analytical treatment of the model, the results of the simulation in the case $k_6 = 0$ are in agreement with eq. (6) above.

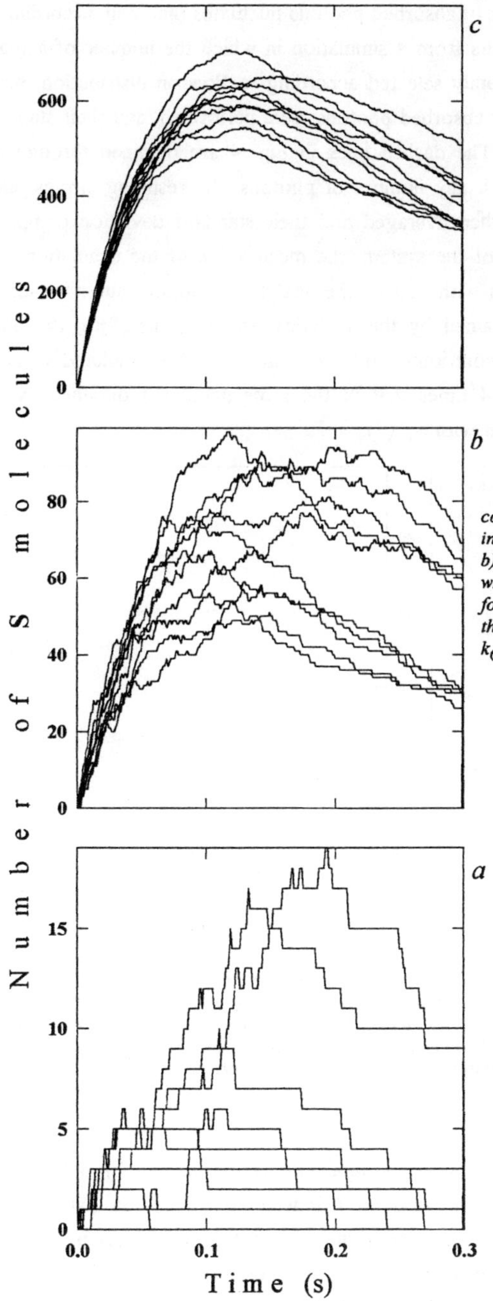

Fig. 2. *Raw traces of the simulated time courses for the production of* S *by the scheme in Fig. 1: a) when a single photon is absorbed; b) when exactly 10 photons are absorbed; c) when exactly 100 photons are absorbed. The following set of parameters was used throughout the simulations:* $k_2 = 143\ s^{-1}$, $k_4 = 14.3\ s^{-1}$; $k_6 = 3\ s^{-1}$.

In a real experiment, the actual number of absorbed photons fluctuates randomly according to Poisson statistics. Fig. 4 reports the results from a simulation in which the number of actually absorbed photons in each trial was randomly selected according to Poisson distribution; many trials (100) with an average number of 10 absorbed photons were performed, and their standard deviation is displayed by the solid line. The dashed lines in Fig. 4 are obtained through the analytical solution of the system, fed with any number of photons; the resulting curves, each multiplied by its Poisson weight, were then averaged and their standard deviation computed accordingly. Due to the linear structure of the system, the mean value of the simulation (not shown in the figure) practically coincides with that of the analytical solution, but its standard deviation is always higher than that obtained by the analytical solution; therefore, the noise intrinsic to chemical transduction actually contributes to the output noise. The simulated standard deviation at its maximum in Fig. 4 is 1.4 times that of the same parameter obtained by the analytical solution, in fairly good agreement with eq. (7).

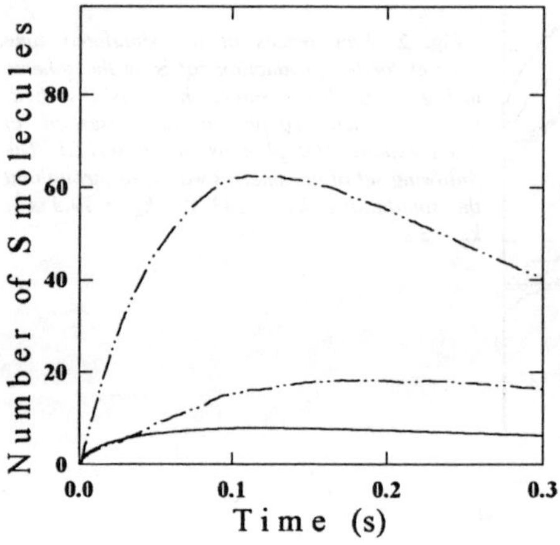

Fig. 3

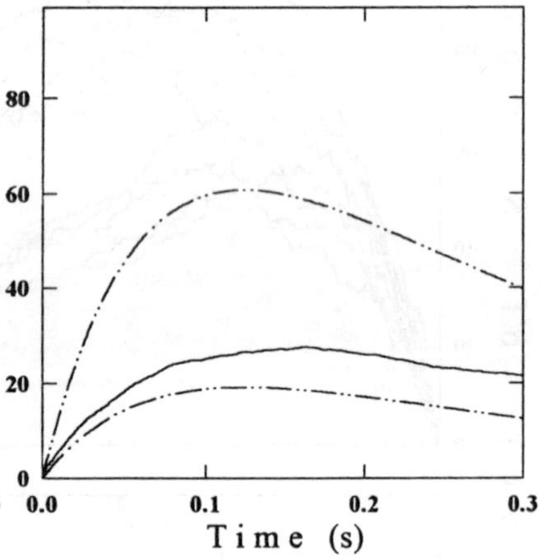

Fig. 4

Fig. 3. *The mean and standard deviation (dashed lines) of 100 raw traces obtained as in Fig. 2 b). The solid line depicts the standard deviation for the same data as expected from eq. (28A) of Borsellino and Fuortes* [3].

Fig. 4. *Time course of S molecules produced by simulation upon the absorption of 10 photons on the average. Dashed lines: mean and standard deviation calculated from the analytical solution of the scheme; solid line: standard deviation for 100 raw traces obtained by Montecarlo simulation.*

B) Interpretation of dose-response curves

Now the following point becomes clear: ***what is important is the number of S molecules produced, not the number of photons which actually do it.*** Indeed, eq. (3) and eq. (5) state that the same number of *S* molecules can be produced (with different probabilities) by different numbers of photons, as can be readily appreciated from Fig. 5. Therefore, the question about the interpretation of dose-response curves can be reshaped as follows.

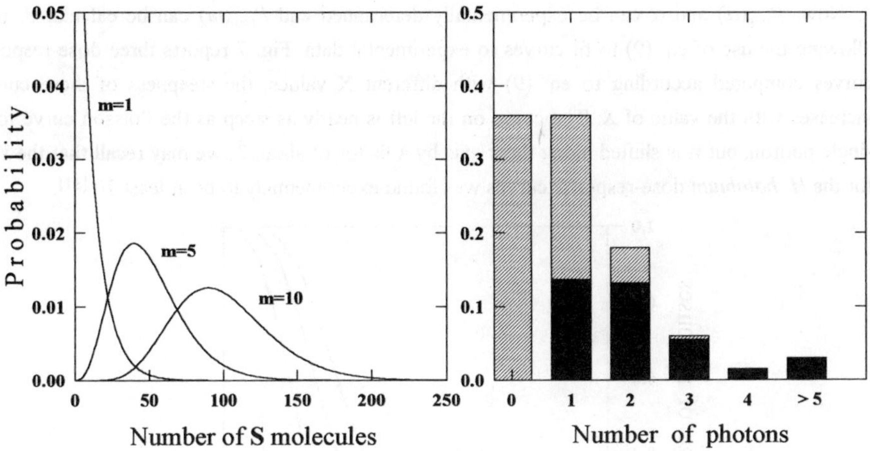

Fig. 5

Fig. 6

Fig. 5. *The probability of producing **n** molecules of S upon the absorption of **m** photons.*

Fig. 6. *The fractions of a cell population absorbing **m** photons, when the average number of photons per flash is unity, are represented by the height of each bar. The fraction of cells responding to light, with an amplifying factor of 10 and provided that S crosses a threshold of 10 molecules, are calculated from eq. (7) and represented by the black area in each bar.*

When a dim light flash is delivered on a cell population, different numbers of photons will be absorbed by each cell. Let us assume that a sharp threshold criterion holds for the response to occur. Upon the absorption of a fixed number of photons (m), the response arises when the number (n) of S molecules crosses a threshold value X. Then, the probability of response (P_{resp}) is given by:

$$P_{resp}(m) = \sum_{n=\mathrm{X}}^{\infty} q^n (1-q)^m \frac{(n+1)(n+2)\cdots(n+m-1)}{(m-1)!} \tag{8}$$

In the bar diagram of Fig. 6, each bar represents a fraction of the cell population (total = 1) according to the number of photons m actually absorbed (the number below each bar). The amplification factor and α values are 10 and 1, respectively, and X = 10. The probability of response calculated from eq. (8) is used to evaluate the percentage of cells that do respond to the light flash (black area in each bar of Fig. 6). It is altogether evident that eq. (8) cannot be used directly to analyse experimental data, when the actual number of absorbed photons by any single cell is unknown. In order to adopt α as a measure of stimulus intensity, eq. (8) has to be combined with the Poisson equation to give:

$$P_{resp}(\alpha) = \sum_{m=1}^{\infty} P_{resp}(m) \frac{\alpha^m}{m!} e^{-\alpha} \tag{9}$$

Now, $P_{resp}(\alpha)$ and α can be experimentally determined and $P_{resp}(m)$ can be calculated, thus allowing the use of eq. (9) to fit curves to experimental data. Fig. 7 reports three dose-response curves computed according to eq. (9) with different X values; the steepness of these curves increases with the value of X. The curve on the left is nearly as steep as the Poisson curve for a single photon, but it is shifted along the *x* axis by a factor of about 3; we may recall that the shift for the *H. halobium* dose-response curves was found experimentally to be at least 10 [11].

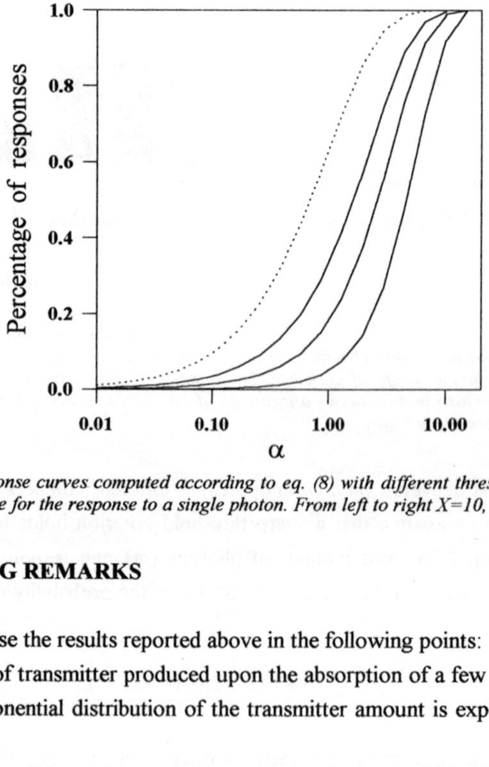

Fig. 7. *Dose-response curves computed according to eq. (8) with different threshold values X. The dashed line is a Poisson curve for the response to a single photon. From left to right X=10, 20,30 with G=10.*

CONCLUDING REMARKS

Let us summarise the results reported above in the following points:

1) the amount of transmitter produced upon the absorption of a few photons is variable and, in particular, an exponential distribution of the transmitter amount is expected upon the absorption of a single photon;

2) the amplitude of the response to a few photons is also variable; for the shape the variability depends on the choice of the system parameters;

3) appreciable noise arises from the fluctuations due to a single-step enzyme amplification and this shot noise is comparable to the input noise, the output variation coefficient increasing by a factor of with respect to the input noise;

4) CSN can partially account for the discrepancy between experimental dose-response curves and the Poisson single-photon curve, because the variability intrinsic to chemical amplification can generate a lower absolute sensitivity, but not so low as to yield a 10- to 20-fold shift along the α axis.

REFERENCES

1. S. Hecht, S. Shlaer and M. H. Pirenne (1942) Energy, quanta and vision, *J. Gen. Physiol.*, *25*, 819-840.
2. G. Fuortes and S. Yeandle (1964) Probability of occurrence of discrete potential waves in the eye of *Limulus*, *J. Gen. Physiol.*, *47*, 443-463.
3. A. Borsellino, and M. G. F. Fuortes (1968) Responses to single photons in visual cells of *Limulus*, *J. Physiol.*, *196*, 507-539.
4. A. Borsellino and M. G. F. Fuortes (1968) Interpretation of responses of visual cells of *Limulus*, *Proc. of the IEEE*, *56*, 1024-1032.
5. P. G. Lillywhite, and S. B. Laughlin (1979) Transducer noise in a photoreceptor, *Nature*, *277*, 569-572.
6. Lederhofer, J. Schnakenberg, H. Stieve (1991) Stochastic treatment of bump latency and temporal overlapping in *Limulus* ventral photoreceptors, *Z. Naturforsch.*, *46c*, 291-304.
7. A. Kirkwood, and J. E. Lisman (1994) Determinants of single photon response variability, *J. Gen. Physiol.*, *103*, 679-690.
8. M. A. Goldring, and J. E. Lisman (1994) Multi-step rhodopsin inactivation schemes can account for the size variability of single photon responses in *Limulus* ventral receptors, *J. Gen. Physiol.*, *103*, 691-726.
9. G. Ratto (1994) Light responsivity in mammalian photoreceptors: dependency on gain and capture efficiency, International School *A. Borsellino and W. Reichardt* "Biophysics of photoreception: molecular and phototransductive events", this volume.
10. D. Petracchi, S. Lucia and G. Cercignani (1994) Photobehaviour of *Halobacterium halobium*: proposed models for signal transduction and motor switching, *J. Photochem. Photobiol. B:Biol.*, *24*, 75-99.
11. W. Marwan, P. Hegemann, and D. Oesterhelt (1988) Single photon detection by an archaebacterium, *J. Mol. Biol.*, *199*, 663-664.
12. T. Takahashi, B. Yan, and J. L. Spudich (1992) Sensitivity increase in the photophobic response of *Halobacterium halobium* reconstituted with retinal analogs: a novel interpretation for the fluence-response relationship and a kinetic modeling, *Photochem. Photobiol.*, *56*, 1119-1128.
13. D. Petracchi, S. Lucia, and G. Cercignani (1994) Fundamental sources of noise in sensory transduction, *Il Nuovo Cimento D*, in press.
14. W. Marwan, W. Schäfer and D. Oesterhelt (1990) Signal transduction in *Halobacterium* depends on fumarate, *EMBO J.*, *9*, 355-362.
15. W. Marwan and D. Oesterhelt (1991) Light-induced release of the switch factor during photophobic responses of *Halobacterium halobium*, *Naturwiss.*, *78*, 127-129.
16. W. Marwan, M. Montrone and D. Oesterhelt (1992) Signal transduction in *Halobacterium halobium* mediated by the switch factor fumarate, in *Structures and Functions of Retinal Proteins*, J. L. Rigaud (ed.), Colloque INSERM, John Libbey Eurotext Ltd., *221*, 313-316.
17. M. Montrone, W. Marwan, H. Grünberg, S. Mußeleck, C. Starostzik and D. Oesterhelt (1993) Sensory rhodopsin-controlled release of the switch factor fumarate in *Halobacterium halobium*, *Mol. Microbiol.*, *10*, 1077-1085.
18. S. Forti, A. Menini, G. Rispoli, and V. Torre (1989) Kinetics of phototransduction in retinal rods of the newt *Triturus cristatus*, *J. Physiol.*, *419*, 265-295.
19. R. A. Bogomolni and J. Spudich (1987) The photochemical reactions of bacterial sensory rhodopsin-I: flash photolysis study in the one microseconds to eight seconds time window, *Biophys. J.*, *52*, 1071-1075.
20. H. Tomioka, T. Takahashi, N. Kamo and Y. Kobatake, Flash spectrophotometric identification of a fourth rhodopsin-like pigment in *Halobacterium halobium*, *Biochem. Biophys. Res. Commun.*, *139* (1986) 389-395.

PHYCOMYCES BLAKESLEEANUS: AN IDEAL ORGANISM TO STUDY PHOTOTROPISM

Chiu-Yuan Chien

Institute of Biological Sciences, National Taiwan Normal University, 88 sect. 4 Ting-chow Road, Taipei, Taiwan, R.O.C.

1. Introduction

Phototropism is caused by an unequal growth of the proximal and distal sides of an illuminated responding element. In order to obtain positive phototropism, the growth of the side distal with respect to the light source must be faster than that of the proximal one. This could be accomplished either by a growth inhibiting or by a growth promoting effect [3-4].

The sporangiophores of *Phycomyces* and *Pilobolus* [1-2,7,15] have been used to a considerable extent as relatively simple systems to evidentiate the effects of external stimuli, particularly light, both on growth and tropic responses [8-10,11,14]. In this paper, the positive phototropism of *Phycomyces* is used to illustrate the light growth response.

2. Materials and Methods

Phycomyces blakesleeanus Burgeff obtained from sheep dungs collected from the Department of Animal Husbandry, National Taiwan University in Taipei, was cultured on potato dextrose agar (PDA, peeled potato 200 g, filtered by cheesecloth after boil, dextrose 20 g, agar 16 g, DEFCO) and malt extract agar (MEA, malt extract 20 g, agar 16 g) and without the supplement of thiamine; the incubation of the plates at room temperature discouraged the growth of contaminating fungi [8].

The large sporangiophores have been favored. Petri dishes (9 cm diam., 6 cm heigh) and glass cylinders (250 ml) sterilized in autoclave and with several mycelial inocula were kept under continuous white light (fluorescent lamp, of about 100 $mW.m^{-2}$, at 6 cm heigh in level) at 24 °C. Sporangiophores induced after inoculation were placed under white light of about 10 $mW.m^{-2}$ for about 24 hr to obtain the initial sporangiophores with growing sporangia.

When the initial sporangiophores grew upward in glass cylinders, some of the cylinders were kept in darkness and some were illuminated by unilateral light; some sporangiophores of the last group of cylinders were completely immersed in liquid paraffin (refractive index $n = 1.45$). All the cylinders have been placed in a dark container with a small hole as a moist chamber (Burder method modified). To record the curvature of the sporangiophores during growth, serial photographs of them were taken at 5 min intervals by a camera (Nikon, Tokyo) with fixed holder and negative film (FX402, Kodak).

3. Results

The sporangiophores grown onto a Petri dish, the growth of which was checked up to 5.6 cm, showed a response to the incident light, bending toward the light. When grown into a glass cylinder in air, if they were in darkness they grew straight upward, while if they were illuminated unilaterally they bent just below the sporangia directing them toward the light.

However, upright sporangiophores completely immersed in liquid paraffin and illuminated unilaterally bent away from the light.

4. Discussion

Over the years, it has been demonstrated that a vertical sporangiophore subjected to unilateral light of wavelength below 520 nm exibit positive phototropism bending toward the light. The curvature results from a faster growth on the side of the elongating region remote from the light with respect to the opposite side.

On the basis of the light growth reaction (transient change in the elongation rate due to a change in fluence rate), we might rather expect that it would be the side nearer the light and presumably more brightly illuminated that would grow faster, resulting in a negative phototropism. However, the unilateral short-wavelength light incident on the sporangiophores (that act as cylindrical lens) is refracted at the cell surface and should concentrate at the distal side [6,12].

If the sporangiophores are immersed in a medium of refractive index different from that of air, the lens effect could be abolished (with n = 1,306) or even inverted [5], thus abolishing or inverting the illumination difference between the two sides of the plasma membrane, which seems to be the site of the photoreceptors [13].

If the cylindrical lens theory is correct, the sporangiophores surrounded by a non-toxic liquid of sufficiently high refractive index should exibit negative phototropism, because under these conditions the rays passing through the growing region would be divergent, not convergent, and the protoplasm of the back half would receive less light than that of the nearer side [3-4,8,14]. In fact, the sporangiophores of *Phycomyces* immersed in paraffin liquid (n = 1,45) bent away from the incident light, exibiting negative phototropism with respect to the light source and giving a positive response to the light stimulus. In such a way, the principle of light growth-response has been verified and confirmed.

Acknowledgements - Thanks are due to the Istituto Italiano per gli Studi Filosofici, Italy, for a partial financial support to attend the "International School on Biophysics of Photoreception: Molecular and Phototransductive Events", and to the National Science Council in Taipei, Taiwan, for the grant no. NSC82-0211-B003-009.

References

[1] Bergman K., P.V. Burke, E. Cerdà-Olmedo, C.N. David, M. Delbruech, K.W. Foster, E.W. Goodell, M. Heisenberg, G. Meissner, M. Zalodar, D.S. Dennison & W. Shropshire Jr (1969) Phycomyces. *Bacteriol. Rev.* 33: 99-157.

[2] Buller A.H.R. (1934) *Researches on Fungi,* Vol. 6, London: Longmans, Gree & Co.

[3] Castle E.S. (1959) Growth distribution in the light-growth response of *Phycomyces. J. Gen. Physiol.* 42: 692-702.

[4] Castle E.S. (1961) Phototropic curvature in *Phycomyces, J. Gen. Physiol.* 45: 743-756.

[5] Fukshansky L. & A.R. Steinhardt (1987) Spatial factors in *Phycomyces* phototropism: analysis of balanced responses. *J. Theor. Biol.* 129:301.

[6] Galland P. & E.D. Lipson (1987) Light physiology of *Phycomyces* sporangiophores. In: *Phycomyces* (E. Cerdà-Olmedo, E.D. Lipson, eds), New York: Cold Spring Harbor, 49-92.

[7] Griffin D.H. (1994) *Fungal physiology*, 2nd edit., New York: Wiley-Liss, 458.

[8] Ingold C.T. (1978) *The biology of Mucor and its allies.* Edward Arnold, Studies in Biology n. 88, 59.

[9] Ingold C.T. & H.J. Hudson (1993) *The biology of Fungi*, 6th edit., London: Chapman & Hall, 224.

[10] Lipson E.D. (1991) Phototropism in Fungi. In: *Biophysics of photoreceptors and photomovements in microorganisms* (F. Lenci, F. Ghetti, G. Colombetti, D.-P. Häder, P.-S. Song, eds), New York and London: Plenum Press, 311-325.

[11] Shropshire W. Jr (1963) Photoresponse of the fungus *Phycomyces, Physiol. Rev.* 43:38-67.

[12] Steinhardt A.R. & L. Fukshansky (1985) Diffusion approximation for scattering in a cylinder: optics of phototropism. *J. Opt. Soc. Am. A.* 2:1725.

[13] Steinhardt A.R., T. Popescu & L. Fukshansky (1989) Is the dichroic photoreceptor for *Phycomyces* phototropism located at the plasma membrane or at the tonoplast? *Photochem. Photobiol.* 49:79.
[14] Webster J. (1980) *Introduction to Fungi*, 2nd edit., New York: Cambridge University Press, 669.
[15] Wolf F.A. & F.T. Wolf (1947) *The Fungi*, Vol. II, New York: John Wiley & Sons, 129-131.

PHOTORECEPTION AND PHOTOTRANSDUCTION IN INVERTEBRATES

Survey of phototransduction in visual cells of invertebrates

Hennig Stieve

Institut für Biologie II, RWTH Aachen, Germany

Abstract

The basic phenomena and the uncontroversial observations on the transduction mechanism of invertebrates are described. Light absorption by a rhodopsin molecule leads via the activation of a complicated enzyme cascade to the formation of intracellular terminal transmitter. Binding of terminal transmitter causes the opening of "light-activated" cation channels in the plasma membrane and thus the receptor current. The enzyme cascade involves the activation of G-protein and the inositol lipid pathway. The nature of the terminal transmitters is not known beyond doubt.

Introduction

Photoreceptor cells act as interpreters which translate the light stimulus into electrical signals, the language of the nervous system. This process is termed visual transduction. Nature has solved the task of visual transduction differently for invertebrate and for vertebrate animals (Fig. 1) [99, 118]. Absorption of a photon by a single rhodopsin molecule leads after large amplification to a transient change in membrane current and by that to a membrane voltage signal. The photoreceptor cells have specialized regions, one of them contains the photopigment rhodopsin. Rhodopsin is an integral membrane protein which resides in the plasma membrane. Visual cells have greatly enlarged rhodopsin-containing membrane areas. These result in a concentration of visual pigment, which ensures a high probability of photon catch. The photosensory membrane contains besides rhodopsin at least two other integral membrane proteins, the light-modulated cation channel and a sodium-calcium exchanger. In photoreceptors of invertebrates like arthropods and mollusks, the enlargement of the photosensory cell membrane is realized by glove finger-like extrusions of the plasma membrane, the so-called microvilli, which are about 0.1 μm in diameter and up to 20 μm in length. Microvilli are supported by a

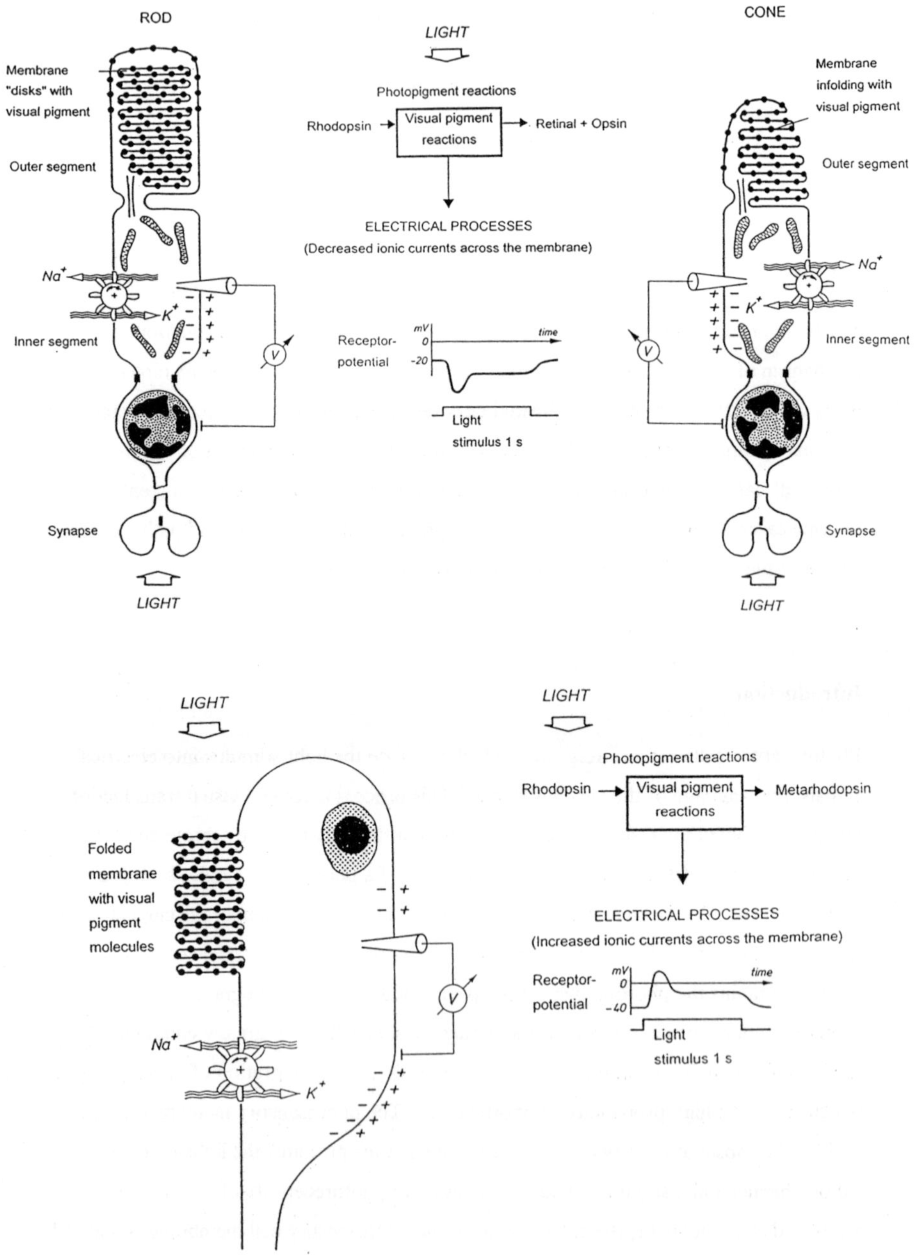

Fig. 1: Functional diagram of visual cells of vertebrates and invertebrates. The paddle wheel represents the active ion transport by a Na, K ATPase (Stieve; 1979).

cytoskeleton consisting of an axial filament and spike-like processes to the cell membrane. Beneath the microvilli is a layer of cisternae, derivatives of the smooth endoplasmic reticulum, called the subrhabdomeric cisternae.

The rest of the visual cell, containing the typical set of cellular organelles, is more or less distinctly separated from the transducing portion and is surrounded by a cell membrane which differs from the photosensory membrane and resembles that of a normal nerve cell. The sensory cell ends in an axon which, after a few micrometers (insects) or centimeters (*Limulus*), forms a synapse with a following nerve cell. The microvilli of many photoreceptor cells of invertebrates are surrounded by an extremely small extracellular compartment.

Single photon response

Photoreceptors are photon counters. If a *Limulus* photoreceptor cell, for instance, is stimulated by extremely dim flashes of light, which are so weak that not each flash elicits a light response, one observes so-called bumps, light responses evoked by the absorption of single photons. These single photon-evoked excitatory events of invertebrates, which were first described by Yeandle and Fuortes [156], vary greatly in delay (latency), time course, and size under identical stimulus conditions [117]. Bump latency, bump size and shape are not correlated, they are independent of each other [51, 117, 122, 133].

The average size of bumps is species-specific. *Limulus* and the cockroach have very large bumps (0.1 to 4 nA). Those of flies are much smaller [4, 53, 107, 154, 155], and in some species, e.g. crayfish and barnacle, they are not observable, presumably because the single events are so small (they probably consist of only a few jointly activated channels each) that they cannot be discovered in the noise.

In a first approximation we can assume that one bump is evoked by the absorption of one photon and that bumps are generated independently of each other [57, 117]. However, there are indications that this is not the whole story [26, 71, 74].

The macroscopic response

The number of bumps evoked by a light flash depends linearly on the energy of the light stimulus, i.e. on the number of absorbed photons [156]. With stronger light stimuli bumps superimpose and fuse to the macroscopic receptor current. Basically the macroscopic receptor current evoked by a brighter light flash is the summation of a volley of bumps. However, the shape of the signal is secondarily modified by bump interactions and feed-back processes [4, 133, 135]. Due to the scattering of bump latencies, the macroscopic receptor current signal is much (several times) longer than a single photon response. The latency of the first bump occurring after the flash determines the latency of the macroscopic light response. The macroscopic response is an average of weighted single photon responses. The latency of the macroscopic response is greatly shortened with increasing stimulus intensity. This is because the probability for bumps with short latencies is raised with increasing number of evoked bumps [125]. In the photoreceptors of vertebrates the latency of the macroscopic response is much less shortened with increasing intensity of the light stimulus since the scatter of the bump latencies is much smaller. In the photoreceptor of the fly still another process contributes to the shortening of the receptor current latency: Bumps with shorter latency are generated when rhodopsin molecules from two neighboring microvilli have absorbed photons from the same flash of light [37, 38].

Intensity dependence

The stimulus energy versus response characteristics is fairly complex. It consists of four regions [13, 125]. At low stimulus intensities it has a linear section (slope about 1) where the size of the response is doubled when the intensity is doubled. In the next section, of an intensity range of less than one logarithmic unit, the curve has a supralinear slope (slope 2-4) where the response size rises more than proportionally to the light intensity. This region is followed by a sublinear section (slope < 0.5), and finally saturation is reached. There are at least two different hypotheses to explain the supralinear slope of the supralinear section: one of them is that in this intensity range the bumps are no longer independent due to overlap of the bump specks [125]. Another explanation comes from a cooperativity which is restricted to one of the three components of the receptor current (see below).

What is a bump:

Bumps are believed to be the result of a concerted opening of many (100 to 1000) light-activated single cation channels. They must be due to a quantal formation or release of excitatory transmitter.

The light-activated cation channels are assumed to be located at the bases of the microvilli. Recordings of light-activated single channel activity in the microvilli range of rhabdomeric photoreceptors is very difficult. Only three laboratories have been successful so far [1, 2, 3, 69, 75, 77, 78, 79].

There is more than one type of light-activated ion channels in the photosensory membrane of one invertebrate photoreceptor cell: three in *Limulus*, at least two in flies, and in the scallop *Pecten*. The light-activated cation channels are all not very selective for Na over K [140, 141]. They have reversal potentials around +10 mV. In *Limulus* they have elementary conductances of about 6, 12, and 30 pS, but only slightly differing ion selectivities [69,70]. In *Drosophila* one of the two light-activated ion channel prefers calcium greatly over sodium [41, 42, 44].

The generation of bumps is probably brought about by a bump speck mechanism [125] where the light activation of rhodopsin leads to the production of transmitter which diffuses under the microvillar membrane to open cation channels in a finite region of a diameter of about 1 to 4 μm (*Limulus*) in the dark-adapted photoreceptor.

Amplification

How is the large amplification realized? Originally it was assumed that rhodopsin itself forms a light-activated cation channel [68]. This was not confirmed. Later it was found for the photoreceptor of vertebrates that rhodopsin acts like a hormone receptor activating an enzyme cascade which leads to the degradation of the terminal transmitter cyclic guanosine monophosphate, cGMP, which opens the ligand-gated cation channel [20, 31]. There are many indications that an analogous mechanism is used by invertebrates. The enzyme cascade in this case leads to the formation of terminal transmitter which opens light-activated cation channels in the photosensory membrane [for review see 127].

Rhodopsin reactions

The rhodopsin molecules in the photoreceptor membrane of invertebrates appear to be virtually immobile [34]. Rhodopsin is a membrane protein consisting of a chain of some 350 to 450 amino acids to which the chromophore (retinal, 3-dehydroretinal, or 3-dehydroxyretinal respectively [33, 147]) is covalently bound. The light-induced rhodopsin reactions, starting with the stereo-isomerization of the 11-cis-chromophore, lead in several reaction steps, after less than 1 to 5 ms, to an activated state metarhodopsin M*. This is inactivated after a certain life time to M_i which is stable for hours [36, 111]. In some species (e.g. *Drosophila* and other flies) the wavelength of the extinction maximum of rhodopsin and metarhodopsin differs greatly, in others (e.g. *Limulus* and the squid *Loligo*) [48] the extinction maxima are almost identical. However, the extinction of metarhodopsin is higher than that of rhodopsin. This ensures the re-conversion to large amounts of transducing rhodopsin even if the eyes are exposed to bright sunlight. Absorption of a second photon causes via intermediate reactions the recovery of transducing rhodopsin. If metarhodopsin has been already inactivated to M_i, light absorption causes first the re-conversion of M_i to non-transducing rhodopsin R_n [35,58].

Too bright illumination with chromatic light which converts rhodopsin to metarhodopsin causes a prolonged depolarizing afterpotential (PDA) [39; 47] which is a prolonged excitation of the photoreceptor which may last for hours after the offset of the light stimulus. It is a saturation phenomenon which does not occur in the normal life of an insect but which is a convenient phenomenon to study the transduction mechanism and to select mutants with transduction defects [66, 83]. The explanation of the PDA was highly controversial for several years [39, 46]. In the meantime it has been settled that it is based on the saturation of the shut-off-mechanism of visual excitation [114].

The recovery of non-transducing rhodopsin, reconverted from inactivated metarhodopsin, takes much more time in the excised eye than in the intact animal [35]. In the excised eye the recovery has a sigmoidal time course indicating that more than one reaction step is involved.

Still somewhat controversial is the mechanism of inactivation of the activated metarhodopsin M*. Whereas Paulsen's lab [8] find that the light-activated metarhodopsin is multiply phosphorylated only after it has bound the protein arrestin II, Selinger and his coworkers report the opposite, that rhodopsin phosphorylation precedes the binding of arrestin [17].

G-proteins

Light activation of G-protein has been proven by measuring the GTPase activity in photoreceptors of the fly *Musca* and *Drosophila* [9, 25]. There is evidence for a light-activated GTP-binding protein in photoreceptors of several invertebrates [18] (*Octopus*); [146] (squid). For some invertebrates a G-protein of the G_q type is identified [54] (*Drosophila*); [95] (Gα, squid *Loligo*); [81] (squid *Allotheutis*); [142] (*Procambarus*); [138] (*Octopus*). The micro-injection of the G-protein inhibitor GDP-ß-S suppresses dose-dependently the light response of the *Limulus* ventral photoreceptor [15, 131].

Is there more than one type of G-protein which can be activated by rhodopsin? Most probably yes: There are reports of more than one G-protein which influence the light response. However, it is not shown yet, whether all of them are light-activated. Robinson et al. [104] find three different light-dependent GTP-binding proteins in the photoreceptors of the squid *Loligo*, Tsuda and Tsuda [144] two light-regulated G-proteins and one which was not influenced by light in the photoreceptors of *Octopus*. Fyles et al. [32] report of at least two light-activated G-proteins in the photoreceptor membrane of *Allotheutis*. Micro-injection of anti-Gα-antibody or other agents which interfere with G-proteins indicate more than one type of G-protein in a single photoreceptor cell of *Limulus* [131].

On the other hand, activation of G-protein by GTP-γ-S or AlF_4 leads to the generation of bumps without any light stimulation [29, 52, 84, 85, 101]. The size distribution of chemically-induced bumps is indistinguishable from the size distribution of bumps which occur spontaneously in the dark. Light-activated bumps are much larger. We assume that the spontaneous and the chemically induced bumps are due to the activation of single G-protein molecules whereas the light-activated bumps are caused by the joint activation of 3 to 12 G-protein molecules [52, 101]. There are indications for a model that reactivation of inactivated meta-rhodopsin may also lead to the generation of bumps without light stimulation [58].

Enzymes of the cascade

Activation of G-protein leads to the initiation of the succeeding steps in the enzyme cascade - the activation of phospholipase C and of other enzymes.

a) The PLC-cascade:
In photoreceptor cells of arthropods, a light-activated enzyme cascade is proven to start from rhodopsin via G-protein, phospholipase C, increase in cytosolic calcium which finally leads via an unknown intermediate to the opening of light-modulated cation channels [11, 25, 27, 30, 86, 114, 139]. The activity of phospholipase C seems to be essential for the transduction process in *Drosophila* [67].

b) Other enzymatic pathways:
There are indications of other light-activated enzymatic branches which lead to the opening of cation conductances. A cGMP-gated cation channel has been cloned from *Drosophila* [5] and there is evidence for a cGMP-induced conductance increase in excised patches of photosensory membrane of *Limulus* [3].
A light-induced change in cyclic GMP level is not shown beyond doubt in photoreceptors of invertebrates: Saibil [105] and Johnson et al. [50] reported an increase in cGMP after illumination of the squid retina. However, no significant change in the cGMP level could be detected under comparable conditions by Brown et al. [16], Seidou et al. [112] and Tsuda [143]. The presence of guanylate and adenylate cyclase (GC & AC) activity is demonstrated histochemically in different locations of the retinula cells of the fly *Calliphora,* AC in the entire microvillar membrane, GC only in the tips of the microvilli [108]. Whether these enzymes CA & CG can be light-activated is not known. According to Robinson and Cote [103] light activation of guanylate cyclase GC could not be demonstrated in the squid retina.

Phosphodiesterase (PDE) activity has been localized histochemically in photoreceptors of the fly *Calliphora* [109]. The distribution of cAMP- and cGMP- PDE activity is different. Whether these phosphodiesterases are light-modulated is not known.

Other indications of more than one light-activated ion channel type will be dealt with in the following lectures. Here is a *terra incognita* of the transduction mechanism of invertebrates which will occupy us much in the further course of this summer school and thereafter.

Adaptation

Adaptation is the change in sensitivity, a feed-back control of the light response. It serves to adjust the intensity versus response curve to the ambient illumination so that the center of the curve corresponds to the average level of illumination of the surroundings [128]. Adaptation can shift the sensitivity by several orders of magnitude (up to 10^6-fold). Light adaptation affects the average bump size in that it becomes smaller both in amplitude and duration, and the latency of the response becomes shortened. After the onset of illumination adaptation reaches a final state in less than one second, whereas dark adaptation, that is the recovery of sensitivity after light adaptation, proceeds in several phases in the order of seconds, minutes, and up to one hour.

Adaptation controls several processes of the transduction. Calcium is the major transmitter for this control, acting on several stages in the enzyme cascade of phototransduction. In the PLC branch of the enzyme cascade calcium probably interacts with the activation of G-protein [32], with the activity of the phospholipase C [98], with the activity of the protein kinase C [45], and with the efficiency of the IP_3-induced calcium release [92].
Besides calcium there are indications for other agents, e.g. cyclic AMP, which contribute to the control of sensitivity [106].

In the following lectures we will learn more about the complex transduction mechanism and its regulation and about the solved and unsolved riddles.

Acknowledgments: I wish to thank Marlies Dorlöchter, Werner von der Heide, Karoly Nagy, and Michael Rack for valuable comments on the manuscript, Bettina Halbe, H.Thomas Hennig, and Irene Wicke for much help with the manuscript; work reported here was supported by the Deutsche Forschungsgemeinschaft (Sti 16/17).

The roles of calcium in phototransduction of invertebrates

Hennig Stieve

Institut für Biologie II, RWTH Aachen, Germany

Abstract:

Calcium ions are very important for the excitability, excitation, and adaptation of visual cells in invertebrates. The extracellular and intracellular calcium concentration determines the functional state of the visual cell. A rise in cytosolic calcium is an essential step in the excitatory transduction chain. Cytosolic calcium is the most important intracellular messenger of adaptation which causes the desensitization of the visual cell in light adaptation.

Introduction:

The extracellular and intracellular calcium concentration determines the state and conformation of proteins and phospholipids and by this many cellular functions. Most enzymatic processes have besides a pH-optimum an optimum of pCa [80].

1. Extracellular calcium ion concentration

Calcium ions in the extracellular saline are essential for the excitability of photoreceptor cells of invertebrates. Reduction of extracellular calcium and magnesium ions to very low values (below 1 μM) causes in *Limulus* photoreceptors an increase in sodium conductance as in the light response and consequently a depolarization of the cell membrane to a similar value as the reversal potential of the light response. In addition, illumination does not evoke a light-induced conductance increase under these conditions. However, additional reduction of the external sodium to 50 mM (about 10 % of its physiological value) causes a recovery of both the dark potential and the electrical light response [124,129]. Under the same low sodium, low calcium and magnesium conditions, the membrane voltage dependence of the peak amplitude of the receptor current is again almost normal as in physiological external calcium and sodium concentrations [120,132].

Measurements of the electroretinogramm of the crayfish retina show corresponding results. However, higher EGTA concentrations (>10 mM) have to be used in order to deplete the small extracellular compartments of calcium [123].

When the extracellular calcium ion concentration $[Ca^{2+}]_{ex}$ surrounding the dark-adapted photoreceptor cell of an invertebrate is lowered to 100 or 250 μM, the electrical light response is enlarged and slowed down, and the latency greatly prolonged. Raising the external calcium has the opposite effect [116,120,128]. Correspondingly, the single photon-evoked events, bumps, become larger and longer, and the histogram of the latency distribution becomes broader and is shifted to longer latencies [121,122, and see below].

Light adaptation, the desensitization of the photoreceptor, is also influenced by the extracellular calcium concentration. A strong conditioning light stimulus causes desensitization, which shifts the intensity dependence of the electrical light response to higher light intensities. This desensitization is the larger the higher the external calcium concentration is, if the photoreceptor cell of *Limulus* is stimulated regularly (every 2 or 3 minutes) by a strong flash of light. Under the same conditions desensitization becomes practically zero if $[Ca^{2+}]_{ex}$ is below 100 μM [130].

This is different for bump adaptation. Under very weak illumination, when single bumps are observed, the light-induced desensitization by a conditioning weak stimulus turned out to be much less dependent on $[Ca^{2+}]_{ex}$ [122, see below].

The site of action of calcium, however, in light adaptation is intracellular [59]. Injection of calcium buffers (EGTA or BAPTA) reduces light adaptation by preventing the light-induced increase in cytosolic calcium concentration [59].

2. Intracellular, cytosolic calcium concentration $[Ca^{2+}]_i$

Cytosolic calcium is a very important transmitter controlling light adaptation [11]; it is also necessary for excitation [152]. The absolute value of the intracellular calcium concentration ot the intact dark-adapted photoreceptor cell of *Limulus* is probably below 100 nM. It is very difficult to determine (see Payne, this issue).

The light-induced cytosolic calcium signal

The receptor potential following a strong flash of light is accompanied by a calcium signal, a transient rise and re-decline in the concentration of intracellular, cytosolic calcium ions. The rise in cytosolic calcium concentration may be locally - in so-called hot spots - even higher than 100 μM [14,56,82, Payne, this issue]. The light-induced cytosolic calcium transient in the *Limulus* photoreceptor originates mainly from intracellular sources which depend on the extracellular calcium concentration and upon illumination (see below).

The cytosolic calcium transient has been measured by the use of the chemilucent protein aequorin, the indicator dye arsenazo, or by fluorescent dyes. I will focus here on the arsenazo measurements. To measure arsenazo signals one needs a measuring light beam which light-adapts the photoreceptor cell to a certain degree. The calcium indicator dye arsenazo can be micro-injected into the *Limulus* photoreceptor cell. Absorption changes in the photoreceptor indicate changes in the cytosolic calcium ion concentration. The arsenazo signal starts several (5 to 15) milliseconds later than the receptor current (or the receptor potential) and lasts several (3 to 12) seconds longer. The time difference of the latencies of both signals is the shorter the higher the stimulus energy [119].

This observation at least does not support the theory that the initial rise of the receptor current in *Limulus* is triggered by a rise in cytosolic calcium. It indicates either that the fast light-induced conductance increase is not caused by the rise in the cytosolic calcium concentration, or that the first channel-opening rise in cytosolic calcium is restricted to such a small area that it does not contribute sufficiently to the arsenazo signal to be detected in the noise.

Intensity dependence of the cytosolic calcium transient

Arsenazo is also, about equally well, an indicator of strontium ions. Since the arsenazo signal is much larger when the external calcium is replaced by strontium [119], the strontium signal can be measured even at lower light intensities than the calcium signal. Therefore we measured the intensity dependence of the arsenazo signal in external strontium saline in which the calcium was completely replaced by strontium [126].

The arsenazo signal is the larger the higher the stimulus intensity. When the maximum of the cytosolic strontium signal and the maximum of the receptor current (or the time integral of the receptor current) are plotted versus the light intensity of the 2 ms light stimulus, in a limited intensity range both signals display a supralinear slope, i.e. a region where the rise of the signal is greater than proportional to the rise of the light intensity. We measured slopes between 1.3 and 2.4 for the receptor current and between 1.2 and 2.0 for the arsenazo signal. The average supralinear slopes of the receptor current and that of the simultaneously measured arsenazo signal are not different, both are about 1.7 [126].

It was an open question whether this supralinearity is due to a cooperative ligand binding to the light-induced cation channels in the plasma membrane or to a cooperativity already at an earlier step. We observed that both the arsenazo signal and the receptor current display the same supralinearity. Therefore the cooperativity must occur earlier than at the light-modulated channel level. We conclude that the transduction chain which leads to calcium release and to the receptor current includes a cooperativity of more than 2 [see also 125]. It may be based either on a cooperativity in the opening of the calcium channels of the submicrovillar cisternae, or/and in a previous step of the transduction chain [126].

The ways of calcium

The cytosolic calcium concentration depends strongly on the extracellular calcium concentration and upon illumination.

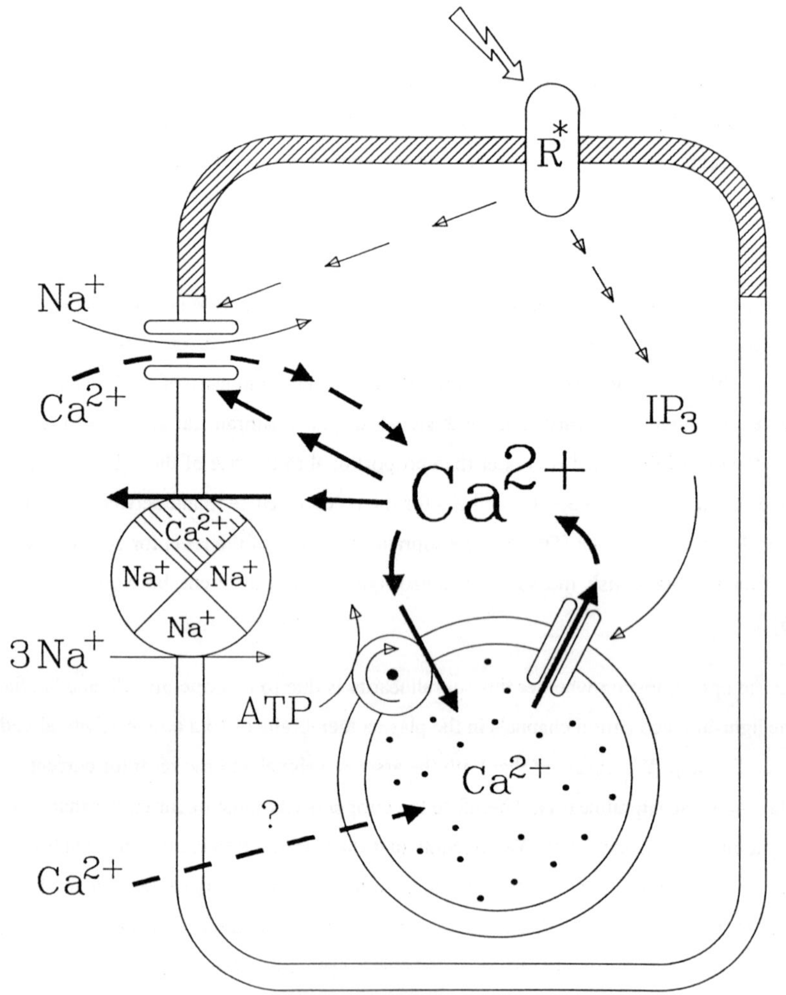

Fig. 1 Calcium connections:

Scheme of the pathways of calcium in the photoreceptor. R* light-activated rhodopsin causes the opening of the "light-activated" cation channels in the plasma membrane via different cascades. Calcium is pumped back into the cisternae by a Ca ATPase and exported out of the cell by a Na-Ca exchanger in the plasma membrane. The cisternae are refilled from extracellular sources - possibly via reverse transport of the Na-Ca exchanger into the cytosol and Ca ATPase-mediated active transport into the cisternae.

a. Calcium transport through the plasma membrane

There are at least two ways linking the extracellular to the cytosolic calcium:

1. an electrogenic sodium-calcium exchanger
2. calcium-permeable cation channels.

ad 1: The exchanger in the plasma membrane seems not to depend upon potassium [82]. The electrogenic sodium-calcium exchange depends upon three driving forces:

1. the calcium gradient
2. the sodium gradient, and
3. the membrane voltage across the plasma membrane.

Under resting conditions the exchanger is in equilibrium and any change in one of the driving forces causes an inward or outward transport of calcium in exchange against sodium across the cell membrane [23].

ad 2: Light-activated cation channels admit besides sodium also calcium ions into the visual cell. In *Limulus* this can be shown by the arsenazo response [see below; 136]. In *Drosophila* the trp channel admits calcium better than sodium [42,43], and a cGMP-gated ion channel cloned from *Drosophila* eyes also admits calcium ions better than sodium ions [113, this meeting].

The cytosolic calcium concentration depends strongly on the extracellular calcium concentration even in the dark, mainly mediated by the sodium-calcium exchanger.

b. Intracellular calcium stores

Calcium is stored in different cell organelles:

1. cisternae from the endoplasmic reticulum
2. mitochondria
3. pigment granules [49,110,145]

Although the calcium in mitochondria and pigment granules can be exchanged with the cytosolic calcium, it seems probable that these two calcium stores serve as calcium buffers and do not directly participate in the process of visual transduction. This is different with the subrhabdomeric cisternae (SRC), differentiations of the endoplasmatic reticulum. They can rapidly release and re-uptake calcium during the light response [23,87,93]. The size of the subrhabdomeric cisternae differs greatly among different species; they are very small in the visual cells of *Sepia* and crayfish, larger in *Limulus* and the blowfly *Calliphora*, and much larger in the honeybee [6,19,49,148,149,150,158]. At least in honeybees the cisternae can release calcium by two different release mechanisms [151]:

1. An IP_3- sensitive release mechanism due to calcium release-controlling IP_3-receptors in the membrane of the cisternae [88,89,91,93] and
2. non-IP_3-sensitive calcium release controlling ryanodine receptors responsible for a calcium-induced calcium release [151]. It seems probable that this non-IP_3-sensitive release mechanisms is also present in *Limulus* photoreceptors.

c. The light-induced cytosolic calcium transient

The light-induced cytosolic calcium transient depends upon the extracellular calcium concentration. However, although the cytosolic calcium concentration directly follows the extracellular calcium changes, the cisternae of the *Limulus* photoreceptor seem not to release and to exchange calcium with the cytosol when the photoreceptor is not exposed to light [12]. However, with repeated illuminations the calcium transient measured by the arsenazo response becomes gradually smaller and finally vanishes [61]. This is different in photoreceptors of crustaceans, those of the barnacle *Balanus* and the crayfish; here the calcium stores seem to loose their calcium already in the dark when the photoreceptors are bathed in saline of reduced calcium concentration [7,12]. In insects, the situation seems to be more similar to *Limulus* than to crustaceans (Minke, personal communication). However, no published results are available. These species differences may be caused by different contribution of the subrhabdomeric cisternae to the cytosolic calcium transient or by differences in the exchange mechanism between cytosolic calcium and calcium in the cisternae.

For refilling the stores illumination seems not to be necessary. If the external calcium concentration is raised, the intracellular calcium stores of the *Limulus* photoreceptor are filled with a higher calcium concentration probably already in the dark [136].

The kinetics of the calcium transient recorded as arsenazo signal is primarily governed by four processes [137]. Regarding the decline of the arsenazo signal one can distinguish two different processes. One of them, the faster decline with a time constant of 1 - 2 s, is not dependent upon the membrane voltage which was varied between -60 and +30 mV [23,134]. A second, slower phase of decline (time constant 6 - 7 s) is strongly membrane voltage-dependent and may even change direction (i.e. displaying a rise of cytosolic calcium) if the membrane potential is more positive than about +10 mV [23]. This indicates that the decline of the cytosolic calcium signal depends on a fast re-uptake of calcium into the cisternae, probably caused by a calcium ATP-ase, and a slower export of calcium through the plasma membrane by the sodium-calcium exchanger.

A similar analysis of the rising phase of the cytosolic calcium signal reveals that it is also due to two processes [137]: a fast (time constant 0.01-0.15 s) calcium rise which is independent of membrane voltage and a slower calcium rise (time constant 0.2-2 s) which becomes the faster the more negative the membrane voltage is. These results are interpreted in that the fast phase of the rise is caused by a release of calcium from the intracellular cisternae whereas the slower phase is due to an influx of calcium through light-activated ion channels (Fig. 1). The arsenazo signal of the *Limulus* photoreceptor can be modeled sufficiently by four exponentials, but less than four do not give a satisfactory description [137].

The intracellular calcium stores are ultimately refilled from extracellular sources. In the *Limulus* photoreceptor there are no indications that the Na-Ca exchanger operating in reverse mode is not sufficient for the observed refill, whereas for the *Drosophila* retinula cell a 'capacity-controlled refill pathway' [96,97], to which the trp calcium channel contributes, is proposed by Hardie and Minke [43].

Bumps are calcium-dependent

Lowering the external calcium concentration from 10 mM to 250 μM has its most conspicuous effect on the bumps of the dark-adapted *Limulus* photoreceptor (without substantially reducing the ability for light adaptation). Amplitude, duration, current integral, and latency of current bumps under voltage clamp conditions are greatly enlarged, and the frequency of light-evoked bumps is raised [133,157]. Raising the external calcium concentration to 100 mM causes the opposite changes, i.e. reduction of bump size on the average to about 60 % and corresponding changes in bump parameters [122].

The calcium dependence of bump adaptation was tested using a conditioning light flash of moderate intensity 2 or 3 seconds prior to the bump-evoking illumination. It turned out that the reduction in bump size, due to such pre-illumination is practically independent of the external calcium concentration between 250 μM and 100 mM calcium in the photoreceptor of *Limulus* [122,157]. Only when the external calcium concentration is very low (below 10 μM), bump adaptation is significantly reduced [122].

The results are interpreted in that due to the lowering of the external calcium concentration the cytosolic calcium concentration is lowered, but still high enough for light adaptation. So the intracellular stores which release calcium into the cytosol for light adaptation can still be refilled in the dark, and the total illumination is not strong enough to exhaust the stores [133,157].

In order to test this assumption, we illuminated cells which were bathed in external calcium lowered to 250 μM by strong flashes of light to exhaust the calcium stores. The results agree with our assumption. The bump size of the dark-adapted photoreceptor under these conditions is slightly reduced and the bump adaptation due to the conditioning pre-illumination is reduced by about half [157].

Injection of 1.2 mM EGTA into the *Limulus* photoreceptor cell causes a reduction in the bump size of dark-adapted cells and an abolishment of bump adaptation, i.e. the conditioning light flash causes no reduction in the average bump size. These results can be explained in that the calcium buffer EGTA prevents the rise in cytosolic calcium and by that reduces light adaptation [157].

Sites of calcium action mediating light adaptation

Calcium is a very important agent for light adaptation. The light-induced cytosolic calcium increase causes a decrease in sensitivity, light adaptation [for review see 11]. Injection of calcium buffers (EDTA, EGTA, or BAPTA), which prevent the light-induced increase in cytosolic calcium concentration, reduces light adaptation [59]. It causes a slowing down and a prolongation of the electrical light response. Another consequence of such an injection is a change in response size. The macroscopic response is transiently enlarged, later diminished. The bump size, however, is greatly reduced, and the bump adaptation following conditioning illumination is diminished [see above; 157].

Calcium acts on several steps of the phototransduction cascade:
1. Probably on the step of G-protein activation by rhodopsin, causing a smaller number of G-proteins to be activated by an active rhodopsin molecule [32].
2. It acts on the IP_3-induced calcium release from the cisternae thus making IP3 less efficient i.e. to release less calcium [55,92].
3. It probably acts on the activity of phospholipase C: The activity of phospholipase C, isolated from outer segments of visual cells of the *Sepia* retina, is strongly dependent on the calcium concentration displaying a bell-shaped curve. The estimated change in cytosolic calcium concentration in light adaptation would reduce the activity of that enzyme by a factor of up to eight [98].
4. It inactivates - at least in Drosophila - the "light-sensitive" cation channels [44].

Calcium seems not to be the only agent controlling the sensitivity in adaptation. Adding cyclic AMP (0. 5 mM dibutyryl-c-AMP or bromo-c-AMP) greatly speeds up the second phase of dark adaptation whereas the first phase of dark adaptation is relatively enhanced by reducing the extracellular calcium concentration [106].

Calcium an excitatory transmitter?

Strong and rapid injections of calcium into the *Limulus* photoreceptor cause a conductance increase similar to the light-induced one [90,91]. Nevertheless, calcium seems not to be a terminal transmitter in the transduction cascade at least in *Drosophila*, since the intracellular photo-release of calcium from a caged compound did not cause an increase in membrane conductance [40].

Acknowledgments: I wish to thank Marlies Dorlöchter, Karoly Nagy, Michael Rack, Georg Stommel, and Weijia Yuan for valuable comments on the manuscript, Bettina Halbe, H. Thomas Hennig, and Irene Wicke for much help with the manuscript. The work reported here was supported by the Deutsche Forschungsgemeinschaft (Sti 16/13).

Phototransduction in *Limulus* photoreceptors and enigmas of the transduction mechanism in photoreceptor cells of invertebrates

Hennig Stieve and Karoly Nagy
Institut für Biologie II, RWTH Aachen, Germany

Abstract

The receptor current of the ventral photoreceptor of *Limulus* is composed of three components which differ in several aspects: Time course of light activation, time course of recovery from light adaptation, reversal potential, and pharmacology. In addition, two different types of single photon-evoked events (bumps) and three different elementary channel conductances are observed in this photoreceptor cell.

These findings indicate that the three receptor current components are controlled by three different enzymatic pathways using three different terminal transmitters. Probably one of the terminal transmitters is cyclic GMP, another one is activated via the IP_3 cascade, the third one might be cyclic AMP.

Introduction

Limulus has three types of photoreceptors: lateral eyes, median ocelli and ventral nerve photoreceptors. The physiological function of the ventral photoreceptors is not known whereas that of the lateral eyes is better understood. In the lateral eye each ommatidium contains up to 20 electrically coupled retinular cells and one response-integrating eccentric cell. The function of the axons of the retinular cells is still a riddle. The median ocelli contain electrically coupled photoreceptor cells for UV light besides those for visible light. The amino acid sequence of rhodopsins from the lateral eye and the median ocelli have been determined by cDNA cloning [115].

The large ventral photoreceptor of *Limulus* is very much suited for electrophysiological investigation. It is, however, not clear how typical it is for general properties of photoreceptors of invertebrates since it has a number of apparently special characteristics:

1. single photon-evoked responses, bumps, are extraordinarily large.
2. intensity-response characteristics are very steep; from threshold to saturation of the receptor potential less than an 100-fold increase in intensity of the light stimulus is needed.
3. the latency is very long, even with the highest stimulus intensity it does not become shorter than 10 ms.

The ventral photoreceptor cell is divided into two parts, an R lobe or rhabdomeric lobe which contains the microvilli with the rhodopsin and an A lobe, or arhabdomeric lobe, which has a smooth surface with no or only very few rhodopsin molecules. From this part the axon originates which leads to the protocerebrum, the distal part of the *Limulus* brain.

Following a suggestion of Kefer Hartline, the ventral photoreceptor cell was first characterized by Milecchia and Mauro [63,64]; their study revealed for this visual cell many characteristic features of an invertebrate photoreceptor cell. We and several other labs chose this photoreceptor as an object assuming that the photoreceptor of an old-timer like *Limulus* would be very simple and easy to understand. This turned out to be an error.

Three phases of dark adaptation

When the course of dark adaptation of a *Limulus* photoreceptor is measured, it is observed that the time course of the recovery of the electrical light response (receptor potential or receptor current) following a desensitizing conditioning light flash occurs in several (at least three) phases. A fast phase with an half-time of about one second is followed by a slower phase (half-time 5 to 15 s) and finally a third phase (half-time > 20 s) [62,106].

The phases of dark adaptation show different sensitivities to calcium and cyclic nucleotides. The first phase of dark adaptation, but not the second phase, is very strongly accelerated when the external calcium concentration is lowered from 10 mM to 250 μM [21,106]; the second phase of dark adaptation, however, is strongly accelerated when dibutyryl-cAMP or bromo-cAMP is added to the external bathing solution [106].

Three components of the receptor current

A detailed analysis reveals that the three phases of dark adaptation are due to three distinct components C1, C2, and C3 of the electrical light response. It can be seen that the different components of the light response recover and saturate after different times of dark adaptation.

These three components of the receptor current have slightly different reversal potentials (C1 around +14 mV; C2 around +7 mV; C3 around +10 mV). Therefore no single voltage is found where the total receptor current of the dark-adapted photoreceptor in response to a strong light stimulus is zero [24].

More than one component of the receptor current with slightly different reversal potentials have also been found by Nasi [77] for the photoreceptor cell of the scallop *Pecten* and by Hardie [41] for the photoreceptor cell of *Drosophila*.

The intensity dependence of the receptor current

The intensity dependence of the light response of the ventral photoreceptor has been measured very accurately [13,125]. For these measurements the stimulus protocol is very important: The photoreceptor is stimulated every 4 minutes by a strong conditioning flash and by two test flashes following 18 and 120 seconds after the conditioning flash to ensure two constant states of adaptation, a relatively light-adapted one (18 s), and a fairly dark-adapted state (120 s) [125]. When such a stimulus protocol is used, the logarithms of the amplitude or of the time integral of the receptor current depends on the logarithm of the stimulus energy in a sigmoidal curve. The dependence starts with a linear slope where the response rises proportionally to the stimulus energy (slope 1); this is followed by a supra-linear slope where the response rises more than proportionally with the stimulus energy (slope 2.7-3.2). This is followed at higher stimulus energies by a curve section where the response rises sub-linearly (slope 0.1-0.3).

Independent bump summation should result in a linear slope of the intensity versus response characteristics. We can explain the supralinear rise by a model of cooperative bump interaction [51,117]. However, it seems probable that the supralinearity is a property of only the second component C2 of the receptor current. It has been shown recently by isolating the different components of the receptor current that the component C1 and C3 have a linear dependence of the stimulus energy, whereas the component C2 depends supralinearly upon the energy of the stimulus [24].

Light adaptation causes the intensity versus response characteristics to be shifted almost in parallel to higher intensities. However, the slope of the supralinear section is somewhat reduced due to the light adaptation [125].

Bumps

Bumps are believed to be the results of concerted openings of 100 -1000 single channels. Up to now single channels could not yet be resolved in bumps of *Limulus*, but in *Pecten* [77]. At least two different types of bumps could be distinguished in *Limulus* ventral photoreceptors [76,100,157]:

1. standard bumps which are the easiest to record; they have amplitudes larger than 0.3 nA. (amplitude averages: 350 pA); they can also be characterized by a shape quotient of (halfwidth / current amplitude) Th/A: < 2 ms/pA ;

2. another type of bumps, "small and long bumps", is a different population mainly characterized by the longer slower decline. On the average they have amplitudes smaller than 0.05 nA and have an average duration of 350 ms (average amplitude: 35 pA; average shape quotient: Th/A: >2 ms/pA).
   Often a small and long bump precedes a standard bump which rides on the tail of the small bump.

There are a number of indications that the standard bumps are the elements which constitute the C2-component. Most probably the small and long bumps are the elements of the component C1 [76]. Up to now we have not identified a third type of bumps responsible for the third component. There is, however, also the possibility that the bumps which underlie the third component are so small that they cannot be detected in the noise. This would correspond to the observation that in the photoresponse of crayfish and barnacle (*Balanus*) no bumps can be recognized.

Bumps can also originate spontaneously in the dark. Most of them (but not all) are standard bumps [Stieve and Hennig, unpublished]; however, their average size is smaller than that of the light-evoked bumps. Spontaneous bumps can be elicited by AlF-activated single G-protein molecules without an activation of rhodopsin. In *Limulus* ventral photoreceptors their average amplitude is 7 to 8 times smaller than that of the light-evoked ones. From this it can be estimated that the bumps of a dark-adapted photoreceptor which vary in size are due to the light-induced activation of 3 to 12 G-protein molecules [52,101].

G-proteins

The action of light-activated rhodopsin to initiate the enzyme cascade is mediated by G-protein. Several types of light-dependent G-proteins have been found in the retina of cephalopods [104,144]. There are indications for more than one G- protein in the *Limulus* ventral photoreceptor. However, it is not yet proven whether they are all light-activated. Injection of GDP-β-S reduces in a dose-dependent manner the size of the receptor current and prolongs the latency [15,131]. This indicates that also in *Limulus* phototransduction, light activation of G-protein is an essential step, as it was shown for the fly photoreceptor [25].

Injection of an antibody against bovine transducin reduces the receptor current without influencing the latency. The exo-toxin C3 of *Clostridium botulinum* inhibits the biological activity of small GTP-binding proteins of the rho-gene family by ADP-ribosylation. Injection of this toxin reduces the receptor current strongly without changing the latency [131]. GDP-β-S is an inhibitor for all G-proteins whereas anti-transducin should be much more specific. The different actions of the three treatments indicate that more than one G-proteins are involved in *Limulus* phototransduction. The effect of the toxin C3 indicates that a small GTP-binding protein plays a role in the excitation cascade or as a pre-requisite for excitability, and maybe also for the refill of the cisternae. Furthermore Petrov et al. [94] report that small GTP-binding proteins interact with light-activated rhodopsin in squid photoreceptors.

Ion channels

Patch clamp measurements to record the activity of light-activated ion channels in the *Limulus* photoreceptor cell are very difficult and have only very rarely been successful [1,2,3,69,75]. Also in other rhabdomeric photoreceptors this patch clamping is very difficult and has not yet been done with success. The only other case where it seems to be easier is the recording from visual cells of the scallops , *Pecten* and *Lima* [77].

In *Limulus* careful measurements in the cell-attached mode show that there is more than one type of light-activated elementary conductance: 6 pS are the rarest events to be seen, 12 pS are the most frequently observed events, and 30 pS events are of an intermediate occurrence [69,75].

The 6 pS events are so rarely observed that they are not yet analyzed in their kinetic behavior. The other two types of events could be analyzed [69]. After the onset of moderate illumination, activity of 12 pS conductance events (of 1.4 ms mean duration) starts with a short delay of 50 - 100 ms after the onset of light, reaches a maximum after about 300 ms and declines to a plateau value within about a second. In a similar illumination the activity of the large 30 pS events (mean open time 5.5 ms) starts with a considerable delay of about 300 ms and increases during the first second of illumination. On turning-off the light stimulus the behavior is corresponding: the 12 pS events activity declines fast (time constant 2.5 s) whereas the activity of the large 30 pS events declines much slower (time constant 13 s); [69]. We assume that the 6 pS events are underlying the current component C1, the 12 pS the component C2 and the 30 pS events the component C3 [24,69].

Also in photoreceptor cells of other invertebrates more than one type of light-activated ion channels have been found: at least two in the fly *Drosophila* and in the mollusk *Pecten* [41,77].

The light-activated cation channels of photoreceptors of invertebrates are all not very selective for Na over K. Their reversal potentials are around +10mV (see above). They probably admit all alkali ions and even large monovalent cations such as choline [140,141]; some of them have a high permeability for calcium [41].

A protein with an amino acid sequence resembling that of the cGMP-gated channel of vertebrate photoreceptors has been cloned from *Drosophila* eyes and was expressed in

Xenopus oocytes [5]. This channel is also very unselective and has a high preference for calcium over sodium [113, this meeting].

Pharmacological characterization of the three components

Three different components (C1, C2, C3) constitute the macroscopic receptor current. As already stated above, these components differ in their recovery time and kinetics after light adaptation, in their reversal potential, and in their dependence on the light intensity [24,70].

For these experiments the stimulus protocol is very important. By applying a special stimulus program (i.e. two identical light flashes separated by about 4 - 5 s which are repeated every 4 - 5 min); [24,70], one can observe two kinds of responses: a response containing all three components evoked by the first flash and a response to the second flash which contains practically only C1 if the time delay between the two flashes is adjusted specifically in each cell. Under these conditions the action of certain drugs and agents microinjected into the photoreceptor cell can be tested. These are the results:

1. Stepwise progressive injection of **neomycine,** an inhibitor of phospholipase C (PLC), and thus of the IP_3 production, blocks progressively and selectively C2; only very large amounts of neomycine reduce C1 and C3 additionally [72].

2. **Heparin,** an inhibitor of the IP_3 receptors in the intracellular calcium stores, blocks C2, reduces C1, and somewhat reduces C3 [22].

3. **BAPTA,** which prevents the increase in cytosolic calcium, blocks C2, reduces C3, and increases C1 [22].

4. **U 73122**, a selective PLC inhibitor, blocks C2 and reduces C1 [102].

5. Injected **l-cis-diltiazem**, a calcium channel blocker which also blocks the cGMP-gated cation channels, blocks C1 and reduces C2 [73].

6. Phosphodiesterase **PDE** (which catalyzes the degradation of both cGMP and cAMP) inhibits C1 and C3 but not C2 [72].

Facit: The different sensitivity to the mentioned drugs indicate that the three components are controlled by three different enzymatic pathways using three different terminal transmitters.

Three transduction pathways, three terminal transmitters

The different behavior of the three components of the receptor current (different time course of light activation and of recovery from light adaptation), the different types of single photon-evoked events, and the different reversal potentials as well as different elementary channel conductances observed, all indicate that there are three different elementary light-activated ion channels which are controlled by different enzymatic systems and different terminal transmitters.

One of the controlling pathways seems to be the IP_3 cascade which leads via an unknown terminal transmitter (probably not calcium) to the C2 current component. Calcium appears not to be a terminal transmitter in the transduction cascade since at least in *Drosophila* intracellular caged calcium did not cause a conductance increase as a light stimulus [40]. However, a strong injection of calcium into *Limulus* photoreceptors causes a conductance increase similar to the light-induced one [90].

The terminal transmitter for the component C1 is probably cyclic GMP. Besides the evidence already mentioned, there is another observation, which may support this hypothesis. The conductance of excised patches of *Limulus* photosensory membrane increased upon exposure to cGMP [3]. We suspect that the component C3 has cyclic AMP as terminal transmitter.

How the complex enzyme cascade leads to the formation of three different types of terminal transmitters is not known yet. Do all the cascade's three branches which control the three channel types start from the same rhodopsin molecule? There are no convincing indications for more than one type of rhodopsin molecule in the *Limulus* photoreceptor from spectroscopic measurements and ERP measurements [28,60]; but this does not exclude differences in the protein composition in domains of the molecule which do not interact with the chromophore. It is unclear how the three branches are interconnected and how they depend upon each other. The two identified different bump types are distinct and seem to be mutually exclusive: no intermediate forms with properties of both

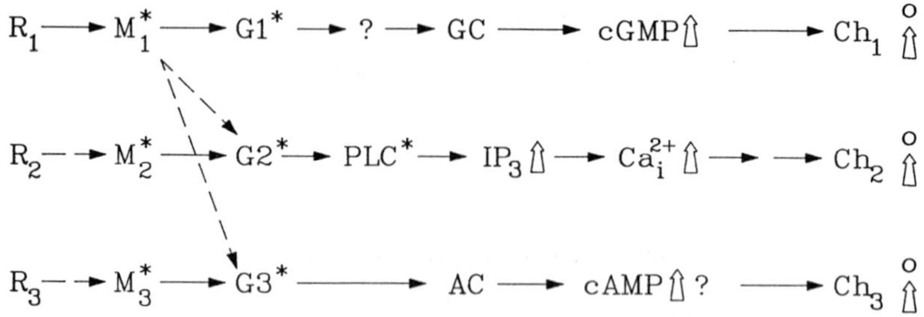

Figure 1. Reaction scheme to discuss the possible mechanism which generates the three observed components of the *Limulus* receptor current.
R_1, R_2, R_3 three (different types? of) rhodopsin molecules, M_1*, M_2*, M_3* (three types? of) light-activated meta-rhodopsin molecules, G1, G2, G3 three different types of G-proteins, GC guanylate cyclase, cGMP cyclic guanosine monophosphate, Ch_1, Ch_2, Ch_3 three different types of "light-activated" cation channels, PLC phospholipase C, IP_3 inositol-trisphosphate, AC adenylate cyclase, cAMP cyclic adenosine monophosphate; asterisks symbolize activated states, open arrows, increases.

types are observed. This indicates that an individual light-activated rhodopsin molecule can initiate only one branch of the cascade and not two at a time [76]. It seems probable that one light-activated rhodopsin molecule is either coupled to one type of G-protein or to another, and there is no promiscuity. It is not yet clear how many types of light-activated G-proteins are in the photoreceptor cell and whether the different rhodopsin molecules are assigned to different G-proteins just by their local neighborhood or whether this is ensured by specific chemical recognition. It is also unclear, how the phospholipase C (which seems to be essential for the whole light response in *Drosophila)*, [66,67] interacts with the two other branches of the transduction cascade.

The scheme (Fig. 1) is meant to show the state of our knowledge. Up to now the enzymes and the pathways of the light-activated cyclic GMP and cyclic AMP branches are not yet demonstrated. However, adenylate- and guanylate cyclase activity could be located histochemically in retinula cells of the fly *Calliphora* [108]. Moreover, the injection of a phosphodiesterase provokes a strong effect on the components C1 and C3 of the *Limulus* photoreceptor response [see above; 72], which suggests that these two components are cyclic nucleotide-dependent.

Right now the state of affairs indicates clearly that there is more than one light-activated transduction pathway. However, there are more open questions now than we thought 5 years ago. The findings reported here suggest that component C1 is controlled by a cyclic GMP- and C3 by a cyclic AMP enzyme pathway, whereas C2 seems to be controlled by the IP_3 pathway. However, its terminal transmitter is not known yet.

Acknowledgments: We wish to thank Marlies Dorlöchter, Klaus Contzen, Michael Rack, and Georg Stommel for valuable comments on the manuscript, Bettina Halbe, H. Thomas Hennig, and Irene Wicke for much help with the manuscript; work reported here was supported by the Deutsche Forschungsgemeinschaft (Na 188 and Sti 16/13).

References

1. J. Bacigalupo and J.E. Lisman (1983), Single channel currents activated by light in *Limulus* ventral photoreceptors. *Nature* 304, 268-270

2. J. Bacigalupo, K. Chinn and J.E. Lisman (1986), Ion channels activated by light in *Limulus* ventral photoreceptors. *J. Gen. Physiol.* 87, 73-89.

3. J. Bacigalupo, E.C. Johnson, C. Vergara and J.E. Lisman (1991), Light-dependent channels from excised patches of *Limulus* ventral photoreceptors are opened by cGMP. *Proc. Natl. Acad. Sci.* 88, 7938-7942.

4. S. Barash and B. Minke (1994), Is the receptor potential of fly photoreceptors a summation of single-photon responses? *Comments Theoretical Biology* 3, 229-263.

5. A. Baumann, S. Frings, M. Godde, R. Seifert and U.B. Kaupp (1994), Primary structure and functional expression of a *Drosophila* cyclic nucleotide-gated channel present in eyes and antennae. *EMBO J.* 13, 5040-5050.

6. O. Baumann and B. Walz (1989), Calcium- and inositol polyphosphate-sensitivity of the calcium-sequestering endoplasmatic reticulum in the photoreceptor cells of the honeybee drone. *J. Comp. Physiol. A* 165, 627-636.

7. U.W. Becker, H. Stieve and E. Wendel (1988), Is desensitization of crayfish retinula due to calcium depletion independent of stimulation?, in: Sense organs (Interfaces between environment and behavior) (N. Elsner and F.G. Barth, eds.). *Proc. 16th Göttingen Neurobiol. Conf.*. Georg Thieme Verlag, Stuttgart, New York, 5290.

8. J. Bentrop, A. Plangger and R. Paulsen (1993), An arrestin homolog of blowfly photoreceptors stimulates visual-pigment phosphorylation by activating a membrane-associated protein kinase. *Eur. J. Biochem.* 216, 67-73.

9. A. Blumenfeld, J.O. Erusalimsky, O. Heichal, Z. Selinger and B. Minke (1985), Light-activated guanosine trisphosphatase in *Musca* eye membranes resembles the prolonged depolarizing afterpotential in photoreceptor cells. *Proc. Natl. Acad. Sci.* 82, 7116-7120.

10. J.E. Brown, A. Combs, K. Ackermann and C.C. Malbon (1991), Light-induced GTPase activity and GTP-γ-S binding in squid retinal photoreceptors. *Vis. Neurosci.* 7, 589-595.

11. J.E. Brown (1986), Calcium and light-adaptation in invertebrate photoreceptors, in: The molecular mechanism of photoreception (H. Stieve, ed.). *Dahlem Konferenzen.* Springer Verlag, Berlin, Heidelberg, New York, Tokyo, 231-240.

12. J.E. Brown and J.R. Blinks (1974), Changes in intracellular free calcium concentration during illumination of invertebrate photoreceptors. Detection with aequorin. *J. Gen. Physiol.* 64, 643-665.

13. J.E. Brown and J.A. Coles (1979), Saturation of the response to light in *Limulus* ventral photoreceptor. *J. Physiol.* 296, 373-392.

14. J.E. Brown, J.A. Coles and L.H. Pinto (1977), Effects of injections of calcium and EGTA into the outer segments of retinal rods of *Bufo marinus. J. Physiol.* 269, 707-722.

15. J.E. Brown, A. Combs, K. Ackermann and C.C. Malbon (1991), Light-induced GTPase activity and GTP-γ-S binding in squid retinal photoreceptors. *Vis. Neurosci.* 7, 589-595.

16. J.E. Brown, M. Faddis and A. Combs (1992), Light does not induce an increase in cyclic-GMP content of squid or *Limulus* photoreceptors. *Exp. Eye Res.* 54, 403-410.

17. T. Byk, M. Bar-Yaacov, Y. N. Doza, B. Minke and Z. Selinger (1993), Regulatory arrestin cycle secures the fidelity and maintenance of the fly photoreceptor cell. *Proc. Natl. Acad. Sci.* 90, 1907-1911.

18. R. Calhoon, M. Tsuda and T.G. Ebrey (1980), A light-activated GTPase from *Octopus* photoreceptors. *Biochem. Biophys. Res. Commun.* 94, 1452-1457.

19. B.G. Calman and S.C. Chamberlain (1982), Distinct lobes of *Limulus* ventral photoreceptors. II. Structure and ultrastructure. *J. Gen. Physiol.* 80, 839-862.

20. M. Chabre and M.L. Applebury (1986), Interaction of photoactivated rhodopsin with photoreceptor proteins: The cGMP cascade, in: The molecular mechanism of photoreception (H. Stieve, ed.). *Dahlem Konferenzen*, Springer Verlag, Berlin, Heidelberg, New York, Tokyo, 51-66.

21. I. Claßen-Linke and H. Stieve (1986), The sensitivity of the ventral nerve photoreceptor of *Limulus* recovers after light adaptation in two phases of dark-adaptation. *Z. Naturforsch.* 41c, 657-667.

22. K. Contzen, K.H. Richter and K. Nagy (1994), Change of the light-activated current components by inhibiting the intracellular calcium release in *Limulus* photoreceptor, in: *17th Ann. Meet. Eur. Neurosci. Ass.*, abstracts 8901, Oxford University Press.

23. A. Deckert and H. Stieve (1991), Electrogenic Na^{+}-Ca^{2+}- exchanger, the link between intra- and extracellular calcium in the *Limulus* ventral photoreceptor. *J. Physiol.* 433, 467-482.

24. A. Deckert, K. Nagy, C.S. Helrich and H. Stieve (1992), Three components in the light-induced current of the *Limulus* ventral photoreceptor. *J. Physiol.* 453, 69-96.

25. O. Devary, O. Heichal, A. Blumenfeld, D. Cassel, E. Suss, S. Barash, C.T. Rubinstein, B. Minke and Z. Selinger (1987), Coupling of photoexcited rhodopsin to inositol phospholipid hydrolysis in fly photoreceptors. *Proc. Natl. Acad. Sci.* 84, 6939-6943.

26. M. Dorlöchter, G. Stommel and H. Stieve (1994), Bump oscillations evoked by local illumination of the *Limulus* ventral nerve photoreceptor, in: Sensory transduction (N. Elsner and H. Breer, eds.). *Proc. 22nd Göttingen Neurobiol. Conf.*. Georg Thieme Verlag, Stuttgart, New York, I, 24.

27. M.N. Faddis and J.E. Brown (1992), Flash photolysis of caged compounds in *Limulus* ventral photoreceptors. *J. Gen. Physiol.* 100, 547-570.

28. A. Fein and R.A. Cone (1973), *Limulus* rhodopsin: rapid return of transient intermediates to the thermally stable state. *Science* 182, 495-497.

29. A. Fein and D.W. Corson (1979), Both photons and fluoride ions excite *Limulus* ventral photoreceptors. *Science* 204, 77-79.

30. A. Fein, R. Payne, D.W. Corson, M.J. Berridge and R. Irvine (1984), Photoreceptor excitation and adaptation by inositol 1,4,5-trisphosphate. *Nature* 311, 157-160.

31. E.E. Fesenko, S.S. Kolesnikov and A.L. Lyubarsky (1984), Induction by cyclic GMP of cationic conductance in plasma membrane of retinal rod outer segment. *Nature* 313, 310-313.

32. J.M. Fyles, J. Baverstock, K. Baer and H.R. Saibil (1991), Effects of calcium on light-activated GTP-binding proteins in squid photoreceptor membranes. *Comp. Biochem. Physiol.* 98B, 215-221.

33. T.H. Goldsmith and G.D. Bernard (1985), Visual pigments of invertebrates. *Photobioch. Photobiol.* 42, 805.

34. T.H. Goldsmith and R. Wehner (1977), Restrictions on rotational and translational diffusion of pigment in the membranes of a rhabdomeric photoreceptor. *J. Gen. Physiol.* 70, 453-490.

35. F.H. Grus and H. Stieve (1991), In vivo and in vitro kinetics of the reactivation of non-transducing-rhodopsin in the median-eye of *Limulus polyphemus* measured with electrophysiological methods, in: Synapse - Transmission - Modulation (N. Elsner and H. Penzlin, eds.). *Proc. 19th Göttingen Neurobiol. Conf.*. Georg Thieme Verlag Stuttgart, New York, 255.

36. K. Hamdorf. (1979), The physiology of invertebrate visual pigments, in: *Handbook of Sensory Physiology*, Comparative Physiology and Evolution of Vision in Invertebrates. A. Invertebrate Photoreceptors (H. Autrum, ed.). Springer Verlag, Berlin, Heidelberg, New York, VII, 6A, 145-224.

37. K. Hamdorf and K. Kirschfeld (1980a), Prebumps: Evidence for double-hits at functional subunits in a rhabdomeric photoreceptor. *Z. Naturforsch.* 35c, 173-174.

38. K. Hamdorf and K. Kirschfeld (1980b), Reversible events in the transduction process of photoreceptors. *Nature* 283, 859-860.

39. K. Hamdorf and S. Razmjoo (1977), The prolonged depolarizing afterpotential and its contribution to the understanding of photoreceptor function. *Biophys. Struct. Mech.* 3, 163-170.

40. R.C. Hardie (1994), Effects of caged Ca^{2+} in *Drosophila* photoreceptors, in: Sensory Transduction (N. Elsner and H. Breer, eds.). *Proc. 22nd Göttingen Neurobiol. Conf.*. Georg Thieme Verlag, Stuttgart, New York, I, 20.

41. R.C. Hardie (1991), Whole-cell recordings of the light-induced current in dissociated *Drosophila* photoreceptors: evidence for feedback by calcium permeating the light-sensitive channels. *Proc. Roy. Soc. Lond. B*. 245, 203-210.

42. R.C. Hardie and B. Minke (1992), The trp gene is essential for a light-activated Ca^{2+} channel in *Drosophila* photoreceptors. *Neuron* 8, 643-651.

43. R.C. Hardie and B. Minke (1993), Novel Ca^{2+} channel underlying transduction in *Drosophila* photoreceptors: Implications for phosphoinositide-mediated Ca^{2+} mobilization. *TINS* 16, 371-376.

44. R. C. Hardie and B. Minke (1994), Calcium-dependent inactivation of light-sensitive channels in *Drosophila* photoreceptors. *J. Gen. Physiol.* 103, 409-427.

45. R.C. Hardie, A. Peretz, E. Suss-Toby, A. Rom-Glas, S.A. Bishop, Z. Selinger and B. Minke (1993), Protein kinase C is required for light adaptation in *Drosophila* photoreceptors. *Nature* 363, 634-637.

46. P. Hillman, S. Hochstein and B. Minke (1983), Transduction in invertebrate photoreceptors. Role of pigment bistability. *Physiol. Rev.* 63, 668-772.

47. P. Hillman, M.E. Keen and J. Winterhager (1977), Discussion of selected topics about the transduction mechanism in photoreceptors. *Biophys. Struct. Mech.* 3, 183-189.

48. R. Hubbard and R.C.C. St. George (1958), The rhodopsin system of the squid. *J. Gen. Physiol.* 41, 501-528.

49. H. Jarminowski, U. Schraermeyer and H. Stieve (1993), Calcium-accumulating organelles in the photoreceptor cells of the crayfish *Orconectes limosus*. *Comp. Biochem. Physiol. B* 105, 389-394.

50. E.C. Johnson, P.R. Robinson and J.E. Lisman (1986), Cyclic GMP is involved in the excitation of invertebrate photoreceptors. *Nature* 324, 468-470.

51. W. Keiper, J. Schnakenberg and H. Stieve (1984), Statistical analysis of quantum bump parameters in *Limulus* ventral nerve photoreceptors. *Z. Naturforsch.* 39c, 781-790.

52. A. Kirkwood, D. Weiner and J.E. Lisman (1989), An estimate of the number of G regulatory proteins activated per excited rhodopsin in living *Limulus. Proc. Natl. Acad. Sci.* 86, 3872-3876.

53. K. Kirschfeld (1966), Discrete and graded receptor potentials in the compound eye of the fly (*Musca*), in: *The functional organization of the compound eye* (C.G. Bernhard, ed.). Pergamon Press, 291-307.

54. Y.J. Lee, M.B. Dobbs, M.L. Verardi and D.R. Hyde (1990), A *Drosophila* gene encoding a visual system specific Gα molecule. *Neuron* 5, 889-898.

55. I. Levitan, P. Hillman and R. Payne (1993), Fast desensitization of the response to $InsP_3$ in *Limulus* ventral photoreceptors. *J. Biophys.* 64, 1354-1360.

56. S. Levy and A. Fein (1985), Relationship between light-sensitivity and intracellular free Ca^{2+}-concentration in *Limulus* ventral photoreceptors. A quantitative study using Ca^{2+}-selective micro-electrodes. *J. Gen. Physiol.* 85, 805-841.

57. P.G. Lillywhite (1977), Single photon-signals and transduction in an insect eye. *J. Comp. Physiol.* 122, 189-200.

58. J.E. Lisman (1985) The role of metarhodopsin in the generation of spontaneous quantum bumps in ultraviolet receptors of *Limulus* median eye. *J. Gen. Physiol.* 85, 171-187.

59. J.E. Lisman and J.E. Brown (1975), Effects of intracellular injection of calcium buffers on light adaptation in *Limulus* ventral photoreceptors. *J. Gen. Physiol.* 66, 489-506.

60. E. Lisman and Y. Sheline (1976), Analysis of the rhodopsin cycle in *Limulus* ventral photoreceptors using the early receptor potential. *J. Gen. Physiol.* 68, 487-501.

61. G. Maaz and H. Stieve (1980), The correlation of the receptor potential with the light-induced transient increase in intracellular calcium-concentration measured by absorption change of arsenazo III injected into *Limulus* ventral nerve photoreceptor cell. *Biophys. Struct. Mech.* 6, 191-208.

62. G. Maaz, K. Nagy, H. Stieve and J. Klomfaß (1981), The electrical light response of the *Limulus* ventral nerve photoreceptor, a superposition of distinct components - observable by variation of the state of light-adaptation. *J. Comp. Physiol. A* 141, 303-310.

63. R. Milecchia and A. Mauro (1969a), The ventral photoreceptor of *Limulus*. II. The basic photoresponse. *J. Gen. Physiol.* 54, 310-330.

64. R. Milecchia and A. Mauro (1969b), The ventral photoreceptor of *Limulus*. III. A voltage-clamp study. *J. Gen. Physiol.* 54, 331-351.

65. B. Minke and Z. Selinger (1991), Inositol lipid pathway in fly photoreceptors: Excitation, calcium mobilization and retinal degeneration, in: *Retinal Research* (Osborne, N.N., G.J. Chader, eds.), Pergamon Press, 99-124.

66. B. Minke and Z. Selinger (1992a), Intracellular messengers in invertebrate photoreceptors studied in mutant flies, in: *Neuromethods* (A. Boulton, G. Baker, C. Taylor, eds.). The Humana Press Inc., 20.

67. B. Minke and Z. Selinger (1992b), The inositol-lipid pathway is necessary for light excitation in fly photoreceptors, in: *Sensory Transduction*. The Rockefeller University Press 12, 202-217.

68. M. Montal, A. Darszon and H.W. Trissl (1977), Transmembrane channel formation in rhodopsin-containing bilayer membranes. *Nature* 267, 221-225.

69. K. Nagy (1990), Kinetic properties of single ion channels activated by light in *Limulus* ventral nerve photoreceptors. *Eur. Biophys. J.* 19, 47-54.

70. K. Nagy (1991), Biophysical processes in invertebrate photoreceptors: recent progress and a critical overview based on *Limulus* photoreceptors. *Quart. Rev. Biophys.* 24, 165-226

71. K. Nagy (1992), Non-independent quantum bumps in *Limulus* ventral photoreceptors - A new insight in the light transduction of invertebrate photoreceptors. *Neurosci. Lett.* 144, 99-102.

72. K. Nagy (1993), Cyclic nucleotides and inositol trisphosphate activate different components of the receptor current in *Limulus* ventral nerve photoreceptors. *Neurosci. Lett.* 152, 1-4.

73. K. Nagy (1994), Inhibition of the first component of the receptor current in *Limulus* photoreceptor. *Neuroreport* 5, 847-849.

74. K. Nagy and K. Contzen (1994), Synchronized activation of quantum bumps in *Limulus* ventral nerve photoreceptor, in: Sensory Transduction (N. Elsner, H. Breer, eds.). *Proc. 22nd Göttingen Neurobiol. Conf.*. Georg Thieme Verlag, Stuttgart, New York, 25.

75. K. Nagy and H. Stieve (1990), Light-activated single channel currents in *Limulus* ventral nerve photoreceptors. *Eur. Biophys. J.* 18, 221-224.

76. K. Nagy, K. Contzen and H. Stieve (1993), Two components of the receptor current are developed from distinct elementary signals in *Limulus* ventral nerve photoreceptor. *Eur. Biophys. J.* 33, 341-350.

77. E. Nasi (1991), Two light-dependent conductances in *Lima* rhabdomeric photoreceptors. *J. Gen. Physiol.* 97, 55-72.

78. E. Nasi and M. Gomez del Pilar (1992), Light-activated ion channels in solitary photoreceptors of the scallop *Pecten irradians*. *J. Gen. Physiol.* 99, 747-769.

79. E. Nasi and M. Gomez (1992), Electrophysiological recordings in solitary photoreceptors from the retina of squid, *Loligo pealei*. *Visual Neurosci.* 8, 349-358.

80. M. Nayal and E. Di Cera (1994), Predicting Ca^{2+}-binding sites in proteins. *Proc. Natl. Acad. Sci.* 91, 817-821.

81. C. Nobes, J. Baverstock and H. Saibil (1992), Activation of GTP-binding protein Gq by rhodopsin in squid photoreceptors. *Biochem. J.* 287, 545-548.

82. P.M. O`Day and M.P. Gray-Keller (1989), Evidence for electrogenic Na^+/Ca^{2+} exchange in *Limulus* ventral photoreceptors. *J. Gen. Physiol.* 93, 473-492.

83. W.L. Pak (1979), Study of photoreceptor function using *Drosophila* mutants, in: *Neurogenetics: Genetic approaches to the nervous system* (X.O. Breakefield, ed.). Elsevier, North Holland, New York, 67-99.

84. R. Payne (1980), Voltage noise accompanying chemically-induced depolarization of insect photoreceptors. *Biophys. Struct. Mech.* 6, 235-251.

85. R. Payne (1982), Fluoride blocks an inactivation step of transduction in a locust photoreceptor. *J. Physiol.* 325, 261-279.

86. R. Payne (1986), Phototransduction by microvillar photoreceptors of invertebrates: mediation of a visual cascade by inositol trisphosphate. *Photobiochem. Photobiophys.* 13, 373-397.

87. R. Payne and A. Fein (1987), Inositol 1,4,5-trisphosphate releases calcium from specialized sites within *Limulus* photoreceptors. *J. Cell Biol.* 104, 933-937.

88. R. Payne and T.M. Flores (1992), The latency of the response of *Limulus* photoreceptors to inositol trisphosphate lacks the calcium-sensitivity of that to light. *J. Comp. Physiol. A* 170, 311-316.

89. R. Payne and B.V.L. Potter (1991), Injection of inositol trisphosphorothionate into *Limulus* ventral photoreceptors causes oscillations of free cytosolic calcium. *J. Gen. Physiol.* 97, 1165-1186.

90. R. Payne, D.W. Corson and A. Fein (1986a), Pressure injection of calcium both excites and adapts *Limulus* ventral photoreceptors. *J. Gen. Physiol.* 88, 107-126.

91. R. Payne, D.W. Corson, A. Fein and M.J. Berridge (1986b), Excitation and adaptation of *Limulus* ventral photoreceptors by inositol 1,4,5 trisphosphate result from a rise in intracellular calcium. *J. Gen. Physiol.* 88, 127-142.

92. R. Payne, T.M. Flores and A. Fein (1990), Feedback inhibition by calcium limits the release of calcium by inositol trisphosphate in *Limulus* ventral photoreceptors. *Neuron* 4, 547-555.

93. R. Payne, B. Walz, S. Levy and A. Fein (1988), The localization of calcium release by inositol trisphosphate in *Limulus* ventral photoreceptors and its control by negative feedback. *Phil. Trans. Roy. Soc. Lond. B*. 320, 359-379.

94. V.M. Petrov, I.D. Artamonov and V.M. Lipkin (1994), Small GTP-binding proteins of squid photoreceptor interaction with photoactivated rhodopsin. *FEBS* 337, 274-276.

95. J.D.D. Pottinger, N.J.P. Ryba, J.N. Keen and J.B.C. Findlay (1991), The identification and purification of the heterotrimeric GTP-binding protein from squid (*Loligo forbesi*) photoreceptors. *Biochem. J.* 279, 323-326.

96. J.W. Putney Jr. (1986), A model for receptor-regulated calcium entry. *Cell Calcium* 7, 1-12.

97. J.W. Putney Jr. (1990), Capacitative calcium entry revisited. *Cell Calcium* 11, 611-624.

98. M. Rack, B. Xhonneux-Cremers, U. Schraermeyer and H. Stieve (1994), On the Ca^{2+}-dependence of inositol-phospholipid-specific phospholipase C of microvillar photoreceptors from *Sepia officinalis. Exp. Eye Res.* 58, 659-664.

99. B. Rayer, M. Naynert and H. Stieve (1990), Phototransduction: Different mechanisms in vertebrates and invertebrates. *J. Photochem. Photobiol. B: Biology* 7, 107-148.

100. H. Reuß and H. Stieve (1993), Two different bump types of the ventral photoreceptor of *Limulus*. *Z. Naturforsch.* 48c, 92-95.

101. H. Reuß, M. Rack and H. Stieve (1992), Spontaneous bumps of the *Limulus* photoreceptor cell are probably triggered by the spontaneous activation of single G-protein molecules. *Z. Naturforsch.* 47c, 932-936.

102. K.H. Richter and K. Nagy (1994), Change of the light-activated current components by inhibiting the intracellular calcium release in *Limulus* photoreceptor, in: *17th Ann. Meet. Eur. Neurosci. Ass.*. abstracts 8902, Oxford University Press.

103. P.R. Robinson and R.H. Cote (1989), Characterization of guanylate cyclase in squid photoreceptors. *Visual Neurosci*. 3, 1-7.

104. P.R. Robinson, S.F. Wood, E.Z. Szuts, A. Fein, H.E. Hamm and J.E. Lisman (1990), Light-dependent GTP-binding proteins in squid photoreceptors. *Biochem. J.* 272, 79-85.

105. H. Saibil (1984), A light-stimulated increase of cyclic GMP in squid photoreceptor. *FEBS Lett.* 12, 168, 213-216.

106. B. Schlösser and H. Stieve (1989), Cyclic nucleotides influence the dark adaptation in *Limulus* ventral photoreceptors, in: Dynamics and plasticity in neuronal systems (N. Elsner and W. Singer, eds.). *Proc. 17th Göttingen Neurobiol. Conf.*. Georg Thieme Verlag, Stuttgart, New York, 93.

107. J. Scholes (1965), Discontinuity of the excitation process in locust visual cells. *Cold Spring Harbor Symp. Quant. Biol.* .XXX, 517-527.

108. U. Schraermeyer, H. Stieve and M. Rack (1994), Cytochemical localization of guanylate cyclase in photoreceptor cells of fly and mouse, in: Sensory transduction (N. Elsner and H. Breer, eds.). *Proc. 22nd Göttingen Neurobiol. Conf.*. Georg Thieme Verlag, Stuttgart, New York, I, 12.

109. U. Schraermeyer, H. Stieve and M. Rack (1993), Cyclic 3′,5′-nucleotide phosphodiesterase: Cytochemical localization in photoreceptor cells of the fly *Calliphora erythrocephala. J. Neurocytol.* 22, 845-853.

110. W. Schröder, D. Frings and H. Stieve (1980), Measuring calcium uptake and release by invertebrate photoreceptor cells by laser microprobe mass spectroscopy. *Scan. Electron Microsc.* 11, 647-652.

111. J. Schwemer (1986), Turnover of photoreceptor membrane and visual pigment in invertebrate, in: The molecular mechanism of photoreception (H. Stieve, ed.). *Dahlem Konferenzen.* Springer Verlag, Berlin, Heidelberg, New York, Tokyo, 303-326.

112. M. Seidou, K. Ohtsu, Z. Yamasita, K. Narita and Y. Kito (1993), The nucleotide content of the *Octopus* photoreceptor cells: No changes in the *Octopus* retina immediately following an intense light flash. *Zool. Sci.* 10, 275-279.

113. R. Seifert, U.B. Kaupp and A. Baumann (1994), Primary structure and functional expression of a *Drosophila* cyclic nucleotide-gated channel.

114. Z. Selinger, Y.N. Doza and B. Minke (1993), Mechanisms and genetics of photoreceptors desensitization in *Drosophila* flies. *Biochim. Biophys. Acta* 1179, 283-299.

115. W.C. Smith, D.A. Price, R.M. Greenberg and B.A. Batelle (1993), Opsins from the lateral eyes and ocelli of the horseshoe crab, *Limulus polyphemus*. *Proc. Natl. Acad. Sci.* 90, 6150-6154.

116. H. Stieve (1981), Roles of calcium in visual transduction in invertebrates, in: *Sense organs* (M.S. Laverack and D.J. Cosens, eds.). Blackie and Son Ltd., Glasgow and London, 163-185.

117. H. Stieve (1986), Bumps, the elementary excitatory responses of invertebrates, in: The molecular mechanism of photoreception (H. Stieve, ed.). *Dahlem Konferenzen*, Springer-Verlag, Berlin, Heidelberg, New York, Tokyo, 199-230.

118. H. Stieve (1990), Microvillar photoreceptor cells of invertebrates, in: *Sensory Transduction* (A. Borsellino et al., eds.). Plenum Press, New York, 109-123.

119. H. Stieve and S. Benner (1992), The light-induced rise in cytosolic calcium starts later than the receptor current of the *Limulus* ventral photoreceptor. *Vision Res.* 32, 403-416.

120. H. Stieve and M. Bruns (1978), Extracellular calcium, magnesium, and sodium ion competition in the conductance control of the photosensory membrane of *Limulus* ventral nerve photoreceptor. *Z. Naturforsch.* 33c, 574-579.

121. H. Stieve and M. Bruns (1980), Dependence of bump rate and bump size in *Limulus* ventral nerve photoreceptors on light-adaptation and calcium concentration. *Biophys. Struct. Mech.* 6, 271-285.

122. H. Stieve and M. Bruns (1983), Bump latency distribution and bump adaptation of *Limulus* ventral nerve photoreceptor in varied extracellular calcium concentrations. *Biophys. Struct. Mech.* 9, 329-339.

123. H. Stieve and I. Claßen-Linke (1980), The effect of changed extracellular calcium and sodium concentration on the electroretinogram of the crayfish retina. *Z. Naturforsch.* 35c, 308-318.

124. H. Stieve and M. Pflaum (1978), Lowering the ratio of extracellular calcium to sodium mimics the effect of light on the photosensory membrane of *Limulus* ventral nerve photoreceptor. *Vision Res.* 18, 883-885.

125. H. Stieve and B. Schlösser (1989), The light energy dependence of the *Limulus* photoreceptor current in two defined states of adaptation. *Z. Naturforsch.* 44c, 999-1014.

126. H. Stieve and G. Stommel (1992), Cooperativity in signal transduction leading to intracellular Ca^{2+} release in *Limulus* ventral nerve photoreceptor, in: Rhythmogenesis in neurons and networks (N. Elsner and D.W. Richter, eds.). *Proc. 20th Göttingen Neurobiol. Conf.*. Georg Thieme Verlag, Stuttgart, New York, 287.

127. H. Stieve (1988), What do we know about the molecular mechanism of visual excitation?. *Naturwissenschaften* 75, 288-296.

128. H. Stieve, M. Bruns and H. Gaube (1983), The influence of the extracellular concentration of calcium, magnesium and sodium on the repolarizing phase of the receptor potential of the *Limulus* ventral nerve photoreceptor. Z. Naturforsch. 38c, 471-483.

129. H. Stieve, M. Bruns and H. Gaube (1983), The intensity dependence of the receptor potential of the *Limulus* ventral nerve photoreceptor in two defined states of light- and dark adaptation. *Z. Naturforsch.* 38c, 1043-1054.

130. H. Stieve, M. Bruns and H. Gaube (1984), The sensitivity shift due to light adaptation depending on the extracellular calcium ion concentration in *Limulus* ventral nerve photoreceptor. *Z. Naturforsch.* 39c, 662-679.

131. H. Stieve, B. Niemeyer, K. Aktories and H.E. Hamm (1992), Disturbing GTP-binding protein function through microinjection into the visual cell of *Limulus*. *Z. Naturforsch.* 47c, 915-921.

132. H. Stieve, M. Pflaum, J. Klomfaß and H. Gaube (1985), Calcium/sodium binding competition in the gating of light-activated membrane conductance studied by voltage clamp technique in *Limulus* ventral nerve photoreceptor. *Z. Naturforsch.* 40c, 278-291.

133. H. Stieve, H. Reuß, H.T. Hennig and J. Klomfaß (1991), Single photon-evoked events of the ventral nerve photoreceptor cell of *Limulus*. *Z. Naturforsch.* 46c, 461-486.

134. H. Stieve, G. Rüsing, W. Yuan and H.T. Hennig (1994), The effects of externally applied lithium and strontium on the arsenazo-monitored cytosolic calcium signal of the *Limulus* ventral photoreceptor. *Z. Naturforsch.* 49c, 372-392.

135. H. Stieve, J. Schnakenberg, A. Kuhn and H. Reuß (1986), An automatic gain control in the *Limulus* photoreceptor, in: Membrane Control (H. Lüttgau, ed.). *Fortschr. d. Zool.*, Gustav Fischer Verlag, Stuttgart, New York, 33, 367-376.

136. G. Stommel and H. Stieve (1995) Effects of reduced external calcium on the light-stimulated photoreceptors. (N. Elsner and R. Menzel, eds.) *Proc. 23rd Göttingen Neurobiol. Conf.*. Georg Thieme Verlag, Stuttgart, New York.

137. G. Stommel, G. Rüsing, W. Yuan and H. Stieve (1994), Two different components of the light stimulated increase in cytosolic calcium or strontium in *Limulus* ventral nerve photoreceptor, in: Sensory Transduction (N. Elsner and H. Breer, eds.). *Proc. 22nd Göttingen Neurobiol. Conf.*. Georg Thieme Verlag, Stuttgart, New York, I, 4.

138. T. Suzuki, K. Narita, K. Yoshihara, K. Nagai and Y. Kito (1993), Immunochemical detection of GTP-binding protein in cephalopod photoreceptors by anti-peptide antibodies. *Zool. Sci.* 10, 425-430.

139. E.Z. Szuts, S.F. Wood, M.S. Reid and A. Fein (1986), Light stimulates the rapid formation of inositol trisphosphate in squid retinas. *Biochem. J.* 240, 929-932.

140. M. Takagi (1994), Light-dependent ion channels in *Octopus* visual cell membranes: the permeability of large organic ions and effects of cyclic GMP. J. Photochem. *Photobiol. B: Biology* 23, 161-177.

141. M. Takagi, K. Azuma, I. Nakagaki and S. Sasaki (1994), Extracellular and intracellular ion concentration of *Octopus* visual cells and ion permeability of the cell membranes. *J. Photochem. Photobiol. B: Biology* 23, 189-200.

142. A. Terakita, T. Hariyama, Y. Tsukahara, Y. Katsukura and H. Tashiro (1993), Interaction of GTP-binding protein Gq with photoactivated rhodopsin in the photoreceptor membranes of crayfish. *FEBS* 33O, 197-200.

143. M. Tsuda (1987), *Octopus* G-protein. A signal coupling protein in invertebrate photoreceptor, in: *Retinal Proteins* (Ovchinnikov, ed.). VNU Science Press, Utrecht, 393-404.

144. M. Tsuda and T. Tsuda (1990), Two distinct light regulated G-proteins in *Octopus* photoreceptors. *Biochim. Biophys. Acta* 1052, 204-210.

145. K. Yu. Ukhanov (1991), Ommochrome pigment granules: a calcium reservoir in the dipteran eyes. *Comp. Biochem. Physiol. A* 98, 9-16.

146. C.A. Vandenberg and M. Montal (1984), Light-regulated biochemical events in invertebrate photoreceptors. 1. Light-activated guanosine-trisphosphate, guanine nucleotide-binding, and cholera toxin catalyzed labeling of squid photoreceptor membranes. *Biochemistry* 23, 2339-2347.

147. K. Vogt and K. Kirschfeld (1984), Chemical identity of the chromophores of fly visual pigment. *Naturwissenschaften* 71, 211-213.

148. J.P. Walrond, E.Z. Szuts (1992), Submicrovillar tubulus in distal segments of squid photoreceptors detected by rapid freezing. *Neurosci.* 12, 1490-1501.

149. B. Walz (1982), Calcium-sequestering smooth endoplasmatic reticulum in retinula cells of the blowfly. *J. Ultrastruct. Res.* 81, 240-248.

150. B. Walz and O. Baumann (1989), Calcium-sequestering cell organelles: in situ localization, morphological and functional characterization, in: *Progress in Histochemistry and Cytochemistry* (W. Grauman, Ph.U. Heitz, Z. Lojoa, A.G.E. Pearse and T.H. Schiebler, eds.). Gustav Fischer Verlag, Stuttgart, New York 20, 1-47.

151. B. Walz, O. Baumann and B. Zimmermann (1994), Ins (1,4,5)P_3-independent Ca^{2+} induced Ca^{2+} release from the endoplasmatic reticulum in honeybee drone photoreceptors, in: Sensory Transduction (N. Elsner and H. Breer, eds.). *Proc. 22nd Göttingen Neurobiol. Conf.*. Georg Thieme Verlag, Stuttgart, New York, I, 3.

152. U. Werner, E. Suss-Toby, A. Rom and B. Minke (1992), Calcium is necessary for light excitation in barnacle photoreceptors. *J. Comp Physiol. A* 170, 427-434.

153. S.F. Wood, E.Z. Szuts and A. Fein (1989), Inositol trisphosphate production in squid photoreceptors. *J. Biol. Chem.* 264, 12970-12976.

154. C.W. Wu and W.L. Pak (1975), Quantal basis of photoreceptor spectral sensitivity of *Drosophila melanogaster*. *J. Gen. Physiol.* 66, 149-168.

155. C.F. Wu and W.L. Pak (1978), Light-induced voltage noise in the photoreceptor of *Drosophila melanogaster*. *J. Gen. Physiol.* 71,249-268.

156. S. Yeandle and M.G. Fuortes (1964), Probability of occurrence of discrete potential waves in the eye of *Limulus*. *J. Gen. Physiol.* 47, 443-463.

157. W. Yuan, H.T. Hennig and H. Stieve (1994), Different calcium sensitivity of the two types of bumps in the ventral photoreceptor of *Limulus*, in: Sensory Transduction (N. Elsner and H. Breer, eds.). *Proc. 22nd Göttingen Neurobiol. Conf.*. Georg Thieme Verlag, Stuttgart, New York, I, 21.

158. A. Ziegler and B. Walz (1989), Analysis of extracellular calcium and volume changes in the compound eye of the honeybee drone, *Apis mellifera*. *J. Comp. Physiol. A* 165, 697-709.

INOSITOL LIPID SIGNALING IN *DROSOPHILA* PHOTOTRANSDUCTION

Zvi Selinger
Department of Biological Chemistry, Life Scieces Institute, and the Minerva Center for Studies of Visual Transduction, The Hebrew University, Jerusalem 91904, Israel

1. Introduction

In the process of phototransduction, photoreceptor cells use chemical reactions to translate and amplify incoming light stimuli to produce finally an electrical signal that can be communicated to other cells of the nervous system. A central question in vision research is therefore the identity of the biochemical pathway that underlies the visual process. A common strategy to unravel such complex pathways has been to dissect the system by single point mutations, each affecting a discrete step of the enzyme cascade. Extensive genetic studies of the fruit fly *Drosophila melanogaster*, initiated at the turn of this century by Morgan and colleagues (1), and since then greatly extended and advanced in many other laboratories have established the fruit fly Drosophila melanogaster as an extremely useful experimental system for genetic analysis of complex biological phenomena. In principal it is possible to systematically screen each of the chromosomes of Drosophila for mutation that produce any identifiable phenotype. Over the years about two dozen different Drosophila visual mutants have been isolated. Recently, molecular, physiological and biochemical characterization of these mutants are producing multifacet insights into the process of invertebrate phototransduction and the mechanisms of its regulation.

2. Isolation of Drosophila visual mutants

The pronounced visual reflexes that dominate the behavior of Drosophila flies enabled several investigators to devise screens for isolation of Drosophila visual mutants. Application of these screens has revealed, however, that they are far too complex for isolation of visual mutants. To get more direct assessment of mutants defective in the visual process, several laboratories have initiated the use of electroretinogram recording (ERG) as a mutant selection scheme (2,3).

Extracellular recording of the ERG measures the current flow into the eye following or during illumination. It is made up of several components not all of which arise from the photoreceptor cells. Nevertheless, the fly's ERG gives a large signal and the method can be easily applied, making it a convenient assay. Despite the need to apply the assay to individual flies, which is extremely labor intensive, the simplicity of the ERG measurement and particularly its direct relationship to the output signal of the photoreceptor cells indeed yielded mutants that were specifically defective in the phototr~n~duction process (2,3). However, even the ERG selection screen did not yield the large number of mutants expected from a multistep and complex process as the phototransduction cascade. Disappointingly, all the above screens produced rather a small number of overlapping mutants. For example, only one mutant that eliminates the light induced electrical response, the no receptor potential A (norpA) mutant have been isolated.

<u>The prolonged depolarizing after potential (PDA) is a useful screen for isolation of visual mutants</u>

Pak who pioneered the use of ERG recording for isolation of visual mutants, has recognized the shortcomings of the X-chromosome linked ERG screen. Seeking a more reliable and better revealing test for isolation of visual mutants Pak introduced the prolonged depolarizing after potential (PDA), screen, and extended his search also to mutants of the second and third chromosomes (4). The PDA response is brought about by genetically removing the red screening pigment and application of blue light which is preferentially absorbed by rhodopsin and thus results in large net conversion of rhodopsin to metarhodopsin. As will be later explained, large net conversion of rhodopsin to metarhodopsin causes disruption of the termination process at the level of the photopigment, which in turn results in sustained excitation long after the light is turned off. Throughout this period the photoreceptors are partially desensitized (inactivated) and are less sensitive to subsequent test lights. Since the PDA test the maximal capacity of the photoreceptor cell to maintain excitation for an extended period and is strictly dependent on the presence of high concentrations of rhodopsin, it detects even minor defects in rhodopsin biogenesis and easily scores deficient replenishment of phototransduction components which results in temporary inactivation of the phototransduction process. Indeed the PDA screen yielded a plethora of new and most interesting visual mutants (reviewed in Ref. 4). Remarkably, to this day the PDA mutants that Pak and his colleagues isolated are the main source of mutants for analysis of the Drosophila visual system. By functional criteria one of the group of the PDA mutants lost all the features of the PDA as they were no longer able to maintain the prolong depolarizing after potential with all its associated features. They were termed nina mutants, which stands for 'neither inactivation nor afterpotential'. Among the nina PDA mutants, ten different allelic groups were isolated (belonging to different mutual complementation groups) designated A-J. All the nina mutants were found to have reduced levels of rhodopsin. The second group of PDA mutants lost the ability to maintain excitation either during intense illumination or in response to an illumination protocol that normally elicits the prolonged depolarizing after potential. These mutants, consisting of six allelic groups, are termed in a A-F, which stands for 'inactivation but not afterpotential'. In contrast to the nina group the ina mutants have close to normal rhodopsin levels.

Photoexcited Rhodopsin Activates a G Protein

In vertebrate rods, photoisomerization of rhodopsin leads to dissociation of the retinal chromophore from the protein opsin. Regeneration of rhodopsin is achieved only after an extended period by a series of chemical reactions, some of which are carried out outside the photoreceptor cell (5). In contrast, in the invertebrates, absorption of light by rhodopsin (R) converts R to the stable metarhodopsin (M), and light absorption by metarhodopsin regenerates M back to R. In flies, R and M have maximal absorptions at ≈ 490 nm and ≈ 570 nm, respectively (6). Thus, by the illumination with either blue (converting) or red (regenerating) lights, the content of M, the active form of the photopigment, can be easily manipulated. We used intense illumination, which brings the photopigment to photoequilibrium between the R and M states, to measure the fractional amount of M at each wavelength of light (the photoequilibrium spectrum of M).

To test which state of the photopigment gives rise to light-dependent GTPase activity, we monitored GTP hydrolysis during illumination with saturating lights of different wavelengths. These studies showed that the action spectrum of the light-dependent GTPase coincides with the photoequilibrium spectrum of metarhodopsin, indicating that metarhodopsin gives rise to the light-dependent GTPase activity.

We also determined the number of G-protein molecules in the photoreceptors by monitoring the time course of [^{35}S]GTPγS binding to eye membrane preparations, preilluminated with blue light. Binding assays over a range of GTPγS concentrations from 10^{-8} M to 10^{-7} M all saturated at about 10 pmoles GTPγS bound per milligram membrane protein, indicating that this value is the total number of binding sites in the membrane. Calculations based on the presence of 1.8×10^{4} peripheral retinula cells in a compound Musca eye, an average yield of 5μg membrane protein per eye and assuming that one GTPγS binds to one G-protein molecule, give a content of 1.6×10^6 molecules of G protein in a peripheral Musca photoreceptor cell.

To identify the G protein, we employed an [α-^{32}P] azidoanilido analog of GTP to label the light-dependent GTP-binding protein component in the Musca eye membranes. Polyacrylamide gel electrophoresis and autoradiograpy revealed in the blue illuminated membranes a labeled protein band of 41 kD that was not seen in the systems illuminated with red light (7).

LIGHT-ACTIVATED PHOSPHOLIPASE C

Evidence for a light-dependent PLC in fly photoreceptors came from our biochemical experiments in membrane preparations of *Musca* and *Drosophila* eyes (7). We have developed membrane preparation in which light-dependent phosphoinositide hydrolysis can be studied under defined conditions and effects of activators and inhibitors can be analyzed. *Musca* eyes or *Drosophila* heads containing intact cells were preincubated with [^{3}H]inositol to label the phosphoinositides. The membrane prepared from these cells was used to analyze the hydrolysis of phosphoinosites. The *Musca* eye membrane preparation responded to illumination by an increase in the accumulation of inositol trisphosphate (InsP3) and inositol bisphosphate ($InsP_2$), the respective products of polyphosphoinositide hydrolysis by a phospholipase-C-type enzyme. In the illuminated system, there is a small increase in the concentration of $InsP_3$ for the first 5 min, then it levels off and decreases slightly at longer incubation periods. The accumulation of inositol phosphate $InsP_1$ (not shown) is characterized by a lag period, indicating that this product may arise by subsequent step-wise enzymatic dephosphorylation of polyphosphoinositols

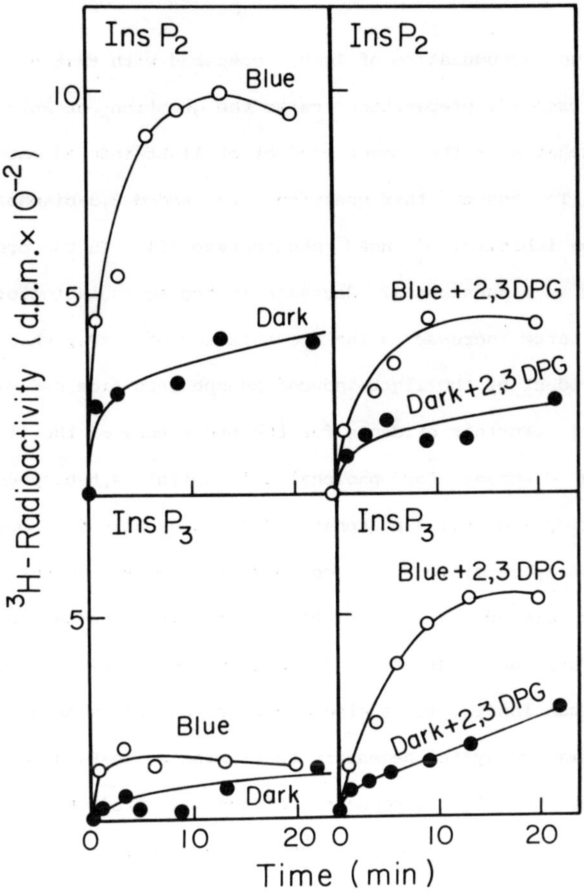

Figure 1 - Light-induced hydrolysis of phosphoinositides in *Musca* eye membranes and effect of 2,3 biphosphoglycerate. Equivalents of 100 *Musca* eyes each cut into two halves were incubated in the dark for 4 hr at 30 °C with [H]inositol. Preparation of eye membranes and measurement of light-dependent phosphoinositide hydrolysis were carried out as described previously (7). The medium was adjusted with Ca-EGTA to give 50 nM free Ca^{2+}. The reaction was initiated by the addition of membranes (500 μg protein/ml). The right column shows the effect of DPG (10 nm) added together with $MgCl_2$ (10 mM) (from 7).

(Fig. 1).

The large accumulation of $InsP_2$ compared with that of $InsP_3$ in the *Musca* and *Drosophila* preparation raise the question of which of the two inositol phosphates is the major product of light-induced phosphoinositide hydrolysis. To answer this question, we added 2,3-bisphosphoglycerate (DPG), a known inhibitor of InsP*3* phosphatase (8). In the presence of 10 mM DPG, there is a substantial decrease in the accumulation of InsP2 and a concurrently large increase in the accumulation of $InsP_3$, which now becomes the major product of the light-induced phosphoinositide hydrolysis (Fig. 1 right panels). Further evidence for the preference of the light-dependent phospholipase C enzyme for phosphatidylinositol 4,5-bisphosphate (PIP_2) over phosphatidylinositol 4-phosphate (PIP) comes also from our observation that, under assay conditions, the concentration of PIP in the fly membranes is five times higher than that of PIP_2, the immediate substrate for $InsP_3$ production. Yet, when hydrolysis of $InsP_3$ is blocked by DPG, more $InsP_3$ is accumulated than $InsP_2$. It is also apparent that fly membranes are endowed with the enzymatic system necessary to eliminate $InsP_3$ after it has been produced. Both the light-dependent preferential production of $InsP_3$ and a turn-off mechanism to stop its action are consistent with an internal messenger role of $InsP_3$ in fly phototransduction.

INSP3 EXCITES PHOTORECEPTORS OF THE FLY

Two criteria of an internal second messenger are (i) introduction of the putative messenger into the cell should reproduce the physiological response; and (ii) agents that inhibit the inactivation of the messenger should augment the effect of the physiological stimulus. These criteria were tested for by light-induced introduction of $InsP_3$ and of DPG into the *Musca* photoreceptor cells. We used pressure injection of $InsP_3$ into the intact eye followed by a 1-min maximal-intensity white light to introduce the $InsP_3$ into the photoreceptors.

The introduction of hydrophilic molecules into fly photoreceptor cells was investigated in detail by Wilcox and Franceschini (8). They showed that lucifer yellow shows light-dependent introduction that stains the whole photoreceptor cell up to its terminal possibly via receptor-mediated pinocytosis. Minke and Stephenson (10) showed that extracellular application of GTPγS and guanosine 5'-[ß,γ-imido]triphosphate had no effect on the membrane potential of *Musca* photoreceptors in the dark but induced noisy depolarization upon illumination in the presence of the nucleotides. In the experiments where white or red lights were used to introduce hydrophilic compounds into the photoreceptors and to stimulate the cells. These lights do not cause prolonged, depolarization (PDA) as that caused by a blue light.

The introduction of $InsP_3$ into the cell by 1-min maximal-intensity white light (Fig. 2) resulted in a large increase in baseline noise in the dark. Subsequent application of an intense 1-sec white test light showed a response with an increased peak amplitude and an extended noisy depolarization that declined in the dark at a slower rate than the control. After $InsP_3$ plus DPG injection. One brief red flash was sufficient to create large voltage fluctuations in the dark similar to quantum bumps. One such flash in the presence of $InsP_3$ alone had no effect. On the other hand, in the presence of both $InsP_3$ and DPG the first maximal-intensity, 1-sec white test pulse elicited a response with an increased peak amplitude and long (20 Min), noisy depolarize. Clearly combination of $InsP_3$ and DPG increased the effect of $InsP_3$ in that much lower light intensities (i.e. less $InsP_3$ in the cell) were needed to induce the excitatory effect, which was much more pronounced and lasted longer. With DPG alone 1 min of

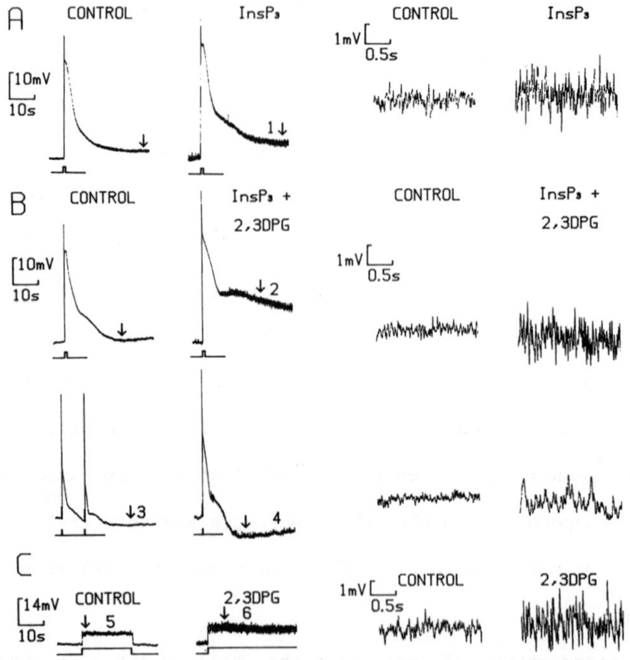

Figure 2 - Excitation of *Musca* photoreceptors by inositol-1,4,5-triphosphate and by 2,3 biphosphoglycerate. The first column from the left shows the control responses before injection, and the second from the left shows responses to identical stimuli in the same cells after pressure injectionns of the indicated test compounds. The third and fourth columns from the left show enlarged segments of the regions of the first and second columns indicated by arrows. Segments from which power spectra were calculated are indicated by numbers. Test compounds were pressure injected in the dark into the retinal extracellular space resulting in a 1:10 dilution (A) InsP3 (1mM) was introduced into the cell by maximal-intensity white illumination (100 W, 12 V halogen lamp; light energy at the level of the eye 20.5 mW cm-2) of 1 min duration. (B, upper) Injection of a solution containing 1mM InsP and 50 mM DPG using a 1 sec white-light stimulation to introduce the compound into the cell. (B, lower) Responses the orange flashes, one short orange flash (150-J photographic flash in conjunction with Scott OG 570 nm edge filter) was sufficient to induce noise. The lower responses preceded the upper responses. (C) Injection of 50 mM DPG solution followed by a 1 min maximal-intensity white light exposure. The responses shown in C are receptor potentials elicited by continuous, dim red light (Schott OG 590 nm edge filter with an intensity of 10.2 mW cm^{-2} attenuated by 4 log units by a neutral-density filter). (Modified from 7).

maximal intensity white light was needed to introduce the DPG into the cell. After penetration of DPG into the cell continuous dim light gave a response with an enhanced amplitude and a higher noise level relative to the control.

Noise analysis suggests that $InsP_3$-induced noisy depolarization is similar to the noisy depolarization induced by dim light and that 2,3-biphosphoglycerate enhances the effects of $InsP_3$ and of dim light suggesting an endogenous production of $InsP_3$ during illumination (not shown).

PHOSPHOLIPASE C ACTIVITY IS CONTROLLED BY A G-PROTEIN

There are a number of indications that PLC activity is controlled by a G-protein (fly, 7) and (squid 11). *In vitro* measurements on the squid retina using exogenous PIP_2 demonstrated that light-activated PLC is GTP dependent (12). In our *in vitro* system omitting GTP from the medium had only little effect on phosphoinositide hydrolysis, possibly is due to the ATP regenerating system needed in our preparation. Therefore, the involvement of a G-protein was tested by addition of GDPßS and GTPγS, the hydrolysis resistant analogs of GDP and GTP, respectively.

Addition of 0.1 mM GDPßS completed inhibited the light-induced hydrolysis of inositol phospholipid. Illumination for 30 sec with blue light increased the rate of inositol phospholipid hydrolysis for ~10 min. However, after illumination for 30 sec with red light, the rate of $InsP_3$ accumulation was not much different from that of a control system incubated in the dark. In contrast, with GTPγS illumination for 30 sec with either blue or red light persistently activated the hydrolysis of inositol phospholipid which was much more pronounced than in the control.

We have also observed an increase accumulation of $InsP_2$ and $InsP_3$ in

Musca eye membranes incubated with fluoride ions (F^-) in the dark which resembles the ability of F^- to bypass the receptor and directly activate the adenylate cyclase and cGMP phosphodiesterase through its actions on the stimulatory G protein and transducin, respectively (13,14). This conclusion was supported by the enhancement of F^- effect by aluminium Bigay et al., 1985). In accord with these findings we have reported that F^- excites the intact fly photorecptors in the dark (10). This excitation can be accounted for by an increased inositol phospholipid hydrolysis elicited by F^-.

Our biochemical and pharmacological experiments strongly suggest that the flow of information in the fly phototransduction cascade is from photo-excited rhodopsin to G-protein and from the G-protein to phospholipase C. We have also found that $InsP_3$ is the major inositolphosphate formed by the light-activated phospholipase C and that $InsP_3$ is rapidly hydrolyzed by specific phosphomonoesterase. Furthermore, introduction of $InsP_3$ into the cell excites the fly photoreceptor indicating that it may function as a second messenger in phototransduction. This has been further corroborated by demonstrating that when 2,3-bisphosphoglycerate, which inhibits $InsP_3$ hydrolysis, is introduced into the cell, it greatly potentiates the excitation by $InsP_3$ as well as augments the physiological excitation by dim light.

References

1. Morgan, T.H., Sturtevant, A.H., Muller, H.J. and Bridges, C.B. (1915) The mechanism of Mendelian heredity, Holt, New York.

2. Pak, W.L., Grossfield, J. and Arnold, K. (1970) Nature 227:518-520.

3. Koenig, J. and Merriam, J. (1977) Drosoph. Inf. Ser. 52:50-51.

4. Pak, W.L. (1979) in Neurogenetics; Genetic Approaches to the Nervous system (Breakfield, X., ed.), pp. 67-99, Elsevier, North Holland.

5. Dowling, J.E. (1960) Nature 188:114.

6. Hamdorf, K. (1979) Handbook of sensory physiology (ed. H. Autumn) vol. 7, p. 145. Springer-Verlag, Berlin.

7. Devary, O., O. Heichal, A. Blumenfeld, D. Cassel, E. Suss, S. Barash, C.T. Rubinstein, B. Minke and Z. Selinger (1987) Proc. Natl. Acad. Sci. 84:6939.

8. Downes, C.P., M.C. Mussat, and R.H. Michell. (1982) Biochem. J. 203:169.

9. Wilcox, M. and N. Franceschini (1984) Science 225:851.

10. Minke, B. and R.S. Stephenson (1985) J. Comp. Physiol. 156:339.

11. Wood, S.F., Szuts, E.Z. and Fein, A. (1989) J. Biol. Chem. 264:12970-12976.

12. Baer, K.M. and Saibil, H.R. (1988) J. Biol. Chem 263:17-20.

13. Northup, J.K., Sternweis, P.C. Smigel, M.D., Schliefer, L.S., Ross, E.M. and Gillman, A.G. (1980) Proc. Natn. Acad. Sci. U.S.A. 77:6516-6520.

14. Bigay, J. Deterre, P., Pfister, C. and Chabre, M. (1985) FEBS Lett. 191:181-185.

THE *TRP* PROTEIN IS A NOVEL Ca^{2+} CHANNEL UNDERLYING TRANSDUCTION IN *DROSOPHILA* PHOTORECEPTORS

Baruch Minke

Department of Physiology and Minerva Center for Studies of Visual Transduction, The Hebrew University - Hadassah Medical School, Jerusalem 91120, Israel

The inositol-lipid signalling cascade mediate phototransduction in *Drosophila* photoreceptors

Invertebrate photoreceptors use the phosphoinositide signalling cascade for phototransduction providing a powerful model system to study this important signalling cascade (Fein et al.,1984; Brown et al.,1984; Devary et al.,1987; Payne et al.,1988; Bloomquist et al.,1988; Minke and Selinger,1991; Hardie and Minke, 1993).

Genetic experiments in *Drosophila* have shown that operation of the phospholipase C enzyme is absolutely required to produce the output signal of depolarization. No parallel pathway that obviates the need for phospholipase C activity exists, as null alleles of the *norpA* locus, that codes the eye specific phospholipase C (Bloomquist et al.,1988) have no measurable electrical response to supersaturating light stimulus (Minke and Selinger, 1992). Furthermore, the $norpA^{H52}$ allele which has a temperature sensitive phospholipase C (Selinger and Minke,1988) looses its response to light seconds after heating to the non-permissive temperature (Deland and Pak,1973). This excludes any conceivable long range and indirect effects on the response to light due to the lack of phospholipase C or its products. Whether the effect of cGMP found in the *Limulus* photoreceptors is an epiphenomenon or whether cGMP is the cellular transmitter that directly opens the light sensitive channels, is not clear at present. The genetic evidence in *Drosophila* indicate, however, that if cGMP is the cellular transmitter of visual transduction, this signal must be generated subsequent to, and is absolutely dependent on, phospholipase C activity. It should also be noted that vertebrate photoreceptors have evolved a transduction mechanism which is uniquely exclusive to these cells, while the invertebrate photoreceptors make use of the inositol lipid signalling system which is ubiquitously present in almost every eukaryotic cell and has been implicated in a plethora of cellular functions (Berridge,1993). It can thus be expected that studies of *Drosophila* photransduction will have wide applications to a variety of activities in other cells.

Since both IP_3 and Ca^{2+} can mimic excitation in *Limulus* and fly (Brown et al. 1984;

Fein et al. 1984; Payne et al. 1986; Devary et al. 1987; Hardie and Minke 1992) it might be supposed that the cascade is completed by Ca^{2+} released from the IP_3-sensitive Ca^{2+} stores (SMC) directly gating cation channels in the plasma membrane. However, at least five lines of arguments have been raised to challenge this simple conclusion. 1) In *Limulus* it has not been possible to block the light induced excitation with the Ca^{2+} chelating agents EGTA or BAPTA (Payne et al., 1986; but see Shin et al., 1993). 2) Light-activated channels in excised patches from *Limulus* are not reported to be gated by Ca^{2+}. The only substance reported to activate such patches is cGMP (Bacigalupo et al., 1991); however, there is no convincing evidence for regulation of cGMP levels by light (Selinger, unpublished). 3) It has not been possible to demonstrate that the light-induced rise in Ca^{2+}_i precedes the electrical response (Stieve and Benner, 1992). 4) Studies in various rhabdomeric photoreceptors (*Limulus*, *Lima* and *Drosophila*) suggest that there may be multiple classes of light-activated channels raising the possibility that these may be activated by different mechanisms and/or messengers (Nagy, 1991; Hardie and Minke, 1992; Nasi, 1991). 5) Rapid concentration jumps of internal Ca^{2+} using caged Ca^{2+} compounds did not excite "blind" mutant photoreceptors of *Drosophila* (Hardie, 1994).

The *trp* protein is necessary for maintained excitation

One problem which faces inositol lipid signaling in cells like the invertebrate photoreceptors is the need for replenishment of Ca^{2+} in the IP_3 -sensitive stores. Once Ca^{2+} is released from these stores in response to an increase in IP_3 concentration, a major part of the released Ca^{2+} is extruded from the cell by the Na^+/Ca^{2+} exchanger and the plasma membrane Ca^{2+} ATPase while only a fraction is taken back into the endoplasmic reticulum vesicles. Continuous operation of the inositol lipid signaling system should thus result in Ca^{2+} deficient intracellular stores, unless a regulatory process is initiated to refill the stores by introduction of extracellular Ca^{2+}. The most difficult problem to explain is how this "reverse transduction" mechanism operates. How the information that the intracellular Ca^{2+} stores are empty is relayed back to the plasma membrane and initiates Ca^{2+} entrance to refill these stores. Experiments on cells other than invertebrate photoreceptors have suggested the possibility that the IP_3 receptor itself may convey the information between the endoplasmic reticulum vesicles and the plasma membrane (Irvine,1991; Berridge,1993; see also Putney,1990). According to this hypothesis the large cytoplasmic domain of the IP_3 receptor communicates between the two Ca^{2+} channels, one in the endoplasmic reticulum (the channel portion of the IP_3 receptor) and the other in the plasma membrane. What hampered further progress in elucidation of the molecular mechanism of this reverse transduction system is the difficulties in reconstitution of the spatial relationship between the endoplasmic reticulum vesicles and the plasma membrane and the lack of a straightforward approach to molecular identification of the plasma membrane channel that responds to the state of filling of the intracellular Ca^{2+} stores. An unbiased approach of Ca^{2+} mobilization in the inositol lipid signalling system came as a result of genetic dissection of the *Drosophila* visual system. A mutant which has been analyzed both at the molecular and at the functional level is the *transient receptor potential* mutant *trp*. The *trp* (Cosens and Manning, 1969) is a spontaneous mutation with a characteristic phenotype, observed by an electroretinogram recording. In contrast to wild type phtoreceptors, the *trp* is unable to maintain sustained excitation in response to continuous intense illumination. Unlike the wild type output signal in which the initial response amplitude declines to a lower steady state level representing adaptation, in the *trp* mutant the response decays back to baseline and consequently the photoreceptor remains inactivated as long as the light stimulus is on (Minke at al.,1975; Minke,1982). On the other hand, the *trp* responses either to a short pulse of light or to a continuous dim light are not much different from the response of wild type photoreceptors indicating that the mutant has all the components necessary to generate the light induced receptor potential but is unable to maintain excitation (Fig 1). Extensive studies by Minke and colleagues and recently by Hardie and Minke (1982) revealed that the *trp* gene product is

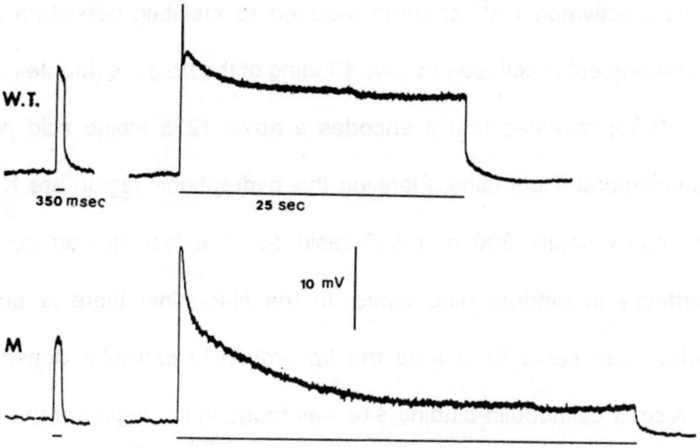

Fig. 1. Intracellularly recorded receptor potentials from peripheral retinular cells of a white-eyed normal fly (W.T.) and from a white-eyed *trp* mutant (M) to monochromatic 540 nm green light. The light intensity was 3.6 x 10^{14} photons $cm^{-2}s^{-1}$ for the upper right trace and was attenuated by 0.5 log unit in the upper left trace and by 2.0 and 2.5 log units in the lower right and left traces respectively. (From B. Minke, Biophys Struct Mech 3:59-64, 1977)

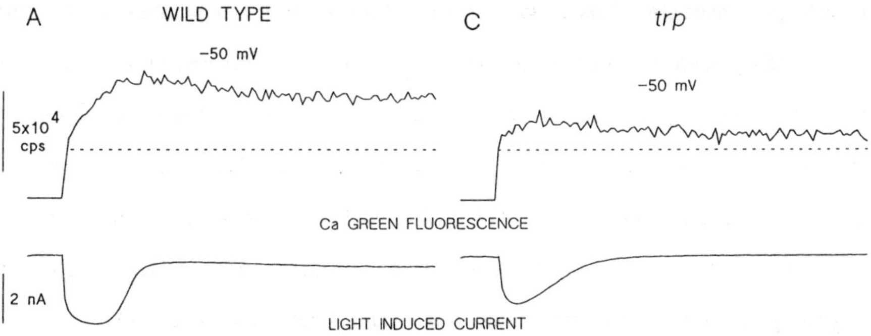

Fig 2. Ca green-5N fluorescence revealed a reduction in the Ca^{2+} signal in *trp* mutant relative to wild-type photoreceptors.

A. Ca green-5N (100μM) fluorescence measured in wild-type (WT) *Drosophila* photoreceptors at -50mV holding potentials.

B. Measurements similar to A performed in the *trp* (adult) mutant. The figure shows that the *trp* mutation largely reduced the Ca signal relative to the Ca signal of WT. (Modified from Peretz et al., 1994, Neuron 12:1257-1267).

probably a light activated Ca^{2+} channel required to maintain persistent excitation of the *Drosophila* photoreceptor cell (see below). Cloning of the *trp* gene (Montell and Rubin, 1989; Wong et al., 1989) revealed that it encodes a novel 1275 amino acid protein with eight putative transmembrane domains. Flanking this hydrophobic region are N- and C-terminal domains of approximately 300 and 600 residues. The last 250 amino acids sequence repeated perfectly in tandem nine times. In the N-terminal there is an ankyrin repeat sequence which can serve to localize the *trp* protein to particular domain in the plasma membrane. Also, a calmodulin-binding site was found in the region of the C terminous.

Light-induced current (LIC) in *Drosophila* photoreceptors

Drosophila has been widely used for studies of phototransduction largely because of the availability of mutant genes for most of the components of the phototransduction cascade many of which have now been cloned (revs. Pak, 1991; Minke and Selinger, 1991; Zuker,1992). This molecular genetic potential has recently been complemented by the possibility of making patch clamp recordings of photoreceptors from a novel preparation consisting of dissociated *Drosophila* ommatidia (Hardie, 1991; Ranganathan et al., 1991). Utilizing the tight-seal whole-cell recording mode (Neher, 1992), this has allowed, for the first time in any insect, rigorous voltage clamp measurements of the light induced current (LIC) which shows a striking dependence on both voltage and Ca^{2+}_o. The response acceleration observed as Ca^{2+}_o is raised, due to influx of Ca^{2+} via the light sensitive channels mediating a sequential positive and negative feedback on the transduction cascade (Hardie, 1991). The negative feedback is likely to play a major role in light adaptation (Hardie, 1991; Hardie and Minke, 1994). It is mediated, in part via Ca^{2+} dependent activation of protein kinase C, since it is reduced in mutants of the *inaC* gene (Hardie et al., 1993a) which encodes an eye-specific protein kinase C (Ranganathan et al., 1991; Smith et al., 1991). Voltage clamp measurements also allow a precise determination of the reversal potential (E_{rev}) of the LIC and its ionic dependence. E_{rev} is found to shift to more positive potentials as Ca^{2+}_o is raised, indicating that the light sensitive channels are permeable to Ca^{2+} (Hardie, 1991; Ranganathan et al., 1991; Hardie and Minke, 1992). Unexpectedly the quantitative dependence of E_{rev} on Ca^{2+}_o indicates that the light sensitive channels are highly selective for Ca^{2+}, (P_{Ca}:P_{Na} = 40-50:1). Formally, therefore phototransduction in *Drosophila* may be described as a process of PI-mediated Ca^{2+} entry. In the standard physiological saline used for these experiments (120 mM Na, 1.5 mM Ca, 8 mM Mg), it can be estimated from constant field theory that ca. 40% of the current is carried by Ca^{2+} ions; the remainder being made up by Mg^{2+} (40%) and Na^+ (20%). This high Ca^{2+} permeability of the light- sensitive channels depends on the *trp* protein (see below). This contrasts with the situation in *Limulus* where the most recent data suggest a very limited or non-existent permeability for Ca^{2+} (Deckert and Stieve, 1991).

To show directly that the *trp* mutation diminishes Ca^{2+} influx into *Drosophila* photoreceptors we measured light induced changes in intracellular Ca^{2+} using fluorescent Ca^{2+} indicators. Photoreceptors of dissociated *Drosophila* retinae were loaded with the fluorescent Ca^{2+} indicators, fluo-3 and Calcium Green-5N. In fluo-3 loaded, wild-type photoreceptors, a rapid increase in fluorescence (Ca^{2+} signal) accompanied the light-evoked inward current. Removal of extracellular Ca^{2+} greatly reduced the Ca^{2+} signal, indicating Ca^{2+} influx as its major cause. In Calcium Green-5N-loaded *trp* mutants, the Ca^{2+} signal was smaller relative to wild-type photoreceptors (Peretz et al.,1994a; Fig 2). This result was verified in intact *Drosophila* using Ca^{2+}-selective microelectrodes (Peretz et al., 1994b).

The *trp* protein is a novel Ca^{2+} channel with unique gating mechanism

Several lines of evidence indicate that the *trp* mutation does in fact represent a defect in one class of light sensitive channel (Hardie and Minke, 1992). 1) The crucial observation is that E_{rev} is at least 10 mV more negative in the *trp* mutant and its dependency on Ca^{2+} greatly reduced, indicating a ca. 10 fold reduction in relative Ca^{2+} permeability. 2) Under conditions where a biphasic reversal potential is observed in WT, there is a unique reversal potential in *trp* flies. 3) La^{3+}, a known Ca^{2+} channel blocker, which closely mimics many other effects of the *trp* mutation (Hochstrate, 1989; Suss-Toby et al., 1991) also quantitatively mimics the effect of the *trp* mutation on reversal potential. These results are most obviously interpreted by assuming that there are two classes of light sensitive channel; and that one of these is highly permeable to Ca^{2+}, blocked by La^{3+} and absent in the *trp* mutant. A recent reanalysis of the *trp* sequence (Phillips et al., 1992) strongly supports the conclusion that *trp* is a structural channel gene since part of it shows extensive homologies to the vertebrate dihydropyridine receptor- or voltage-gated Ca^{2+} channel. Phillips et al., (1992) also sequenced a *trp* homologue gen (*trpl*) expressed in *Drosophila* photoreceptors and suggested that this may represent a second class of light-sensitive channel. However, since both sequences only represent a single subunit of a putative tetrameric channel protein, it is also possible that *trp* and *trpl* represent different subunits of the same channel. Intriguingly, both *trpl*, and *trp* contain calmodulin binding sites suggesting a role for Ca^{2+}-calmodulin in channel activation or regulation. The decay of the light response in *trp* during continuous illumination represents the exhaustion of the excitatory process, and not excessive adaptation, which on the contrary is also largely blocked in the *trp* (Minke, 1982; Suss et al., 1989). Minke and Selinger (1991) suggested that the vital factor which is exhausted might be cellular Ca^{2+}, and that the *trp* protein might represent a plasma membrane Ca^{2+} transporter essential for replenishing stored Ca^{2+}. An obvious and self consistent explanation for the *trp* phenotype is as follows: 1) the non-*trp* channels remaining in the mutant are activated (directly, or via an intermediate) - by the Ca^{2+} released from these stores (and in wild type, possibly also directly by the influx of Ca^{2+} via the *trp* dependent channels). 2) these stores (SMC) are of limited capacity; 3) the SMC require influx of Ca^{2+} via the *trp* dependent Ca channels to be replenished, and consequently in the mutant become rapidly depleted. Such arguments suggest that the non-*trp* channels require Ca^{2+} for activation, but what about the *trp* dependent channels? Since there is absolutely no response to light in the PLC mutant *norpA*^{H44} (Minke and Selinger, 1992), it seems clear that these channels also require production of the IP_3. Two lines of evidence suggest that Ca^{2+} is also required for their activation. 1) During a developmental study of the light response we discovered a critical stage when responses to light can *only* be evoked if the cells are first provided with micromolar concentrations of Ca^{2+} by the whole-cell recording pipette (Hardie et al., 1993b). 2) It is possible to greatly reduce the light response in *Drosophila* by bathing photoreceptors in Ca^{2+}-free Ringer and buffering intracellular Ca^{2+} below 100 nM with EGTA in the electrode (Hardie and Minke, 1992). If the more powerful Ca^{2+} buffer BAPTA is used, it is in fact possible to completely abolish the light response (Hardie and Minke, in preparation). In either case, if Ca^{2+} is provided either in the bath, or in the electrode, responses rapidly recover (Hardie and Minke, 1992; Hardie unpublished). Ca^{2+} therefore appears to be required for the activation of at least two classes of light-activated channels in *Drosophila* photoreceptors. One obvious mechanism, already suggested (e.g. Payne et al., 1986) is that the Ca^{2+} released from the SMC, activates the light-sensitive channels, either directly or via an intermediate. We suggest that this can account for the activation of non-*trp* dependent channels. The *trp*-dependent channels may be activated by a depletion of the IP_3-sensitive Ca^{2+} stores.

REFERNCES

Bacigalupo J, Johnson E, Veraga C, and Lisman JE (1991) Light-dependent channels from excised patches of *Limulus* ventral photoreceptors are opened by cGMP. Proc. Natl. Acad. Sci. USA 88:7938-7942.

Berridge MJ (1993) Inositol trisphosphate and calcium signalling. Nature 361:315-325.

Bloomquist BT, Shortridge RD, Schneuwly S, Perdew M, Montell C, Steller H, Rubin G, Pak WL (1988) Isolation of a putative phospholipase C gene of Drosophila, norpA, and its role in phototransduction. Cell 54:723-733.

Brown JE, Rubin LJ, Ghalayini AJ, Tarver AP, Irvine RF, Berridge MJ, Anderson RE (1984) myo-Inositol polyphosphate may be a messenger for visual excitation in Limulus photoreceptors. Nature 311:160-163.

Cosens DJ, Manning A (1969) Abnormal electroretinogram from a Drosophila mutant. Nature 224:285-287.

Deckert A, Stieve, H (1991) Electrogenic Na^{+}-Ca^{2+} exchanger, the link between intra-and extracellular calcium in the Limulus ventral photoreceptor. J Physiol 433:467-482.

Deland MC, Pak WL (1973) Reversible temperature sensitive phototransduction mutant of *Drosophila melangoaster*. Nature New Biol 244:184-186.

Devary O, Heichal O, Blumenfeld A, Cassel D, Suss E, Barash S, Rubinstein CT, Minke B, Selinger Z (1987) Coupling of photoexcited rhodopsin to inositol phospholipid hydrolysis in fly photoreceptors. Proc Natl Acad Sci U S A 84:6939-6943.

Fein A, Payne R, Corson DW, Berridge MJ, Irvine RF (1984) Photoreceptor excitation and adaptation by inositol 1,4,5-trisphosphate. Nature 311:157-160.

Hardie RC (1991) Whole-cell recordings of the light induced current in dissociated Drosophila photoreceptors: evidence for feedback by calcium permeating the light-sensitive channels. Proc R Soc Lond B 245:203-210.

Hardie RC, Minke B (1992) The *trp* gene is essential for a light-activated Ca^{2+} channel in *Drosophila* photoreceptors. Neuron 8:643-651.

Hardie RC, Minke B (1993) Novel Ca^{2+} channels underlying transduction in *Drosophila* photoreceptors: implications for phosphoinositide-mediated Ca2+ mobilization. Trends Neurosci 16:371-376.

Hardie RC, Peretz A, Pollock JA and Minke B (1993b) Ca^{2+} limits the development of the light response in *Drosophila* photoreceptors. Proc R Soc Lond B 252:223-229.

Hardie RC,Peretz A, Suss-Toby E, Rom-Glas A, Bishop SA, Selinger Z, and Minke B (1993a) Protein kinase C is required for light adaptation in *Drosophila* photoreceptors. Nature 363:634-637.

Hardie RC, Minke B (1994) Calcium-dependent inactivation of light sensitive-channels in

Drosophila photoreceptors. J Gen Physiol 103:409-427.

Hardie RC (1994) Photolysis of caged Ca^{2+} facilitates and inactivates, but does not directly gate light-sensitive channels in *Drosophila* photoreceptors. J Neurosci (in press).

Hochstrate P (1989) Lanthanum mimics the trp photoreceptor mutant of Drosophila in the blowfly *Calliphora*. J Comp Physiol A 166:179-187.

Irvine RF (1991) Inositol tetrakisphosphate as a second messenger: confusions, contradictions, and a potential resolution. Bioessays 13:419-427.

Minke B (1982) Light-induced reduction in excitation efficiency in the trp mutant of Drosophila. J Gen Physiol 79:361-385.

Minke B, Selinger Z (1991) Inositol lipid pathway in fly photoreceptors: excitation, calcium mobilization and retinal degeneration. In: Progress in retinal research (Osborne NA, Chader GJ eds), pp 99-124. Pergamon Press Oxford.

Minke B, Wu C, Pak WL (1975) Induction of photoreceptor voltage noise in the dark in *Drosophila* mutant. Nature 258:84-87.

Minke B, Selinger Z (1992) Inositol lipid pathway is necessary for light excitation in fly photoreceptors. In: Sensory Transduction. D.P. Corey and S.D. Roper, eds. Rockefeller University Press, New York, pp. 201-217.

Montell C, Rubin GM (1989) Molecular characterization of the Drosophila trp locus: a putative integral membrane protein required for phototransduction. Neuron 2:1313-1323.

Nagy K (1991) Biophysical processes in invertebrate photoreceptors. Recent progress and a critical review based on *Limulus* photoreceptors. Q Rev Biophys 24:165-226.

Nasi E (1991) Electrophysiological properties of isolated photoreceptors from the eye of *Lima scabra*. J Gen Physiol 97:17-34.

Neher E (1992) Ion channels for communication between and within cells. Neuron 8:605-612.

Pak WL (1991) The molecular biology of the retina, basic and clinically relevant studies: Molecular genetic studies of photoreceptor function using *Drosophila* mutants. Prog.

Stieve H, Benner S (1992) The light-induced rise in cytosolic calcium starts later than the receptor current of the *Limulus* ventral photoreceptor. Vision Res 32:403-416.

Suss Toby E, Selinger Z, Minke B (1991) Lanthanum reduces the excitation efficiency in fly photoreceptors. J Gen Physiol 98:849-868.

Suss E, Barash S, Stavenga DG, Stieve H, Selinger Z, Minke B (1989) Chemical excitation and inactivation in photoreceptors of the fly mutants *trp* and *nss*. J Gen Physiol 94:465-491.

Wong F, Schaefer EL, Roop BC, LaMendola JN, Johnson Seaton D, Shao D (1989) Proper function of the Drosophila trp gene product during pupal development is important for normal visual transduction in the adult. Neuron 3:81-94.

Zuker CS (1992) Phototransduction in *Drosophila*: a paradigm for the genetic dissection of sensory transduction cascades. Curr Opin Neurobiol 2:622-627.

TURNING OFF THE SIGNAL: A REGULATORY ARRESTIN CYCLES SECURES THE FIDELITY AND MAINTENANCE OF THE FLY PHOTORECEPTOR CELLS

Zvi Selinger

Department of Biological Chemistry, Life Scieces Institute, and the Minerva Center for Studies of Visual Transduction, The Hebrew University, Jerusalem 91904, Israel

INTRODUCTION

The use of illumination protocol that results in prolonged depolarizing after potential (PDA) for isolation of *Drosophila* visual mutants has been very successful (1) but the underlying mechanism for the PDA response remained obscure for many years. When we recently examined again the mechanism that generates the PDA, one of the first things that we did was to give it a new and more meaningful name rather than the descriptive name of prolonged depolarizing after potential. Since the PDA results in sustained excitation long after the light is turned off another way of looking at this phenomenon is as an experimental protocol which disrupts the termination process. The temporal resolution required by the visual process dictates that each step in the transduction pathway should have not only an efficient "turn on" mechanism, but also an equally effective 'turn off' mechanism. Conceivably termination of one of the steps in transducton pathway is disrupted upon generation of the PDA response and we seeked to identify this step. The first clue to the step in the phototransduction cascade which is disrupted during the PDA response came as a result of studies of the light dependent GTPase activity. It was found that the in-vitro assay of light dependent GTPase activity closely resembles the PDA phenominology, indicating that the reactions that give rise to the PDA must take place at the level of the interaction between metarhodopsin and the G-protein. Further analysis was therefore concentrated on this step.

RESULTS

Phosphorylation and Dephosphorylation of Rhodopsin.

A characteristic of fly photoreceptors is that their photopigment is thermostable and photoreversible.

Blue light (<90 nm) converts rhodopsin to metarhodopsin (80%), and orange light (>580 nm) regenerates it to rhodopsin (100%).

Drosophila eye membranes, preilluminated with blue light, revealed substantial phosphorylation of metarhodopsin. This phosphorylation was strictly dependent on conversion of rhodopsin to metarhodopsin, as no phosphorylation of the photopigment took place in membranes that had been preilluminated with orange light. The highest extent of metarhodopsin phosphorylation was obtained in the presence of the Ca^{2+} chelator EGTA, whereas Ca^{2+} considerably reduced the extent of metarhodopsin phosphorylation. This effect was not dependent on calmodulin, as an efficient peptide inhibitor of calmodulin, M5, did not change the extent of metarhodopsin phosphorylation. Ca^{2+} was required for the dephosphorylation reaction, but the rate of dephosphorylation was much faster in membranes in which phosphorylated metarhodopsin had been first converted to phosphorylated rhodopsin by illumination with orange light and then subjected to dephosphorylation in the presence of Ca^{2+}. Parallel assays were conducted on *Drosophila* eye preparations of the retinal degeneration C (rdgC) mutant (Fig. 1). Recent cloning of the rdgC gene showed sequence similarity to mammalian serine/threonine protein phosphatases with an appended domain containing direct Ca^{2+} binding sites (4). Optimal dephosphorylation conditions resulted in removal of 95% of the phosphate from wild-type rhodopsin, whereas little if any dephosphorylation of the rdgC rhodopsin was observed, indicating that phosphorylated rhodopsin is a major substrate for the rdgC protein phosphatase (Fig. 1). The much higher extend of 49-kDa arrestin phosphorylation in the rdgC preparation was due to higher content of Ca2in this preparation, as in the presence of 0.1 mM Ca^{2+} an equal extent of 49-kDa arrestin phosphorylation was obtained in both the wild-type and the rdgC preparations (data not shown). Because arrestin dephosphorylation in wild-type preparations was not Ca^{2+}-dependent, it was not further studied.

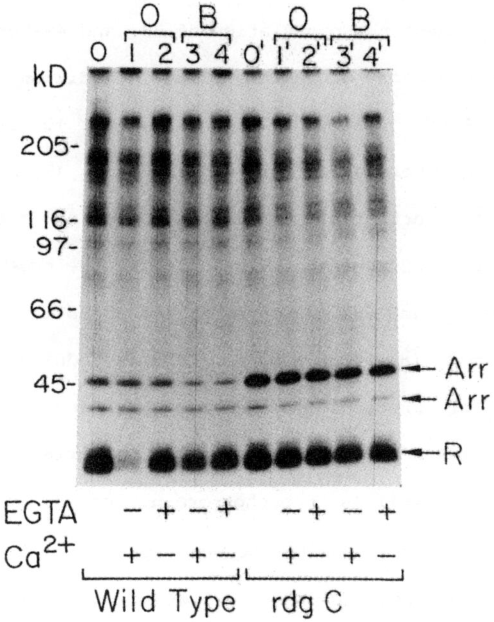

Figure 1 - Dephosphorylation of proteins in wild-type and rdgC eye membranes. Membranes. Membranes were derived from blue-illuminated eyes and phosphorylated in the presence of 10pM EGTA, followed by addition of unlabeled ATP to 0.5 mM and centrifugation. The membrane pellets were suspended in homogenization buffer and illuminated with orange (O) or blue (B) lights. Dephosphorylation was carried out for 5 min at 25°C, with the indicated additions. Lanes in radiogram: O and 0', phosphoproteins before dephosphorylation; 1, 1', 3, and 3', dephosphorylation in the presence of 0.1 mM Ca^{2+}; 2, 2', 4, and 4', dephosphorylation in the presence of 1 mM EGTA. Arr, arrestin; R, rhodopsin.

PHOSPHORYLATION OF ARRESTIN AND ITS INTERACTION WITH THE MEMBRANE.

Phosphorylation of *Musca* eye cytosol derived from dark-adapted flies revealed Ca^{2+}-dependent phosphorylation of 39- and 49-kDa proteins that was blocked by the presence of excess EGTA. Addition of the calmodulin inhibitor M5 at 1 µM greatly diminished, and at 10 µM almost completely abolished, the phosphorylation of the 39- and 49-kDa proteins in the presence of Ca^{2+}. Furthermore, addition of bovine calmodulin in excess over M5 completely restored and even increased the phosphorylation of the 39- and 49-kDa proteins, indicating that the arrestins in the eye cytosol are phosphorylated *in vitro* by a Ca^{2+}/calmodulin dependent kinase. Phosphorylation of proteins in membranes derived from blue-illuminated *Musca* eyes revealed a phosphoprotein of 31 kDa corresponding to *Drosophila* R1-6 rhodopsin, as well as phosphoproteins of 39 and 49 kDa. These proteins remained associated with the membrane upon subsequent illumination with blue light. On the other hand, illumination with orange light, which regenerates all the photopigment into rhodopsin, resulted in an essentially quantitative release of the 49-kDa protein to the medium. The 49-kDa protein was the major phosphoprotein that was released into the medium upon illumination with orange light (Fig. 2). Quantitation of the 49-kDa arrestin with antibodies raised against a peptide sequence of *Drosophila* 49-kDa arrestin showed that the 49-kDa arrestin was associated with membranes containing metarhodopsin and that it was quantitatively released into the medium upon regeneration of metarhodopsin to rhodopsin.

ARRESTIN BINDING INHIBITS DEPHOSPHORYLATION OF METARHODOPSIN

As shown in Fig. 1, both an increase in free Ca^{2+} and regeneration of p-M to p-R are required to obtain substantial dephosphorylation of the photopigment. These findings raised the question whether the resistance of p-M to phosphatase activity is due to the metarhodopsin conformation or

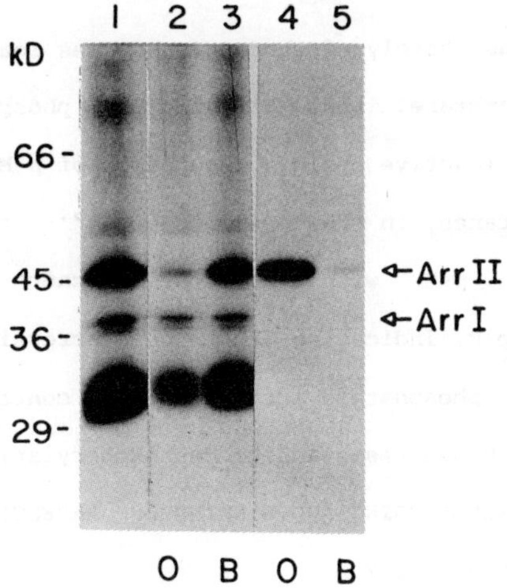

Figure 2 - Effect of illumination on the release of phosphorylated proteins from *Musca* eye membranes: (A) Membranes derived from blue-illuminated eyes were phosphorylated in the presence of 0.1 mM Ca^{2+}. At the end of the phosphorylation reaction, unlabeled ATP was added to 0.5 mM, and membranes were precipitated by centrifugation (14,000 x g,10 min), suspended in 100 pl of homogenization buffer, and illuminated for 15 sec with orange (O) or blue (B) light. Membranes were separated from the supernatant by centrifugation (14,000 x g,10 min) and each fraction was analyzed by SDS/PAGE and radiography. Lanes: 1, phosphorylated membranes before centrifugation; 2 and 3, membrane fractions obtained by centrifugation following illumination with orange or blue light, respectively; 4 and 5, supernatant fractions obtained by centrifugation following illumination with orange or blue light, respectively. Positions of arrestin (Arr) are indicated.

whether a protein component binds to p-M, thus making it inaccessible to the phosphatase. Indeed, it has been demonstrated that bovine arrestin inhibits dephosphorylation of freshly bleached bovine rhodopsin by protein phosphatase 2A. As the photolyzed rhodopsin becomes inactive, arrestin is released from the membrane, thus allowing the phosphatase to remove phosphate from the now inactive opsin. Incubation of p-M, which resides in arrestin-depleted membranes, in the presence of Ca^{2+}, resulted in partial dephosphorylation of p-M with no significant difference from dephosphorylation of p-R, indicating that once arrestin is removed, p-M becomes accessible to phosphatase activity. In contrast, in arrestin-containing membranes p-M was resistant to dephosphorylation.

PHOSPHORYLATION OF METARHODOPSIN AND BINDING OF ARRESTIN TO THE MEMBRANE TERMINATE METARHODOPSIN ACTIVITY

Membranes depleted of or containing 49-kDa arrestin were prepared from eyes that had been illuminated with orange or blue light, respectively. To test the ability of metarrhodopsin to sustain GTPase activity after illumination, each set of membranes was illuminated again at zero time, with blue light for 10 sec, followed by incubation in the dark in the presence of MgATP. Assay of GTPase activity of these membranes in the dark gave a measure of the functional metarhodopsin in the system. Membranes depleted of 49-kDa arrestin showed an undiminished ability of metarhodopsin to catalyze GTPase activity. The GTPase activity was considerably higher in the presence of Ca^{2+} than in the presence of EGTA. In contrast, membranes that contained 49-kDa arrestin showed a progressive decline in the ability of metarhodopsin to sustain GTPase activity after a flash of blue light, with little difference between the systems incubated either in the presence or in the absence of Ca^{2+}. Addition of partially purified 49-kDa arrestin to eye membranes that were depleted of 49-kDa arrestin caused a progressive decrease in the ability of metarhodopsin to activate the G protein both in the presence and in the absence of Ca^{2+}.

DISCUSSION

ARRESTIN SECURES THE FIDELITY OF PHOTOTRANSDUCTION

The fly visual system accommodates both highly active rhodopsin kinase and phosphatase, yet inactivation of the photopigment goes to completion, and reversal of the inactivation reaction does not reinitiate transduction in the dark. The present results indicate that 49-kDa arrestin is the crucial component that prevents reinitiation of phototransduction in the dark, by inhibiting dephosphorylation of metarhodopsin. Inactivation of p-M by 49-kDa arrestin, however, is not a dead-end reaction. Once the p-M/49-kDa arrestin complex is hit by another photon, p-M regenerates back to p-R with concomitant release of 49-kDa arrestin. The p-R intermediate then becomes an efficient substrate for rhodopsin phosphatase, which safely reintroduces it to the rhodopsin pool, ready for the next round of photoexcitation. A similar mechanism whereby binding of arrestin to the p-M2 intermediate protects it from dephosphorlylation by phosphatase 2A has been described for vertebrate rods (5). However, in this system it is not know how the latent activity of phosphatase 2A is regulated nor is it clear whether dephosphorylation for the photopigment is important for the maintenance of the photoreceptor cell (Fig. 3).

DISRUPTION OF THE TERMINATION PROCESS

Analysis of the PDA mutant revealed that reduction of the photopigment level in the photoreceptor cells, either by mutations in a number of different genes or by vitamin A deprivation, completely

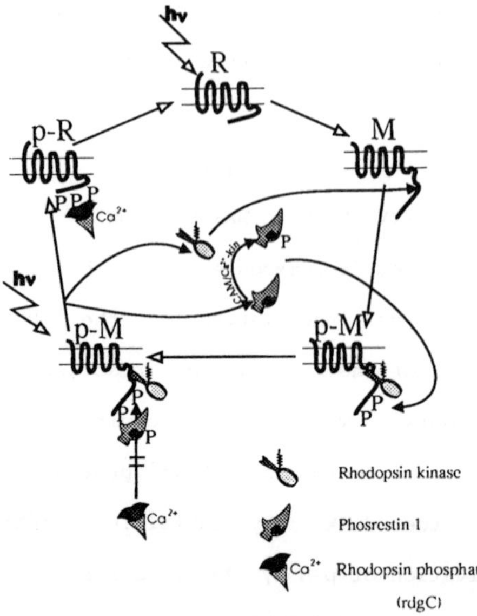

Figure 3 - Regulatory arrestin cycle secures the fidelity of phototransduction. The fly photoreceptor accommodates both a highly active rhodopsin kinase and an equally effective rhodopsin phosphatase, yet inactivation of the photoexcited metarhodopsin goes to completion and the inactivated photopigment is efficiently regenerated to the excitable rhodopsin state without initiation of phototransduction in the dark. This is accomplished by concerted operation of the following reactions. Photoconversion of rhodopsin (R) to metarhodopsin (M) results in rapid phosphorylation of M by rhodopsin kinase to give p-M followed by binding of 49 kDa phoerestin 1. Phosphorylation of H decreases its ability to activate the G protein. Binding of phosrestin 1 further quenches the ability of M to activate G protein and protect it from phosphatase activity. Absorption of a photon by the p-M/phosrestin 1 complex regenerates it to p-R with concomitant release of phoerestin 1. The p-R then becomes a substrate for rdg C rhodopsin phosphatase that reintroduces it to the excitable rhodopsin pool without reinitiation of phototransduction in the dark. Binding and release of phoarestin 1 to p-M and p-R, respectively, confines the rhodopsin phosphatase activity only to p-R, thereby securing the fidelity of phototransduction (see Refs. 12-14 for experimental details).

eliminated the PDA response (6). The PDA response thus depends on the net amount of conversion, rather than on the fractional conversion, of rhodopsin to metarhodopsin. Taken together, these experiments are consistent with a mechanism in which a component that is present at a concentration lower than that of the photopigment inactivates an equivalent amount of metarhodopsin, leaving the excess metarhodopsin in a persistently active state. The present work corroborates this mechanism by demonstrating that membranes depleted of 49-kDa arrestin show undiminished GTPase activity after a flash of blue light, whereas membranes containing 49-kDa arrestin demonstrate a progressive decline in GTPase activity.

RELEVANCE TO RETINAL DEGENERATION

Genetic analysis of the *Drosophila* mutant rdgC revealed that retinal degeneration is dependent on high levels of activated rhodopsin and placed the site of action of the rdgC gene product, before phospholipase C (7). More recently, molecular cloning of the rdgC gene showed sequence similarity with mammalian serine/threonine phosphatase and identified an appended domain containing putative direct Ca^{2+} binding sites (4). Our findings that phosphorylated rhodopsin is a major substrate for the rdgC phosphatase and that dephosphorylation of rhodopsin in wild-type flies requires Ca^{2+} but not calmodulin are in perfect agreement with the above studies by O'Tousa and colleagues (4). While the full range of substrates for the rdgC phosphatase has still to be established, these results suggest a similar mechanism of retinal degeneration in yet another *Drosophila* visual mutant. Studies of the no receptor potential A (norpA) mutant revealed that while norpA alleles at 2 days of age have lose to normal rhodopsin levels, they later develop a substantial age-dependent decrease in rhodopsin concentration and retinal degeneration. Alleles with complete elimination of the receptor potential exhibit the largest decrease in rhodopsin levels and, astonishingly, this decrease was enhanced by exposure to light (8). In view of the knowledge that norpA mutants are defective in an eye-specific phospholipase C (9) and the present findings of Ca^{2+} dependent dephosphorylation of rhodopsin, an analogy can be drawn between the norpA and the rdgC mutants. We suggest that in both mutants phosphorylation of the photopigment generated by photoconversion of rhodopsin to metarhodopsin is not adequately followed by dephosphorylation. In the rdgC mutant this is due to deficiency in rhodopsin phosphatase, and in the norpA mutant, due to a failure to generate the Ca^{2+} signal required for activation of rhodopsin phosphatase. Since arrestin phosphorylation does not take place in norpA mutants *in vivo* (10), excessive phosphorylation of arrestin as the cause of retinal degeneration of norpA can perhaps be excluded. Whether a protein phosphatase also plays a role in vertebrate retinal degeneration is unknown, nor is it known what are the physiological regulators of the latent protein phosphatase 2A that dephosphorylates rhodopsin. In view of the large number of mutations in the human opsin gene causing retinitis pigmentosa and the remarkable differences in the phenotypic expression among individuals afflicted by the same mutations (11), the possibility that rhodopsin dephosphorylation may play a direct or indict role in vertebrate retinal degeneration should be considered.

References

1. Pak, W.L. (1979) in Neurogenetics: Genetic Approaches to the Nervous system (Breakfield, X., ed), pp. 67-99, Elsevier, North Holland.

2. Blumenfeld, A. Erusalimsky, J., Heichal, O., Selinger, Z. and Minke, B. (1985) Proc. Natl. Acad. Sci. USA 82, 7116-7120

3. Hamdorf, K. (1979) Handbook of sensory physiology. Vol. VII/6A, pp. 145-224

4. Steele, F. Washburn, T., Rieger R. and O'Tousa, J.E. (1992) Cell 69, 669-676.

5. Palczewski, K., McDowell, J.H., Jakes, S., Ingebritsen, T.S. & Hargrave, P.A. (1989) J. Biol. Chem 264, 15770-15773

6. Stark, W.S. and Zsitzmann, W.G. (1976) J. Comp. Physiol. 105, 15-27.

7. Steele, F.R. and O'Tousa, J.E. (1990 Neuron 4, 883-890

8. Meyertholen, E.P., Stein, P.J., Williams, M.A. and Ostroy, S.E. (1987) J. Comp. Physiol. 161, 793-798.

9. Bloomquist, B.T., Shortridge, R.D., Schneuwly, S. Perdew, M., Montell, C., Steller, H., Rubin, G. and Pak, W.L. (1988) Cell 54, 723-733.

10. Matsumoto, H. and Yamada, T. (1991) Biochem. Biophys. Res. Commun. 177, 1306-1312.

11. Applebury, M. (1992) Curr. Biol. 2, 113-115.

12. Byk, T., Bar-Yaacov, M., Doza, Y.N., Minke, B. and Selinger, Z. (1993) Proc. Natl. Acad. Sci. 90, 1907-1911.

13. Doza, Y.N., Chorev, M. Minke, B. and Selinger Z. (1992) Eur. J. Biochem. 209, 1035-1040.

14. Selinger, Z., Doza, Y.N. and Minke, B. (1993) Biochimic. et Biophysica Acta 1179, 283-299.

AN EYE SPECIFIC PKC IS REQUIRED FOR LIGHT ADAPTATION IN *DROSOPHILA* PHOTORECEPTORS

Baruch Minke

Department of Physiology and Minerva Center for Studies of Visual Transduction, The Hebrew University - Hadassah Medical School, Jerusalem 91120, Israel

Properties of the single photon responses (quantum bumps)

In invertebrate photoreceptors the gating mechanism of the light-activated channels is still unknown. Although there is still a dispute about the identity of the final excitatory messenger, a large body of evidence indicates that transduction in invertebrate photoreceptors is mediated by the ubiquitous phosphoinositide (PI) signaling cascade (revs. Selinger and Minke, 1988; Payne et al.,1988; Nagy, 1991; Minke and Selinger, 1991; Zuker, 1992). Phospholipase C (PLC) a key enzyme in the PI cascade of *Drosophila* (Bloomquist et al, 1988), generates one soluble intemal messenger, 1,4,5 IP_3 and one membrane bound messenger, diacylglycerol (DAG). DAG is presumed to facilitate Ca^{2+}-activation of protein kinase C (rev. Berridge 1993), involved in regulation of the cascade (Smith et al., 1991; Hardie et al., 1993). The major known function of IP_3 is to release Ca^{2+} from intracellular stores (rev. Berridge, 1993), which in rhabdomeric photoreceptors are represented by specialized endoplasmic reticulum abutting the base of the microvilli - the submicrovillar cisternae (SMC) (Walz, 1982).

In the dark adapted state invertebrate photoreceptors respond to dim light with discrete electrical signals, each initiated by the absorption of a single photon of light. The ability to respond to a single photon is achieved by the production of a highly amplified single photon responses called quantum bumps. Each quantum bump represents a discrete voltage fluctuation generated by the concerted opening of many of ionic channels (Stieve, 1986). Upon increasing the light intensity, the bumps sum up to produce the macroscopic photoreceptor potential (Dodge et al.,1968). The mechanisms underlying the various bump parameters such as latency, rise time, peak amplitude and area size are largely unknown. Results obtained from fluctuation analysis (also called shot noise analysis) of the light induced macroscopic currents have led Knight and colleagues (Wong et al.,1982) to suggest the "adapting bump model". This model which is based on a synthesis between quantitative theory and actual experiments suggests that the major effect of light adaptation is to reduce the size of the quantum bumps by several orders of magnitude and to shorten their duration by a smaller factor (Wong et al.,1982). Adaptation thus prevents the receptor potential from reaching saturation and produces large reduction in the noise created by individual bumps. As an increase in cellular Ca^{2+} either by light or by injection of Ca^{2+} into the *Limulus* ventral photoreceptors causes adaptation to subsequent test lights (Lisman and Brown, 1972) and injection of the Ca^{2+} chelator EGTA reduces light adaptation it is now generally accepted that Ca^{2+} is the internal transmitter that mediates light adaptations (Lisman and Brown, 1972; Stieve, 1986).

Experiments in *Limulus* (Fein and Corson,1981) and Flies (Minke and Stephenson,1985; Suss et al.,1989) have shown that introduction of either the metabolically stable GTP analog GTPγS or IP_3 both excite and adapt the photoreceptor cells. In all of these studies GTPγS and IP_3 produced quantal bumps similar in shape (but different in amplitude) to those produced by light.The current knowledge about inositol lipid signaling, indicates that IP_3 acts on a subpopulation of the endoplasmic reticulum vesicles, that constitute intracellular Ca^{2+} stores, causing them to release their Ca^{2+} content into the cytosol. It thus follows that the structural elements responsible for generation of quantum bumps and consequently also the site of light adaptation are possibly the submicrovillar cisternae (SMC) that constitute the intracellular Ca^{2+} stores of the invertebrate photoreceptor cells (Walz, 1982). Whereas the response to IP_3 injection and the light response are composed of noisy sequence of transient events, the response to Ca^{2+} injection was a smooth inward current indicating that Ca^{2+} excites the cell subsequent to the site which generates the IP_3 induced unitary events.

Adaptation decreases the size rather than the frequency of single photon responses.

Summation of large and fast single photon responses that form the receptor potential gives rise to two undesirable functional features: 1) They produce large fluctuations, thereby resulting in poor signal to noise ratio. 2) Saturation of the light response is expected already at dim light due to the limited voltage range which is available in every excitable cell. These problems are solved in an ingenious way by the mechanism of light and dark adaptation. Adaptation is therefore a gain control mechanism operating by a negative feedback loop that reduces the amplification of one or more steps of the phototransduction cascade. The molecular mechanism of this gain control is still unknown but protein kinase C (PKC) may be involved (Hardie et al.,1993).

Recent cloning of an eye specific protein kinase C (eye-PKC) and immunological analysis of its presence in mutants eye preparations indicated that the *inactivation but no afterpotential C* (*inaC*) locus encodes the eye-PKC enzyme (Smith et al.,1991). Final identification was achieved by demonstration that introduction of the cloned eye-PKC gene into an *inaC* host completely rescued the *inaC* phenotype (Smith et al.,1991). These findings provided an excellent opportunity to study the physiological functions of PKC in a well-defined system as the fly's phototransduction cascade and to study its effects on light excitation and adaptation.

Initial analysis led to the conclusion that the role of PKC is distinct from light adaptation.

Functional analysis of a null *inaC* mutant *inaC*^{P209}, which completely lacks the PKC enzyme, using whole cell recording, showed that the *inaC* demonstrates much slower termination of the response to short pulse of light. Furthermore sustained light stimuli did not elicit in the *inaC* mutant the rapid decrease from peak amplitude to steady state plateau, a characteristic manifestation of light adaptation in wild type flies (see Fig 1a). As these effects were strongly influenced by external Ca^{2+} (Lisman and Brown, 1972) the effects of the *inaC* mutation on light adaptation was tested. In this effect, two weak test pulses of light were interspaced by a pulse of strong adapting light and adaptation was scored as the ratio of the second to the first test lights. This analysis gave apparent "hyperadaptation" in the *inaC* mutant (Smith et al.,1991). It thus paradoxically follows that in the wild type fly, light induced inositol lipid signaling and thereby activation of the PKC enzyme should prevent light adaptation. To reconcile these peculiar observations with the known effect of light on adaptation Smith et al. (1991) normalized their results to the response of the adapting light and obtained no substantial difference between the apparent adaptation of *inaC* and the wild type fly, concluding that the role of PKC in the fly photorectors is distinct from light adaptation (Smith et al.,1991). Unfortunately, these initial studies are problematic having several serious conceptual and experimental shortcomings. In particular, when testing light adaptation in a mutant which is known to undergo light-dependent response inactivation (i.e. exhaustion of excitation, see Fig 1a) it is inadequate to use as an indicator of adaptation, the only parameter which is common to both adaptation and response inactivation (i.e. the decrease in the response to test light after application of strong adapting light). Furthermore, the use of EGTA in excess over Ca^{2+} in the recording pipette is hazardous as this strong Ca^{2+} buffering precludes the effects caused by an increase in cellular Ca^{2+} and may lead to depletion of the internal Ca^{2+} stores.

Absence of eye-PKC results in pleiotropic manifestations, each revealed under particular experimental conditions.

The effect of the *inaC* mutation on light adaptation was recently re-examined under physiological conditions of intracellular recording from the intact fly and by whole cell recording but in this case, using physiological concentrations of Na^+ and K^+ with minimal concentration of EGTA (Hardie et al., 1993). These conditions yielded results which were entirely different from those obtained by Smith et al. (1991) and led to opposite conclusions. The intracellular recording was used to test the criteria that discriminates between adaptation and response inactivation. As these required long periods of recording and multiple stimulations they are not easily applicable to whole cell recording from dissociated ommatidia. Using the criteria which has been developed to discriminate between light adaptation and exhaustion of excitation in the *trp* mutant (Minke, 1982), it was apparent that adaptation is

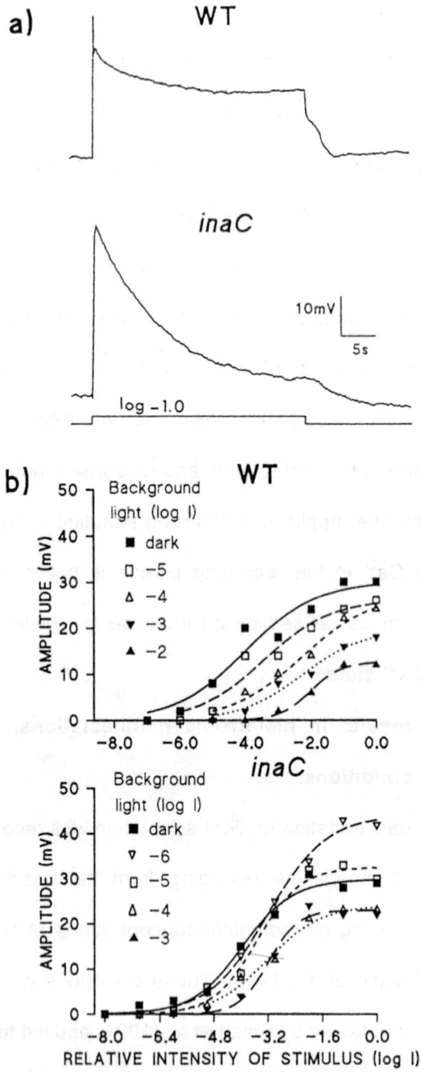

Fig. 1. Adaptation is blocked in *inaC* while bright lights or low Ca^{2+} induce inactivation. a) Voltage responses to prolonged intense light in WT and *inaC* photoreceptors (OG 570 edge filter attenuated by log -1). In WT the peak response adapts rapidly to a maintained plateau. In *inaC* there is no rapid adaptation, but the receptor potential decays near to baseline within 30 seconds. b) Response intensity functions (V/log -1) determined at different background intensities (OG 570 edge filter) in WT and *inaC* by intracellular voltage recordings in intact *Drosophila.* Voltage increments above steady state are plotted. In WT all backgrounds induce a lateral shift in V/logI curve. In *inaC* a backgrounds of log -6, actually induces facilitation and there is no shift observed at log -5. With a brighter (>log -4.0) background there is mainly response compression. Data fitted (least square deviations) with curves of the form:

$$V/V_{max} = I^n / (I^n + \sigma^n)$$

where I is the light intensity, n is a constant (0.52±0.16) and σ is the light intensity which elicits half of the maximal response (V_{max}). (Modified from Hardie et al.,1993, Nature 363:634-637).

severely reduced in the *inaC* mutant (Fig 1b) and as the background light intensities increased, response inactivation of the *inaC* photoreceptors was evident. The combination of intracellular and whole cell recordings revealed the following cluster of phenotypic expressions which characterize the *inaC* phenotype: 1) Decay of the response to baseline upon continuous intense illumination. 2) Supersensitivity to dim light. 3) Decreased sensitivity to strong light. 4) Supersensitivity to response inactivation upon removal of external Ca^{2+}. 5) Slow termination of the response to short pulse of light. This last feature was previously described by Ranganathan et al. (1991) and Smith et al. (1991) but the mechanism that underlies the slow termination of the light response remained obscure.

In the absence of eye-PKC individual bumps fail to terminate.

Whole cell voltage clamp analysis of the light induced current at the level of single photon elicited quantum bumps revealed the primary defect caused by the absence of eye-PKC (Hardie et al.,1993). In the *inaC* mutant individual quantum bumps fail to terminate normally, each having continuous slow phase of decay accompanied by channel noise (Fig 2). This is in contrast to the quantum bumps of wild type fly which have a sharp rise time and a fast decay. On the assumption that quantal Ca^{2+} released from the intracellular stores by IP_3 is represented by the individual single photon events and that Ca^{2+} is needed for excitation, it was possible to account for all the pleotropic manifestations of the *inaC* phenotype, as arising from one primary defect (i.e. failure of individual quantum bumps to terminate). The supersensitivity to dim light takes place when the Ca^{2+} stores are still full and the failure of the released Ca^{2+} to terminate results in an increased release per each photon and hence the supersensitivity. On the other intense sustained illumination leads to depletion of the Ca^{2+} stores, and thereafter results in less Ca^{2+} released per each photon which gives rise to decrease sensitivity and eventually to response inactivation. The same arguments hold true for the decrease in sensitivity, after strong adapting light, which presumably depletes the *inaC* Ca^{2+} stores as well as for the slow termination of the light response which

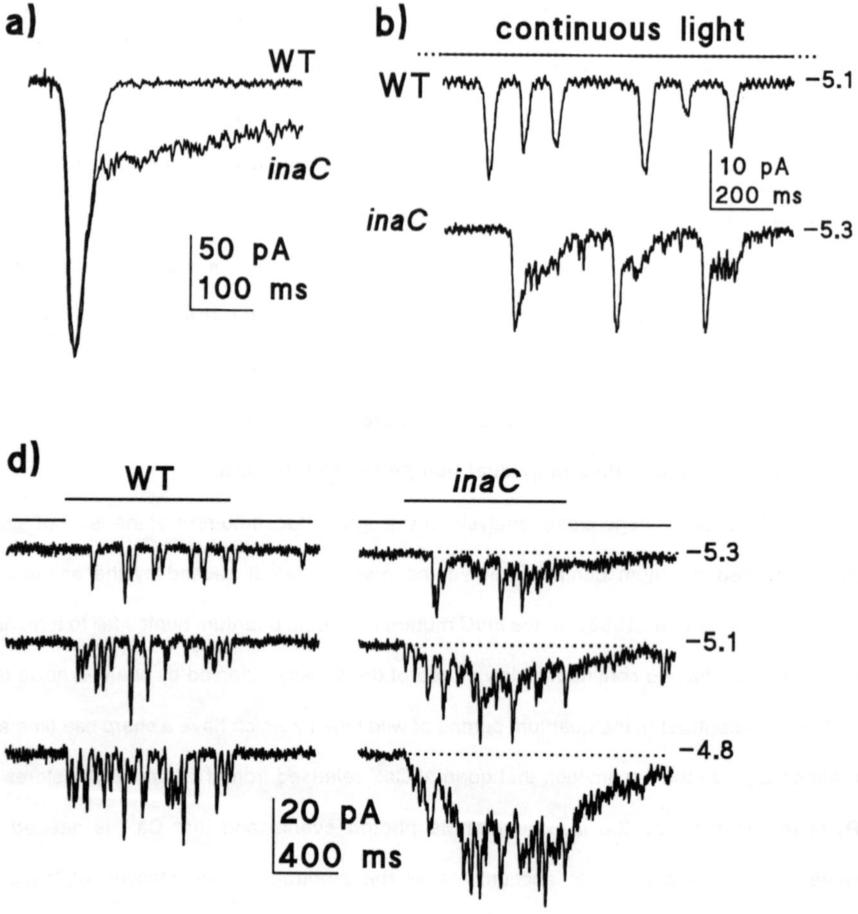

Fig. 2. Whole-cell current responses in *inaC* and wild type photoreceptors. a) Responses to 10 ms flashes (relative intensity log -2.5) in adult *inaC* and WT. Initially responses overlap; however, the *inaC* response has a characteristic slowly decaying tail with high frequency noise. b) Quantum bumps recorded in response to continuous dim illumination (p15 pupae). In *inaC* although bump rise and initial decay appear similar to wild type there is again a conspicuous tail. c) In WT photoreceptor (adult) steps of light evokes discrete quantum bumps. These sum to produce a noisy steady-state response (reached in less than 100ms) as the intensity is increased (log I indicated). In *inaC,* summation of the incompletely terminated bumps results in a slowly rising ramp of current. (Modified from Hardie et al.,1993, Nature 363:634-637).

is due to slow termination of individual quantal responses. Finally, upon removal of Ca^{2+} from the external medium dim lights which in presence of physiological extracellular Ca^{2+} concentrations causes an increased response, now causes response inactivation. This is due to the absence of external Ca^{2+} which must enter the cell to refill the stores (see below). Under these conditions internal Ca^{2+} stores are rapidly emptied and the response is inactivated but could be restored by readdition of Ca^{2+} to the medium. While these results strongly suggest that eye-PKC is an important player in light adaptation it is probably not the only one. Both Ca^{2+} -induced Ca^{2+}- release and Ca^{2+} inhibition of Ca^{2+} release from the intracellular stores have been described in other cells and are likely to be contributing factors also in the photoreceptor cell, accounting for the observations of facilitation and adaptation respectively. Likewise, an effect on the light sensitive channels is possible. If the other *ina* mutants are also defective in Ca^{2+} homeostasis, analysis of the remaining *Drosophila ina* mutants will yield a comprehensive picture of light adaptation in the not too distant future.

Ca^{2+} influx is enhanced in the *inaC* mutant

Unlike the *trp* mutation, which diminishes Ca^{2+} influx into the photoreceptors (Hardie and Minke,1992), we expect that in the dark adapted state the *inaC* mutation will *enhances* Ca^{2+} influx. To verify this expectation we measured $\Delta[Ca^{2+}]_o$ using Ca^{2+}-selective microelectrodes in the *inaC* mutant. In wild type (WT) flies illumination induced a decrease in extracellular $[Ca^{2+}]$ with kinetics and time-dependent magnitude that changed with light intensity. Compared to wild type, the light-induced decrease in $\Delta[Ca^{2+}]_o$ (the Ca^{2+} signal) was diminished in *trp* (Peretz et al.,1994a) but significantly enhanced in *inaC* (Fig 3). The enhanced Ca^{2+} signal was also diminished in the double mutant *inaC;trp* indicating that the effect of the *trp* mutation overrides the enhancement observed in the absence of eye-PKC (Peretz et al.,1994a). We suggest that the reduction in $[Ca^{2+}]_o$ reflects light-induced Ca^{2+} influx into the photoreceptors and that the *trp* mutation blocks a large fraction of this Ca^{2+} influx. In contrast the absence of eye specific PKC led to enhancement of light-induced Ca^{2+} influx.

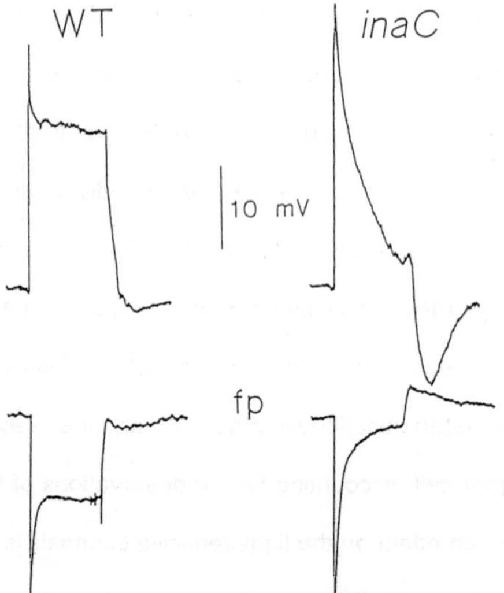

Fig. 3. Light induced $\Delta[Ca^{2+}]_0$ changes in the retina of WT as compared to the retina of the white-eyed *inaC* mutant. The upper row is intracellular recordings of the receptor potential. The middle row (fp) is the extracellular recordings of the receptor potential and the bottom row is ΔE_{ca}. recorded from the Ca_{2+} selective barrel of the same double barrel electrodes used to record the middle traces. The relative light intensity was -logI=O. The figure shows that in contrast to WT the receptor potential of *inaC* declines towards baseline (inactivates) during illumination. It furthermore shows that $\Delta[Ca^{2+}]_0$ is significantly larger (and slower) in *inaC* relative to WT. The large hyperpolarization of the receptor potential of *inaC* at light off most likely reflects activation of the Na^+-K^+ pump.

This suggestion was supported by measurements of light-induced increase in cellular Ca^{2+} using patch clamp whole-cell perfusion technique with a fluorescent Ca^{2+} indicator. In Ca-green-5N-loaded WT and *inaC* photoreceptors (in isolated ommatidia) a ~3 fold larger increase cellular Ca^{2+} was found in *inaC* over WT. Our observations are consistent with the hypothesis that TRP is a light activated Ca^{2+} channel with activation level depended on PKC.

REFERENCES

Berridge MJ (1993) Inositol trisphosphate and calcium signalling. Nature 361:315-325.

Bloomquist BT, Shortridge RD, Schneuwly S, Perdew M, Montell C, Steller H, Rubin G, Pak WL (1988) Isolation of a putative phospholipase C gene of *Drosophila*, norpA, and its role in phototransduction. Cell 54:723-733.

Dodge FA, Jr., Knight BW, Toyoda J (1968) Voltage noise in *Limulus* visual cells. Science 160:88-90.

Fein A, Corson DW (1981) Excitation of *Limulus* photoreceptors by vanadate and by a hydrolysis-resistant analogue of guanosine triphosphate. Science 212:555-557.

Hardie RC, Peretz A, Suss-Toby E. Rom-Glas A, Bishop SA, Selinger Z, and Minke B (1993) Protein kinase C is required for light adaptation in *Drosophila* photoreceptors. Nature 363:634-637.

Hardie RC, Minke B (1992) The *trp* gene is essential for a light-activated Ca^{2+} channel in *Drosophila* photoreceptors. Neuron 8:643-651.

Lisman JE, Brown JE (1972) The effects of intracellular iontophoretic injection of calcium and sodium ions on the light responses of *Limulus* ventral photoreceptors. J Gen Physiol 59:701-719.

Minke B (1982) Light-induced reduction in excitation efficiency in the trp mutant of

Drosophila. J Gen Physiol 79:361-385.

Minke B, Stephenson RS (1985) The characteristics of chemically induced noise in *Musca* photoreceptors. J Comp Physiol 156:339-356.

Minke B, Selinger Z (1991) Inositol lipid pathway in fly photoreceptors: excitation, calcium mobilization and retinal degeneration. In: Progress in retinal research (Osborne NA, Chader GJ eds), pp 99-124. Pergamon Press Oxford.

Nagy K (1991) Biophysical processes in invertebrate photoreceptors. Recent progress and a critical review based on *Limulus* photoreceptors. Q Rev Biophys 24:165-226.

Payne R, Corson DW, Fein A and Berridge MJ (1986) Excitation and adaptation of *Limulus* ventral photoreceptors by inositol 1,4,5-trisphosphate result from a rise in intracellular calcium. J Gen Physiol 88:127-142.

Payne R, Walz B, Levy S, Fein A (1988) The localization of calcium release by inositol trisphosphate in *Limulus* photoreceptors and its control by negative feedback. Philos Trans R Soc Lond Biol 320:359-379.

Peretz A, Sandler C, Kirschfeld K, Hardie RC, Minke B (1994a) Genetic dissection of light-induced Ca^{2+} influx into *Drosophila* photoreceptors. J Gen Physiol (in press)

Peretz A, Suss-Toby E, Rom-Glas A, Arnon A, Payne R, Minke B (1994b) The light response of *Drosophila* photoreceptors is accompanied by an increase in cellular calcium: effects of specific mutations. Neuron 12:1257-1267.

Ranganathan R, Harris GL, Stevens CF, Zuker CS (1991) A *Drosophila* mutant defective in extracellular calcium-dependent photoreceptor deactivation and rapid desensitization.

Selinger Z, Minke B (1988) Inositol lipid cascade of vision studied in mutant flies. Gold Spring Harbor Symposium on Quantitative Biology 53:333-341.

Smith DP, Ranganathan R, Hardy RW, Marx J, Tsuchida T and Zuker CS (1991) Photoreceptor deactivation and retinal degeneration mediated by a photoreceptor-specific protein kinase C. Science 254:1478-1484.

Stieve H (1986) Bumps, the elementary excitatory responses of invertebrates, in *The Molecular Mechanism of Photoreception* (Stieve H, Ed.), Springer-Verlag, Berlin, pp. 199-230.

Suss E, Barash S, Stavenga DG, Stieve H, Selinger Z, Minke B (1989) Chemical excitation and inactivation in photoreceptors of the fly mutants *trp* and *nss*. J Gen Physiol 94:465-491.

Walz B (1982) Ca^{2+}-sequestering smooth endoplasmic reticulum in an invertebrate photoreceptor. II. Its properties as revealed by microphotometric measurements. J Cell Biol 93:849-859.

Wong F, Knight BW, Dodge FA (1982) Adapting bump model for ventral photoreceptors of *Limulus*. J Gen Physiol 79:1089-1113.

Zuker CS (1992) Phototransduction in *Drosophila:* a paradigm for the genetic dissection of sensory transduction cascades. Curr Opin Neurobiol 2:622-627.

USE OF FLUORESCENT DYES TO MEASURE INTRACELLULAR CALCIUM ION CONCENTRATION IN MICROVILLAR PHOTORECEPTORS

Richard Payne and Kyrill Y. Ukhanov, *Dept. of Zoology, University of Maryland, College Park MD 20742, U.S.A.*

Introduction

Phototransduction by the microvillar photoreceptors of invertebrate species is accompanied by an increase in intracellular free calcium concentration (Ca_i)[1]. The increase in Ca_i appears to be mediated through activation by rhodopsin of the phosphoinositide (PI) pathway [2]. During illumination, calcium ions enter the cytosol either from intracellular stores or from the extracellular space. The relative contribution of these sources seems to vary from species-to-species. For Limulus ventral photoreceptors, calcium is released from storage through inositol 1,4,5 trisphosphate ($InsP_3$)-activated channels in the endoplasmic reticulum [3, 4]. In barnacle and Drosophila photoreceptors, calcium ions enter the cell from the extracellular space through calcium channels activated by an unknown messenger also perhaps produced via the PI pathway [1, 5].

This paper is a discussion of the use of fluorescent indicators of calcium ion concentration to monitor the light-induced elevation of Ca_i in microvillar photoreceptors. Fluorescent calcium indicators offer some advantages over other methods used to monitor Ca_i in microvillar photoreceptors. Their speed of response, approximate linearity and range of sensitivities should enable quantification of localized elevations of Ca_i through ratio fluorescence microscopy [6]. Confocal imaging of the fluorescence using a laser-scanning microscope allows measurement of Ca_i to be made in small volumes of cytosol [7]. In addition, the equipment used to monitor dye fluorescence can also be modified to enable the instantaneous release of messengers from "caged compounds" using UV-illumination [8].

The use of fluorescent indicators in photoreceptors has allowed the first measurements of Ca_i in intact rods under physiological conditions [9], in intact Drosophila ommatidia [5,10] and in bee photoreceptors [11]. The work in our laboratory has focused on Limulus ventral photoreceptors, where comparisons may be made with results previously obtained using other methods of measuring Ca_i in a large cell whose morphology is well-described [12 - 14]. This has enabled us to test the behaviour of the dyes before proceeding to apply them to confocal imaging and new preparations, such as Drosophila ommatidia. What follows is a essentially a progress report on this work.

Special considerations when using fluorescent dyes in photoreceptors.

The use of fluorescent calcium indicators in microvillar photoreceptors suffer from all of the general problems associated with the calibration of the dyes. These include autofluorescence, estimation of the effect of protein and ionic strength on the dye's performance and the measurement of the signals associated with fully saturated or unsaturated dye when performing in vivo calibrations [6]. However, in addition, two specific problems must be especially considered when using the dyes in microvillar photoreceptors.

Excitation of the photoreceptor by the measuring light.

The most obvious limitation in using fluorescent dyes is that the photoreceptors are themselves sensitive to the wavelengths of light used to excite the dyes that are currently commercially available. In Limulus ventral photoreceptors, for instance, adequate signals from the dye fura-2 require light intensities that photoisomerise approximately 10^7 rhodopsin molecules per second. As a result, Ca_i can never be measured other than during intense stimulation of the transduction cascade. This is a severe limitation if one wishes to investigate the action in darkness of intermediates in the visual cascade, such as inositol 1,4,5 triphosphate. However, there are ways around this limitation.

a) In slowly-responding photoreceptors, such as Limulus ventral photoreceptors, calcium can be estimated during the latent period of the response to the measuring flash. This period, 15-50 ms, depending on temperature and light intensity, is an estimate of the "dark level" of Ca_i taken before Ca_i begins to increase as a result of the measuring light. A series of brief test flashes at different intervals after an event could, in theory, be used to piece together the time course of the "dark level" of Ca_i after an event such as the injection of $InsP_3$ or a dim light flash.

b) The efficiency with which the rhodopsin is excited relative to the fluorescent dye can be reduced. For Limulus ventral photoreceptors, chemically bleaching rhodopsin using hydroxylamine reduces quantum efficiency by 2-3 log units [15]. This is clearly not enough to enable excitation of the dye without appreciable photoisomerisation of rhodopsin. However, hydroxylamine treatment, together with a lowered temperature and lowered extracellular calcium concentration, may extend the latent period of the light response to 100ms or more and reduce the adapting effect of brief measuring flashes used to measure levels of Ca_i during the latent period.

For Drosophila photoreceptors, mutations can be engineered that shift the spectrum of rhodopsin away from that of the dye [10], that reduce pigment content (**ninaE**;[16, 17]), or that completely eliminate the light-induced elevation of Ca_i by eliminating phospholipase-C (**norpA**;[18]). These mutations enable measurements of Ca_i to be made with minimal background excitation of rhodopsin by the measuring light.

2) Large local gradients of Ca_i may exist in photoreceptors.

Local gradients of calcium concentration are present in the cytosol of many cells [19]. Their presence complicates the interpretation of optical signals such as fluorescence, which are average measurements made within the illuminated volume. Photoreceptors are particularly likely to display large gradients of Ca_i for the following reasons. First, the phototransduction apparatus is highly concentrated in one region of the photoreceptor to maximise the absorption of light. Second, phototransduction results from the sum of highly amplified quantal events, whose influence may spread only a few microns from the site of a photoisomerisation. Third, excitation and adaptation are initiated faster than calcium can diffuse within the cell, increasing the localization of calcium ions during the period of their greatest physiological effect. These factors result in the possibility of very high concentrations of calcium transiently localized to the sites of calcium release or influx. Even in response to dim lights, transient gradients may exist due to the amplified quantal nature of the response.

Implications of calcium gradients for dye measurements.

Dye signals measure a spatial average of Ca_i

It is important to recognize that estimates of Ca_i using fluorescent dyes reflect the spatial average of Ca_i within the volume from which fluorescence is collected. Both light-induced calcium-release and calcium influx first elevate Ca_i immediately beneath the microvillar plasma-membrane. In the case of calcium influx through light-sensitive channels in the plasma membrane, it is obvious why this should be. For calcium release, localization is achieved either by virtue of the limited diffusion of messengers released by light, such as inositol 1,4,5 trisphosphate, or due to the localization of calcium stores just beneath the plasma membrane. It may take several hundred ms for calcium or $InsP_3$ to diffuse into the body of the cell. During this time, it is probable that gradients of calcium ions, bound and unbound dye will exist within the cell, so that any dye signal collected from the entire cell, or a large portion of it will underestimate the value of Ca_i at the microvillar membrane, where calcium ions act to influence visual transduction. Confocal measurements will reduce this averaging effect, since calcium is sampled in a smaller volume which can be placed close to the microvillar membrane. However, even within volumes of a few cubic microns, large calcium gradients may exist around the mouths of open calcium channels and a steady state calcium concentration may take several ms to build up or dissipate due to diffusion within the volume.

<u>Saturation of indicator dyes</u>

We cannot avoid underestimation of dye signals due to spatial averaging. However, there is an additional, avoidable, systematic error in estimating the spatial average of Ca_i if the fluorescent dye becomes saturated with calcium at any point within the cell. Saturation of dye within the measurement volume as light-induced elevation of calcium proceeds will greatly reduce the apparent rate of rise of the estimate of Ca_i and will retard the time to peak. The calibrated dye signal will begin to measure the true spatial average of Ca_i only when Ca_i at the site of calcium release falls below the level which saturates the dye [20]. As a result, high affinity dyes such as fura2 should not be used to estimate levels of Ca_i during the initial rise or peak of Ca_i during illumination, when local saturation of the dye may occur. The signal from a low affinity dye, such as mag-fura2 [21] or calcium-green 5N [22] should be compared to that of the high affinity indicator if saturation of the dye is suspected.

Application of fluorescent indicators to Limulus ventral photoreceptors.

General considerations

A recent paper [23] summarizes the expected messenger roles of calcium and $InsP_3$. Free calcium ions are expected to be at high concentration (>100 μM) at the mouths of ion channels [24]. However, free calcium ions probably travel only very short distances (approx. 0.1μm) from their site of influx into the cytosol before being bound by immobile or slowly diffusing calcium buffers. Calcium ions are expected to be buffered by intracellular binding sites within less than 100μs of entering the cytosol. Thus only events immediately adjacent to a source of calcium ions may be mediated by free calcium ions and then only while the source is active.

Once calcium ions are buffered, diffusion is much slower and free calcium ion concentrations are lowered by a factor of 10 or more, into or below the micromolar range. Removal of calcium from the cytosol by sequestration into stores is expected to limit the lifetime of buffered calcium to a few seconds. As a result buffered calcium ions are thought to be able mediate signals over distances of a few μm.

By contrast, $InsP_3$ does not appear to be bound by the cytosol so that its diffusion is as rapid as that of free calcium. $InsP_3$ is expected to survive for a few seconds before hydrolysis by 5-phosphatase activity so that the effects of $InsP_3$ are expected to be exerted over distances of tens of μm following its release from the plasma membrane.

Anatomical considerations

The <u>Limulus</u> ventral photoreceptor is a giant cell, some 150 μm in length and 60 μm in diameter. The arhabdomeral (A)-lobe contains the cell's nucleus and is the point of departure of the axon. The rhabdomeral, or R-lobe is the only part of the cell that contains microvilli with rhodopsin in their plasma membrane. These microvilli cover the plasma membrane of the photoreceptor, including occasional invaginations known as "internal rhabdom" [12,13]. Underlying the microvilli and extending as a dense mesh throughout the R-lobe is a network of smooth endoplasmic reticulum [14]. This meshwork of endoplasmic reticulum appears to be continuous with a less dense network in the A-lobe. Thus the microvilli in the R-lobe are the probable site of $InsP_3$ production, while the endoplasmic reticulum network in the R-lobe is the probable site of calcium release by $InsP_3$ [25]. The dense network of endoplasmic reticulum in the R-lobe may result in the entire R-lobe cytoplasm being within 1 μm of an $InsP_3$-activated calcium channel. Thus if the signal to release calcium is spread by $InsP_3$, very high calcium concentrations might be possible almost everywhere within the cytoplasm of the R-lobe. $InsP_3$ appears to be less effective in releasing calcium from endoplasmic reticulum in the A-lobe [26], which probably explains the much smaller and delayed light-induced elevation of Ca_i in the A-lobe [27].

Summary of previous studies.

There is a rich history of attempts at measuring Ca_i in Limulus ventral photoreceptors [28], providing plenty of comparisons for results obtained using fluorescent dyes. Briefly, the previous measurements indicate that a rapid calcium elevation is initiated in the R-lobe within the first hundred milliseconds of illumination as the result of $InsP_3$-induced release of calcium from intracellular stores. Calcium concentrations of at least tens of micromolar are achieved in the R-lobe at the peak of the elevation [27, 29]. Within a few seconds after the onset of bright illumination, calcium concentrations have declined to the micromolar range (25, 26]. The magnitude of both the large initial elevation of calcium and the final lower level are graded with light intensity.

Studies to test the use of fluorescent dyes in Limulus ventral photoreceptors.

We have concentrated our studies on events within the R-lobe at early times in the photoexcitation process. The large gradient of calcium elevation between the R-lobe and the A-lobe has been well described by previous calcium-electrode and aequorin studies. The arsenazo III study of Harary and Brown [30] makes it clear that, during the first few hundred ms of the light response, calcium spreads from a localized site of release, but that study did not quantify the release or localize it relative to the plasma membrane of the R-lobe.

We began our studies by using a conventional microfluorimeter (Deltascan, Photon Technology International) in which the photoreceptor is uniformly illuminated by the light used to excite the dye, but a mask placed in a focal plane of the image ensures that only a portion of the cell (i.e. the R-lobe) is viewed by the photomultiplier that counts emitted light. Given the large depth of field of the 20x lens used to image the photoreceptor, the measurement volume is therefore the entire it-lobe of the cell. What follows is a summary of the completed study, which may be found described fully elsewhere [31].

Fura-2 provides an estimate of resting Ca_i in dark-adapted photoreceptors.

We started by testing the use of fura-2, a comparatively high affinity indicator with a dissociation constant for calcium binding of 0.64 μM under our conditions. Photoreceptors were injected with 10-100 pl of a solution containing 500 μM fura-2.

The photoreceptors were then dark-adapted for up to 20 mins, until single photon events of amplitude 1-5 Mv were observed. A 4.6s step of rapidly alternating 340/380 nm light, producing approximately 4 x 107 photoisomerisations/s, was then delivered to the cell. Following the onset of this illumination, fluorescence to 380 nm light dropped and fluorescence to 340 nm light increased as the electrical response of the cell to the flash proceeded. Figure 2 shows an attempt to calibrate the calcium signal, so as to estimate the average Ca_i within the R-lobe. The increased noise of the Ca signal at the peak of the response is a result of dye saturation. During the latent period of the electrical response to light, between 10 and 50 ms after the shutter opened, Ca_i remained at approximately 0.4 μM.

We assume that this value also represents the dark-adapted level of Ca_i.

The comparatively slow rise of the calibrated fura2 signal in fig.1 is in contradiction to expectations from aequorin measurements, which suggest that calcium should reach concentrations of above 5 μM and saturate fura-2 within 200 ms of the onset of bright illumination [291.

We attributed the slow rise of the fura-2 signal to indicate that we were not measuring the true spatial average of Ca_i within the R-lobe, but rather the slower diffusion of saturated dye from release sites close to the plasma membrane into the centre of the R-lobe. We therefore decided to compare our measurements with those using the less sensitive indicator mag-fura2.

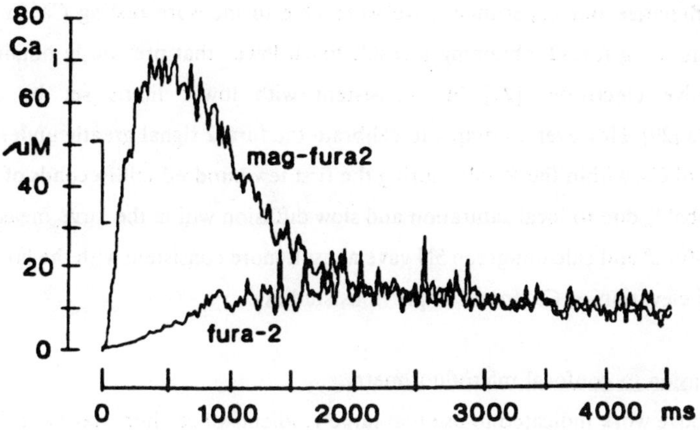

Figure 1 - Comparison of estimates of the spatial average of Ca_i within the R-lobe of a <u>Limulus</u> ventral photoreceptor as obtained using fura2 and mag-fura2.

Mag-fura2 has a much lower Kd than fura-2 (approximately 56 µM under our conditions). On the assumption that magnesium does not change during the light response, a calibrated result also shown in fig.1 was obtained. We checked the result by using a different low affinity dye, calcium-green 5N. Since calcium-green 5N does not change its excitation or emission spectrum upon binding calcium, we co-injected a calcium-insensitive dye, ANTS (8-aminonaphthalene-1,3,6-trisulfonic acid), with the calcium-green 5N and used the ratio between calcium-green 5N and ANTS fluorescence to calibrate Ca_i. The resulting estimation of the spatial average of Ca_i within the R-lobe was very similar to that obtained using mag-fura2.

In order to investigate the differences in amplitude and time course of the fura2 and mag-fura2 signals in fig.1, we constructed a simple model, in which calcium was released only at the outer rim of a sphere representing the R-lobe (for details see [31]).

The results of the model indicated that the large underestimation of Ca_i by fura-2 could be plausibly explained by an immediate initial saturation of fura-2 at the release site, followed by the diffusion of a wave-front of saturated dye and calcium into the centre of the R-lobe.

We concluded from the model that the mag-fura2 signal was likely to be a much better estimate of the spatial average of Ca_i within the R-lobe.

To summarize our experiments, we were able to measure resting Ca_i in Limulus photoreceptors using fura-2, obtaining a result much lower that previously obtained with calcium-selective electrodes [27] but consistent with lower limits set by aequorin measurements [29]. However, attempts to calibrate the fura-2 signal greatly underestimate the elevation of Ca_i within the R-lobe during the first few hundred milliseconds of the light response, probably due to local saturation and slow diffusion within the large measurement volume. Mag-fura2 and calcium-green 5N gave a result more consistent with the kinetics and magnitude of elevation of Ca_i found in previous studies.

Possible extension to confocal microfluorimetry.

The above work indicated to use that large gradients of calcium may exist within the R-lobe during the initial few hundred ms of the light response [30]. These gradients severely distort the signals generated by high affinity dyes at early times after the onset of the light response. We have therefore initiated a study using a confocal microscope to measure Ca_i within small volumes close to the microvillar membrane and to scan calcium concentrations along lines within the R-lobe. So far, we have obtained evidence that, as expected, the rate of elevation of calcium close to the plasma membrane is much more rapid than indicated by spatially averaged recordings from the whole R-lobe and that the elevation spreads throughout the R-lobe over the first few hundred ms of the light response. Peak local concentrations exceed 100μM. The results suggest that the kinetics of dye signals from the whole R-lobe reflect the kinetics of the mixing of calcium, dye and $InsP_3$ within the larger volume, rather than the kinetics of calcium elevation at its site of action. Overall our work on Limulus photoreceptors so far indicates the importance of localized measurement of calcium with low affinity indicators.

Measurement of Ca_i in Drosophila photoreceptors

We have collaborated with Dr. Baruch Minke and his colleagues to measure calcium in Drosophila photoreceptors [5]. The high affinity dye, fluo3, and the low affinity dye, calcium-green 5N, have been used to show that, unlike in Limulus, the elevation of Ca_i appears to arise largely through influx through light-activated channels in the plasma membrane. Just as in limulus photoreceptors, the PI pathway appears to mediate the elevation of Ca_i through the enzyme phospholipase-C, but the messenger that links this pathway to the opening of channels in the plasm membrane is unknown. As with our initial studies using Limulus, we collected fluorescence from a large volume of the Drosophila photoreceptor. We cannot, therefore, easily interpret the timecourse of the signal or calibrate the signal. We expect that our results will be superseded by confocal imaging of much more rapidly changing and larger elevations of Ca_i localized to the cytoplasm beneath the rhabdom.

The use of fluorescent dyes to measure Ca_i in Limulus and Drosophila opens up new several new areas in the study of phototransduction in microvillar photoreceptors. What is the relative timing of the electrical response of Limulus photoreceptors and calcium release as indicated by local measurement of Ca_i close to the microvilli? What novel biochemical or biophysical processes might be initiated in the R-lobe by calcium concentrations of tens or even hundreds of micromolar? How does the PI-pathway activate the opening of calcium channels in the microvillar plasma membrane of Drosophila?

References

1. J.E. Brown and J.R. Blinks (1974) Changes in intracellular free calcium during illumination of invertebrate photoreceptors. Detected with aequorin. *J. Gen. Physiol.* 64, 643-665.

2. S. Yarfitz and J.R. Hurley (1994) Transduction Mechanisms of Vertebrate and Invertebrate Photoreceptors. *J. Biol. Chem.* 269, 14329-14332.

3. R. Payne, D.W. Corson, A. Fein, and M.J. Berridge (1986). Excitation and adaptation of Limulus ventral photoreceptors by inositol 1,4,5 trisphosphate result from a rise in intracellular calcium. *J. Gen. Physiol.* 88, 127-142.

4. T.M. Frank and A. Fein (1991) The role of inositol phosphate cascade in visual excitation of invertebrate microvillar photoreceptors. *J. Gen. Physiol.* 97, 697-723.

5. A. Peretz, E. Suss-Toby, A. Rom-Glas, A. Arnon, R. Payne and B. Minke (1994) The light response of Drosophila photoreceptors is accompanied by an increase in cellular calcium. Effects of specific mutations. *Neuron* 12, 1257 - 1267

6. R.Y. Tsien (1989) Fluorescent probes of cell signalling. *Annu. Rev. Neurosci.* 12, 227-253.

7. A. Hernandez-Cruz, F. Sala and P.R. Adams (1990) Subcellular calcium transients

visualized by confocal microscope in a voltage-clamped vertebrate neuron. *Science*, 247, 858-862.

8. S.R. Adams and R.Y. Tsien (1993) Controlling cell chemsitry with caged compounds. *Annu. Rev. Physiol.* 55, 755-784.

9. G.M. Ratto, R. Payne, W.G. Owen and R.Y. Tsien (1988). The concentration of cytosolic free calcium in vertebrate rod outer segments measured with Fura-2. *J. Neurosci.* 8, 3240-3246.

10. R. Ranganathan, B.J. Backsai, R.Y. Tsien and C.S. Zuker (1994) Cytosolic calcium transients: spatial localization in Drosophila photoreceptor cell function. *Neuron* 13, 837-848.

11. B. Walz, B. Zimmerman and S. Seidl (1994) Intracellular Ca_{2+} concentration and latency of light-induced Ca^{2+} changes in photoreceptors of the honrybee drone. *J. Comp. Physiol.* 174, 421-431.

12. R. Millecchia and A. Mauro (1969). The ventral photoreceptor cell of Limulus I. The microanatomy. *J. Gen. Physiol.*, 54, 289-309.

13. B.G. Calman and S.C. Chamberlain (1982). Distinct lobes of Limulus photoreceptors II. Structure and ultrastructure. *J. Gen. Physiol.* 80, 839-862.

14. J.J. Feng, J.H. Carson, F. Morgan, B. Walz and A. Fein (1994) Three dimensional organisation of endoplasmic reticulum in the ventral photoreceptors of Limulus. *J. Comp. Neurol.* 341, 172-183.

15. M.N. Faddis and J.E. Brown (1993) Flash photolysis of caged compounds in Limulus ventral photoreceptors. *J. Gen. Physiol.* 100, 547-570.

16. J.E. O'Tousa, W. Baehr, R.L. Martin, J. Hirsh, W.L. Pak and M.L. Applebury (1985) The Drosophila ninaE gene encodes an opsin. *Cell*, 40, 839-850.

17. C.S. Zuker, A.F. Cowman, and G.M. Rubin (1985) Isolation and structure of a rhodopsin gene from D. melanogaster. *Cell*, 40,851-858.

18. B.T. Bloomquist, R.D. Shortridge, S. Schneuwly, N. Pedrew, C. Montell, H. Steller, G. Rubin and W.L. Pak (1988). Isolation of a putative phospholipase C gene of Drosophila norpA and its role in phototransduction. *Cell* 54, 723-733.

19. G.J. Augustine and E. Neher (1992) Neuronal Ca^{2+} signalling takes the local route. *Curr. Opin. Neurobiol.* 2, 302-307.

20. D. Tillotson and E. Nasi (1988) Ca^{2+} diffusion in the cytoplasm of Aplysia neurons: its relationship to local concentration changes. In Grinnel A.D. and Armstron D. (eds) *Calcium and ion channel modulation.* New York: Plenum Press, 1988.

21. B. Raju, E. Murphy, L.A. Levy, R.D. Hall and R.E. London (1989) A fluorescent indicator for measuring cytosolic free magnesium. *Am. J. Physiol.* 256, C540-C548.

22. R.P. Haugland (1992). Handbook of Fluorescent Probes and Research Chemicals. 5th. Ed. Molecular Probes Inc., Eugene, Oregon.

23. N.L. Allbritton, T. Meyer and L. Stryer (1992) Range of messenger action of calcium ion and inositol 1,4,5-trisphosphate. *Science* 258, 1812-1815.

24. S.M. Simon and R.R. Llinas (1985) Compartementalization of the submembrane calcium activity during calcium influx and its significance in transmitter release. *Biophys. J.* 48, 485-498.

25. R. Payne, B. Walz, S. Levy, and A. Fein (1988). The localization of calcium release by inositol trisphosphate in Limulus ventral photoreceptors and its control by negative feedback. *Phil. Trans. Roy. Soc. B.* 320, 359-379.

26. R. Payne and A. Fein (1987). Inositol 1,4,5 trisphosphate releases calcium from specialized sites within Limulus photoreceptors. *J. Cell Biol.* 104, 933-937

27. S. Levy and A. Fein (1985). Relationship between light-sensitivity and intracellular free calcium in Limulus ventral photoreceptors. *J. Gen. Physiol.* 85, 805-841.

28. K. Nagy, Biophysical processes in invertebrate photoreceptors: recent progress and a critical overview based on Limulus photoreceptors. *Quart. Rev. Biophys.* 24, 165-226.

29. P.M. O'Day, M.P. Gray-Keller (1989) Evidence for electrogenic Na^+/Ca exchange in Limulus ventral photoreceptors. *J Gen Physiol.* 93, 473-492.

30. H.H. Harary and J.E. Brown (1984) Spatially nonuniform changes in intracellular calcium ion concentrations. *Science* 224, 292-294.

31. K. Ukhanov, T.M. Flores, C.H. Pitts, P. Mohapatra, H. Sheng-Hsiao, and R. Payne (1995) Measurement of intracellular calcium in Limulus ventral photoreceptors using fluorescent dyes. *J. Gen. Physiol.* (in press)

INOSITOL 1,4,5 TRISPHOSPHATE AND VISION IN LIMULUS PHOTORECEPTORS

Richard Payne and Kyrill Y. Ukhanov, *Dept. of Zoology, University of Maryland, College Park, MD 20742, U.S.A.*

The electrical response of most photoreceptors in invertebrate species is very different from that of the rods and cones of vertebrates (1). Following illumination, invertebrate photoreceptors depolarize due to the opening of cationic channels in the plasma membrane, rather than hyperpolarize as do rods and cones. Over the past several years, the involvement of the intracellular messenger d-myo-inositol 1,4,5 trisphosphate ($InsP_3$) in visual transduction in the ventral photoreceptors of the horseshoe crab (Limulus polyphemus) has been the subject of detailed investigation. Interest in $InsP_3$ stems from three observations. First, release of calcium from intracellular stores occurs upon illumination of Limulus photoreceptors (2 and Fig. 1). Second, $InsP_3$ is -

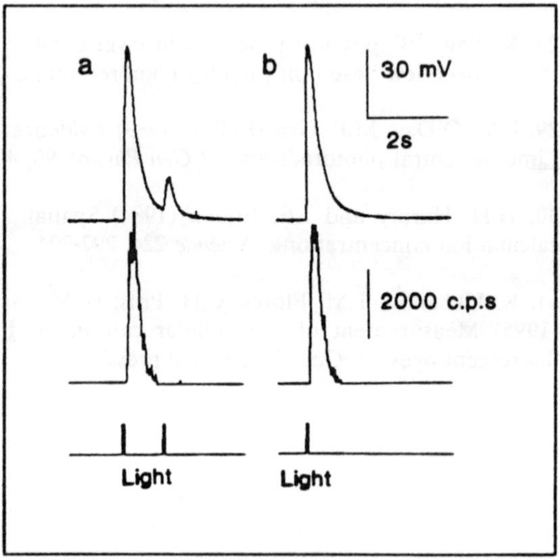

Figure 1.
Light-induced depolarization and calcium release in Limulus ventral photoreceptors.
a) Membrane potential (top trace) and aequorin luminescence (middle trace; indicating intracellular Ca^{2+} concentration) following two consecutive light flashes (bottom trace). Adaptation greatly reduces the depolarization and calcium elevation produced in response to the second flash.
b) Recovery from adaptation of light-induced depolarization (top trace) and calcium elevation (middle trace), recorded 100s after the traces in (a). For experimental details see ref. 2

accepted as a messenger of calcium release in transduction pathways that are activated by rhodopsin-like receptor proteins (3). Third, intracellular injection of $InsP_3$ mimics some aspects of the light response (fig. 2 and refs. 4,5). Injection of $InsP_3$ depolarizes the photoreceptors and desensitizes (adapts) the cell to subsequent light flashes. The investigation of the role played by $InsP_3$ in Limulus photoreceptors has been accompanied by progress in research on the photoreceptors of other invertebrate species (6) as well as on olfaction (7,8). The present article is an evaluation of the extent to which the actions of $InsP_3$ "explains" phototransduction in the photoreceptors of Limulus.

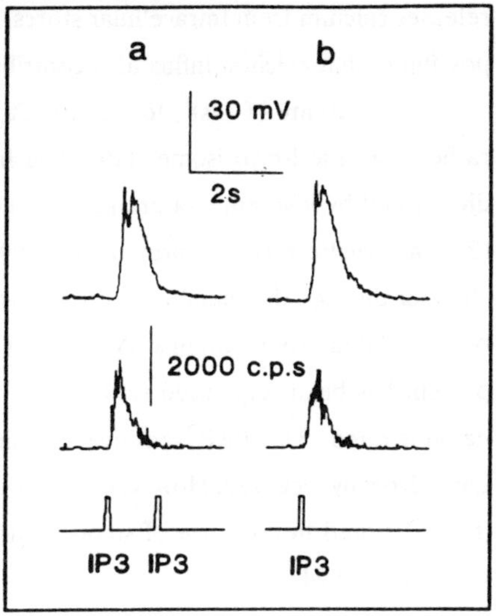

Figure 2.
$InsP_3$-induced depolarization and calcium release in Limulus ventral photoreceptors.
a) Membrane potential (top trace) and aequorin luminescence (middle trace; indicating intracellular Ca^{2+} concentration) following two consecutive pulsed-pressure injections of 100 μM $InsP_3$ (bottom trace). Feedback desensitization greatly reduces the depolarization and calcium elevation produced in response to the second injection.
b) Recovery from desensitization of $InsP_3$-induced depolarization (top trace) and calcium elevation (middle trace), recorded 100s after the traces in (a). For experimental details see ref. 18

$InsP_3$ is generated following a light flash.

Following a light flash, Limulus photoreceptors generate $InsP_3$ from the lipid, phosphatidylinositol bisphosphate (PIP_2) (4). With the exception of rhodopsin (9), the enzymes responsible for the production of $InsP_3$ are not yet characterized in Limulus photoreceptors. In other invertebrate species, enzymes for the production of $InsP_3$, namely phospholipase-C (10) and a GTP-binding protein of the G_q subfamily (11,12), have been sequenced and localized to the photoreceptors. Regulatory enzymes are now emerging such as arrestin (13-15), which inhibits the activation of the GTP-binding protein by rhodopsin and may play a role in terminating the light response and in adaptation.

The only known action of $InsP_3$ in Limulus photoreceptors is to release calcium from intracellular stores

Intracellular injection of $InsP_3$ into Limulus ventral photoreceptors results in a rapid, transient elevation of cytosolic free calcium ion concentration, Ca_i, (fig. 2). The $InsP_3$-induced elevation of Ca_i has been detected by injection of the luminescent Ca-sensitive photoprotein aequorin (16,18) and by impaling them with a Ca-selective micro-electrode (17). Because the ability of $InsP_3$ to elevate Ca_i in Limulus photoreceptors persists in the presence of nominally Ca-free extracellular solutions (16,18), it is assumed that $InsP_3$ releases calcium from intracellular stores. This observation does not, however, rule out the possibility that calcium influx also contributes a small fraction to the elevation of Ca_i.

The ability of $InsP_3$ to elevate Ca_i is stereospecific, being confined to the d-myo- rather than the l-myo isomer (19). The $InsP_3$-induced elevation of Ca_i is also not greatly diminished by low temperature (20). It seems therefore likely that, as in mammalian cells (21), a specific receptor protein for $InsP_3$ exists, which acts as an $InsP_3$-gated calcium channel through the membrane of an intracellular calcium store. No such protein has yet been isolated from Limulus. A Drosophila homolog of the mammalian $InsP_3$ receptor protein has been sequenced (22), but its localization within photoreceptors has not been demonstrated. The $InsP_3$-sensitive calcium store in Limulus photoreceptors is also not unequivocally localized. However, by analogy with bee photoreceptors, one probable source is Ca^{2+} stored in cisternae of smooth endoplasmic reticulum beneath the photoreceptive microvilli (23,24).

Calcium release following an injection of $InsP_3$ is rapid. The latency observed before the calcium elevation is detected by aequorin is less than 56 ms at room temperature (20). The aequorin signal reaches its peak within 500 ms, and declines to baseline within 5s (fig.2 and refs. 16,18). The magnitude of the peak $InsP_3$-induced elevation of Ca_i, is unknown because the aequorin signals are uncalibrated. Ca-selective microelectrodes are too slow to catch the peak elevation of Ca_i, but they indicate a maximum $InsP_3$-induced elevation of Ca_i of 1-2 μM (17), so presumably the peak concentration exceeds this range. A small sustained elevation of Ca_i, detectable by Ca-selective microelectrodes but not by aequorin, lingers at the injection site for tens of seconds after an injection of $InsP_3$ (17). This lingering elevation of Ca_i probably mediates feedback desensitization of the $InsP_3$ receptor (see below). By comparison, recent measurements with fluorescent Ca-sensitive dyes indicate peak Ca_i elevations of more than 60 μM following bright light flashes (25).

All the effects of injected $InsP_3$ can be mimicked by Ca^{2+} injection and can be reversed by injection of calcium chelators (16,18). Therefore, in Limulus ventral photoreceptors, the known actions of $InsP_3$ all appear to result from released calcium. Thus, the role $InsP_3$ plays in phototransduction therefore depends on the role played by calcium.

Roles of calcium in visual transduction in Limulus.

a) The activation of a cation conductance resembling the light-activated conductance.

Injection of $InsP_3$ or calcium activates a cation conductance in the plasma membrane with a similar reversal potential to that activated by light (4,5,18,26). This conductance is responsible for the $InsP_3$- or Ca2+-induced depolarization of the photoreceptor. While the similarity of reversal potential encourages the belief that calcium and light activate the same ion channels, the final proof requires identification of the calcium-activated channels by patch-clamp analysis. No calcium-activated channels have yet been observed in patches of membrane removed from the photoreceptor (27), suggesting that calcium may stimulate the production of another intracellular messenger that directly opens the channels (see below for possibilities). The ability of Ca^{2+} to stimulate production of this messenger is presumably lost when the patches are removed from the cell and exposed to artificial salines. Calcium-activated channels in cell-attached patches of membrane have also not been observed Thus the pathway by which calcium (and therefore $InsP_3$) excites the photoreceptor and the exact nature of the channels opened is unknown.

b) Feedback inhibition of further calcium release by $InsP_3$

In addition to activating a cationic conductance, injection of calcium desensitizes Limulus photoreceptors to subsequent injections of $InsP_3$ (28,29). This desensitization constitutes a negative feedback loop that rapidly terminates calcium release. Desensitization lasts for several tens of seconds and is accompanied by a lingering elevation of Ca_i. Rapidly lowering Ca_i during this period by injection of small amounts of BAPTA or EGTA restores a large fraction of the cell's sensitivity to $InsP_3$ (17).

c) Adaptation of the light response

In addition to the feedback desensitization of the response to $InsP_3$, Ca2+ injection also desensitizes the cell to subsequent light flashes (30). This has long been proposed as the likely mechanism of light adaptation. Ca^{2+} released by light feeds back to desensitize the cell. This desensitization process may, in part, involve feedback inhibition of $InsP_3$-induced elevation of calcium as described above.

Evidence that $InsP_3$ acts together with another messenger to produce the electrical response to light.

The work described above suggests the elements of the visual transduction pathway described in fig. 3. Note that the scheme is dependent on calcium as the mediator of both excitation and adaptation (feedback inhibition). The next section of this paper reviews the evidence that this pathway, while it is likely to operate in the photoreceptors, fails to fully explain visual transduction in Limulus photoreceptors and that another pathway for visual excitation may exist. The argument presented in this review hinges on the failure of pharmacological treatments expected to eliminate the light-induced elevation of Ca_i, to also eliminate light-induced depolarization of the photoreceptor. Other studies have reached a similar conclusion based on a detailed analysis of the complex dynamics of the electrical response to light (31).

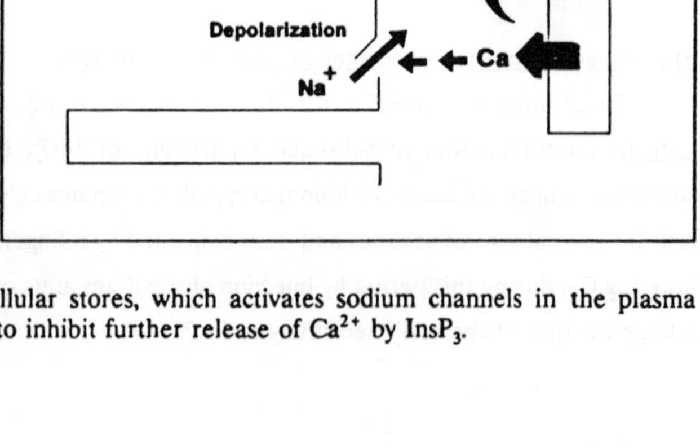

Figure 3. Simple model of visual transduction in Limulus photoreceptors. Photoisomerisation of rhodopsin (Rh) in the photoreceptive microvilli initiates a cascade that activates a GTP-binding protein (G_q) and phospholipase-C (PLC) which catalyses the production of diacylglycerol (DAG) and $Ins(1,4,5)P_3$ from phosphatidylinositol bisphosphate (PIP_2). $InsP_3$ releases Ca^{2+} from intracellular stores, which activates sodium channels in the plasma membrane and feeds back to inhibit further release of Ca^{2+} by $InsP_3$.

The ability of light (and $InsP_3$) to elevate Ca_i in Limulus photoreceptors has been attacked in three different ways.

1) To prevent Ca^{2+} from reaching its site of action, the calcium chelators EGTA and BAPTA have been injected into photoreceptors at a high concentration (32-35)

2) To prevent $InsP_3$ from acting at its receptor site, a competitive inhibitor of the $InsP_3$-receptor, heparin, has been injected into photoreceptors (32,33)

3) To discharge the $InsP_3$- and light-sensitive calcium stores, a Ca-ionophore, 4-bromo-A23187 has been applied to the photoreceptors (K. Yu. Ukhanov and R. Payne, unpublished observations)

It has been suggested that in the case of injection of calcium chelators, the persistence of the light response after treatment might arise from saturation of the chelators by light-induced calcium release, rather than from a calcium-independent transduction pathway (34). However, this argument is less easy to apply to the latter two treatments. In these cases, the abolition of the light-induced elevation of Ca_i by heparin injection or by Ca-ionophore treatment was verified by the intracellular Ca-indicators Arsenazo III (33) or fura-2, yet visual transduction persisted - bright illumination induced large depolarizations or inward currents.

All three of the treatments that eliminate the light-induced elevation of Ca_i also greatly modify the sensitivity and kinetics of the transduction process. The sensitivity of the cell to dim flashes of light is reduced 10-100 fold by the above treatments and the response timecourse is greatly slowed. As a result, responses to dim flashes of light and to single photons become undetectable above the recording noise level. It is therefore apparent that light-induced elevation of Ca_i is required to maintain the sensitivity and speed of the electrical response to light. Bright flashes, however, can still induce large, slow depolarizations or inward currents even in the absence of a detectable elevation of Ca_i.

Non-metabolizable analogues of $InsP_3$ do not mimic the response to steady light.

Another set of experiments casts doubt on the ability of $InsP_3$ to fully mimic the light response. After the initial large transient depolarization, the photoreceptor responds to bright light with a sustained depolarization of 10-30 mV. One might expect the response to sustained bright light to be mimicked by large injections of the non-metabolizable $InsP_3$ analogue, inositol 1,4,5 trisphosphorothioate ($InsP_3S_3$). However, such injections do not produce a sustained, steady depolarization, but instead induce dramatic oscillations of membrane potential caused by bursts of calcium release occurring every 10-20s (19).

Although one might argue that the differences between the actions of $InsP_3S_3$ and steady light result from a pharmacological difference between the actions of $InsP_3$ and $InsP_3S_3$ at the $InsP_3$-receptor, we also observe transient bursts of depolarization, rather than a steady, sustained depolarization following large injections of $InsP_3$ (29) and other analogues (R. Payne and B.V.L. Potter, unpublished observations). It may be that some other product of illumination suppresses the oscillations of calcium release that would otherwise result from production of $InsP_3$. In support of this idea, we have observed that dim illumination suppresses the $InsP_3S_3$-induced bursts of depolarization (19).

Alternatives to $InsP_3$ as a messenger of visual excitation.

All of the above results lend credibility to a search for an alternative pathway of visual excitation in Limulus photoreceptors, one that does not utilize $InsP_3$ or calcium release. This alternative pathway is likely to act in concert with $InsP_3$ so as to excite the cell. What are the probable messengers?

cyclic guanosine monophosphate (cGMP)

cGMP is the only messenger so far shown to open cation channels in excised patches of the plasma membrane of Limulus photoreceptors (27). These channels have similar electophysiological properties to the light-sensitive channels observed in cell-attached patches. The problem with the acceptance of cGMP as a messenger of visual transduction in Limulus lies in the lack of a link between cGMP production and the phosphoinositide cascade. Recently, however, a head-specific putative guanylate cyclase, the enzyme required to synthesize cGMP, has been sequenced in Drosophila (36). Perhaps some aspect of the regulation of this enzyme will offer the missing link.

Diacylglycerol

Diacylglycerol (DAG) has the advantage that it is almost certainly produced following a light flash. Diacylglycerol is the companion product, along with $InsP_3$, of the action of phospholipase-C on PIP_2. The problem with its acceptance as a messenger for excitation lies in the lack of any demonstration that injection of DAG analogues is able to depolarize the photoreceptor. However, an involvement in light-adaptation is suggested in Drosophila photoreceptors, where mutation of an eye-specific protein kinase-C, a target protein that is activated by DAG and calcium, results in deficiencies in adaptation and in the turn-off of individual quantal responses (37,38). Malfunction in the metabolism of DAG has also recently been implicated as a possible cause of retinal degeneration in rdgA and rdgB mutant Drosophila. The rdgA gene is a diacylglycerol kinase, an enzyme responsible for metabolism of DAG (39). The degeneration of rdgB mutants is accelerated by activators of protein kinase-C (40). The rdgB gene is a phosphatidylinositol transfer protein (41). Both rdgA and rdgB may therefore act to ensure that DAG is rapidly recycled to PIP_2. DAG has also been proposed as a stimulus to the daily turnover of the photoreceptive membrane in crab eyes (42).

Emptying of calcium stores

An additional message that might be sent during transduction, that also requires no further enzymes for its generation, is the emptying of Ca2+ stores by $InsP_3$. It has been proposed in other cell types that the emptying of calcium stores triggers the opening of channels in the plasma membrane that allow calcium to enter the cell and refill the stores (43). A direct protein-to protein contact between the Ca2+-channel in the plasma membrane and the $InsP_3$-receptor in the endoplasmic reticulum has been proposed (44). This transduction pathway has been advanced as an explanation for the ability of light to open a class of calcium-permeable channels in Drosophila photoreceptors (the trp gene product) (45).

Acknowledgements This review is an abridged version of a review by the same authors which will appear in Sensory Systems, Plenum Press. N.Y. This work was supported by the National Eye Institute, EY 07743.

References

1. B. Rayer, M. Naynert and H. Stieve, Phototransduction: Different mechanisms in vertebrates and invertebrate, J. Photochem. Photobiol. B. 7, 107-148 (1990).

2. J.E. Brown and J.R. Blinks, Changes in intracellular free calcium concentration during illumination of invertebrate photoreceptors: detection with Aequorin, J. Gen. Physiol. 64, 643-665 (1974).

3. M.J. Berridge, Inositol trisphosphate and cell signalling, Nature 361, 315-325 (1993).

4. J.E. Brown, L.J. Rubin, A.J. Ghalayini, A.L. Tarver, R.F. Irvine, M.J. Berridge and R.E. Anderson, myo-Inositol polyphosphate may be a messenger for visual excitation in Limulus photoreceptors, Nature 311, 160-162 (1984).

5. A. Fein, R. Payne, D.W. Corson, M.J. Berridge and R.F. Irvine, Photoreceptor excitation and adaptation of Limulus ventral photoreceptors by inositol 1,4,5 trisphosphate, Nature 311, 157-160 (1984).

6. D.P. Smith, M.A. Stamnes and C.S. Zuker, Signal transduction in the visual system of Drosophila, Ann. Rev. Cell. Biol. 7, 161-190 (1991).

7. H. Breer, and I. Boekhoff, Second messenger signalling in olfaction, Curr. Opinion. Neurobiol. 2, 439 - 443 (1992).

8. D.A. Fadool and B.W. Ache, Plasma membrane inositol 1,4,5 trisphosphate-activated channels mediate sensory transduction in lobster olfactory receptor neurons, Neuron 9, 907-918 (1992).

9. C.W. Smith, D.A. Price, R.M. Greenberg and B.-A. Battelle Opsins from the lateral eyes and ocelli of the horseshoe crab Limulus polyphemus, Proc. Natl. Acad. USA 90, 6150-6154 (1993).

10. B.T. Bloomquist, R.D. Shortridge, S. Schneuwly, M. Pedrew, C. Montell, H. Steller, G. Rubin, and W.L. Pak, Isolation of a putative phospholipase-C gene of Drosophila, norpA, and its role in phototransduction, Cell 54, 723-733 (1988).

11. Y.J. Lee, M.B. Dobbs, M.L. Verardi, and D.R Hyde, dgq a Drosphila gene encoding a visual system-specific G alpha molecule, Neuron 5, 889-898 (1990).

12. C. Nobes, J. Baverstock and H. Saibil, Activation of the GTP-binding protein G_q by rhodopsin in squid photoreceptors, Biochem. J. 287, 545-548 (1992).

13. P.J. Dolph, R. Ranganathan, N.J. Colley, R.W. Hardy, M. Socolich and C.S. Zuker, Arrestin function in inactivation of G-protein coupled receptor rhodopsin in vivo, Science **260**, 1910-1916 (1993).

14. D.R. Hyde, K.L. Mecklenberg, J.A. Pollock, T.S. Vihtelic and S. Benzer, Twenty Drosophila visual system cDNA clones One is a homolog of human arrestin, Proc. Natl. Acad. Sci. 87 1008 - 1012 (1990).

15. H. Matsumoto, T. Yamada, Phosrestin I and II: arrestin homologs which undergo differential light-induced phosphorylation in the Drosophila photoreceptor in vitro. Biochem. Biophys. Res. Comm. 177 1306-1312 (1991)

16. J.E. Brown, and L.J. Rubin, A direct demonstration that inositol trisphosphate induces an increase in intracellular calcium in Limulus photoreceptors, Biochem. Biophys. Res. Comm. 125, 1137-1142 (1984).

17. S. Levy and R. Payne, Inhibition of IP3-induced Ca release in Limulus ventral photoreceptors requires the continuous presence of elevated resting Ca_i, J. Gen. Physiol. 101, 67-84 (1993).

18. R. Payne, D.W. Corson, A. Fein and M.J. Berridge, Excitation and adaptation of Limulus ventral photoreceptors by inositol 1,4,5 trisphosphate result from a rise in intracellular calcium, J. Gen. Physiol. 88, 127-142 (1986).

19. R. Payne and B.V.L. Potter. Injection of inositol trisphosphorothioate into Limulus ventral photoreceptors causes oscillations of free cytosolic calcium, J. Gen. Physiol. 97, 1165-1186 (1991).

20. R. Payne and T.M. Flores, The latency of the response of Limulus photoreceptors to inositol trisphosphate lacks the calcium-sensitivity of that to light, J. Comp. Physiol.170, 311-316 (1992).

21. T. Furuichi, S. Yoshikama, A. Miyawaki, K. Waden, N. Maeda and K. Mikoshiba, Primary structure and functional expression of the inositol 1,4,5 trisphosphate-binding protein P_{400}, Nature 342, 32-38 (1989).

22. S. Yoshikawa, T. Tanimura, A. Miyawaki, M. Yuzaki, T. Furuichi and K. Mikoshiba, Molecular cloning and characterization of the inositol 1,4,5 trisphosphate receptor from Drosophila melanogaster, J. Biol. Chem. 267, 16613-16619 (1992).

23. O. Baumann, B. Walz, A.V. Somlyo, and A.P. Somlyo, Electron probe microanalysis of calcium release and magnesium uptake by endoplasmic reticulum in bee photoreceptors, Proc. Natl. Acad. Sci. USA 88, 741-744 (1991).

24. J.E. Lisman and J.A. Strong, The intiation of excitation and light adaptation in Limulus ventral photoreceptors, J. Gen. Physiol. 73, 219-243 (1979).

25. K. Yu. Ukhanov, T.M. Flores, C.H. Pitts, P. Mohapatra, H. Sheng-Hsiao and Payne R. Measurement of intracellular calcium in Limulus ventral photoreceptors using fluorescent

dyes, J. Gen. Physiol. 105, 95-116 (1995).

26. R. Payne, D.W. Corson and A. Fein, Pressure injection of calcium both excites and adapts Limulus ventral photoreceptors, J. Gen. Physiol. 88, 107-126 (1986).

27. J. Bacigalupo, E.C. Johnson, C. Vergara, and J.E. Lisman, Light-dependent channels in excised patches of Limulus ventral photoreceptors are opened by cyclic GMP, Proc. Natl. Acad. Sci. U.S.A. 88, 7938-7942 (1991).

28. R. Payne, T.M. Flores and A. Fein, Feedback inhibition by calcium limits the release of calcium by inositol trisphosphate in Limulus ventral photoreceptors, Neuron 4, 547-555 (1990).

29. R. Payne, B. Walz, S. Levy and A. Fein, The localization of calcium release by inositol trisphosphate in Limulus photoreceptors and its control by negative feedback, Phil. Trans. Roy. Soc. (Lond) B 320, 359-379 (1988).

30. J.E. Lisman and J.E. Brown, The effect of intracellular iontophoretic injection of calcium and sodium ions on the light response of Limulus ventral photoreceptors, J. Gen. Physiol. 59, 701-719 (1972).

31. K. Nagy, K. Contzen and H. Stieve, Two components of the receptor current are developed from distinct elementary signals in Limulus ventral nerve photoreceptor, Eur. Biophys. J. 22, 341-350 (1993).

32. M.N. Faddis and J.E. Brown Intracellular injection of heparin and polyamines: Effects on phototransduction in Limulus ventral photoreceptors. J. Gen. Physiol. 101, 909-931 (1993).

33. T.M. Frank and A. Fein, The role of the inositol phosphate cascade in visual excitation of invertebrate microvillar photoreceptors, J. Gen. Physiol. 97, 697-723 (1991).

34. J.E. Lisman and J.E. Brown, Effects of intracellular injection of calcium buffers on light adaptation in Limulus ventral photoreceptors, J. Gen. Physiol. 66, 489-506 (1975).

35. J. Shin, E.A. Richard and J.E. Lisman, Ca^{2+} is an obligatory intermediate in the excitation cascade of Limulus photoreceptors, Neuron 11, 845-855 (1993).

36. S. Yoshikawa, I. Miyamoto, J. Aruga, T. Furuichi, H. Okano and K. Mikoshiba, Isolation of a Drosophila gene encoding a head-specific quanylyl cyclase, J. Neurochem. 60, 1570-1573 (1993).

37. R.C. Hardie, A. Peretz, E. Suss-Toby, A. Rom-Glas, S.A. Bishop, Z. Selinger and B. Minke, Protein kinase C is required for light-adaptation in Drosophila photoreceptors, Nature 363, 634-637 (1993).

38. R. Ranganathan, G.L. Harris, C.F. Stevens and C.S. Zuker, A Drosophila mutant defective in extracellular calcium-dependent photoreceptor deactivation and rapid desensitization, Nature 354, 230-232 (1991).

39. I. Masai, A. Ozazaki, T. Hosoya and Y. Hotta, Drosophila retinal degeneration A gene encodes an eye-specific diacylglycerol kinase with cysteine-rich zinc-finger motifs and ankyrin repeats, Proc. Natl. Acad. Sci. USA 90, 11157-11161 (1993).

40. B. Minke, C.T. Rubinstein, I. Sahley, S. Bar-Nachum, R. Timberg and Z. Selinger, Phorbol ester induces photoreceptor-specific degeneration in a Drosophila mutant, Proc. Natl. Acad. Sci. USA 87, 113-117 (1990).

41. T.S. Vihteic, M. Goebl, S. Milligan, J.E. O'Tousa and D.R. Hyde, Localization of Drosophila retinal degeneration B: a membrane-associated PI transfer protein, J. Cell. Biol. 122, 1013-1022 (1993).

42. A.D. Blest, S. Stowe and A. Delaney, Diacylglycerols as putative second messengers that regulate phototransductive membrane turnover by arthropods, J. Comp. Physiol. A. 173, 57-63 (1993).

43. H. Takemura, A.R. Hughes, O. Thastrup and J.W. Putney Jr., Activation of calcium entry by the tumor promotor thapsigargin in parotid acinar cells, J. Biol. Chem. **264**, 12288-12271 (1989).

44. R.F. Irvine, Quantal Ca^{2+} release and the control of Ca^{2+} entry by inositol phosphates, Febs. Lett. 263, 5-9 (1990).

45. B. Minke and Z. Selinger, Inositol lipid pathway in fly photoreceptors: Excitation, calcium mobilization and retinal degeneration, In Progress in Retinal research vol. 11, 99-124. Osborne N.N. and Chader G.J. (eds) Pergamon Press, NY (1992).

VISUAL PIGMENT REGENERATION IN INVERTEBRATES. RETINAL PHOTOISOMERASE

Isidoro M. Pepe and C.Cugnoli

Istituto di Biofisica, Facolta' di Medicina, Universita' di Genova and Istituto di Cibernetica e Biofisica del C.N.R., Genova, Italy.

Introduction

In visual systems of invertebrates, rhodopsin is not hydrolyzed by photon absorption but is converted to a stable metarhodopsin, which can be reconverted to rhodopsin by light (1,2). This fast process, called photoregeneration, plays an essential role in maintaining the rhodopsin content, and thus the receptor sensitivity, at a high level (3-5). Visual pigments of invertebrates are also regenerated by a slow process of membrane renewal quite similar to that occurring in vertebrates (6-9).

Studies on the visual system of the blowfly have been of fundamental importance in clarifying the possible pathways of rhodopsin regeneration in invertebrates. During continuous illumination a photoequilibrium is reached between rhodopsin and metarhodopsin: the ratio between the concentration of rhodopsin and that of metarhodopsin in the fly compound eye depends on the spectral composition of the environmental light and on the respective extinction coefficients of the two pigments. The Kinetics of the photoreactions is dependent on the light intensity (3,5).

The alternation of the two states rhodopsin and metarhodopsin, which are caused by photoabsorption, is not of unlimited duration: in the ommatidia there is continuous renewal which implies degradation and resynthesis of molecules. Rhodopsin degrades with a mean lifetime of about 5 days, while metarhodopsin decays 60 fold more rapidly (10).When the animals are kept in alternation of light and darkness, rhodopsin content remains constant because of photoregeneration and biosynthesis processes. In contrast, when the animals are kept in the dark for days, rhodopsin content decays exponentially, and is restored to the original content by injecting 11-cis retinal in their eyes. Alternatively, the flies were tranferred from dark to light and their rhodopsin content was found to recover to normal values rapidly with a rate constant very close to the value obtained for the biosynthesis of rhodopsin in the dark following injection of 11-cis retinal into the eyes of vitamin A-deprived flies (11). So these experiments clearly demonstrate that retinal isomerization in the fly ommatidia is a light- dependent process. In fact the maximum effect was reached keeping the animals in light of the blue-violet range (about 440 nm). This may seem strange: blue-violet light is scarcely absorbed by free-retinal which absorbs maximally in the U.V range (380 nm). However, all-trans retinal released in the cell by the metarhodopsin degradation cannot stay free because of its high reactivity, so that possible binding with a protein may have the twofold effect of shifting its absorbance to the visible range and making highly stereospecific photoisomerization possible. Indeed, retinal can bind to an amino group of protein residue via a protonated Schiff-base linkage, thereby shifting its absorbance maximum from 380 nm to about 440 nm. Moreover, the protein structure around the binding site could favour the photoconvertion of all-trans retinal into 11-cis retinal, which is the only isomer able to combine with opsin to give rhodopsin.

Retinal-photoisomerase from honeybee retina

A pigment isolated in the honeybee compound eye seems to have the required characteristics of photoisomerase activity. This pigment is made up of a protein which was binds all-trans retinal as chromophore.

Retinal-photoisomerase was extracted from homogenized honeybee heads in Tris-glycine buffer, incubated with tritiated retinol, isolated by preparative electrophoresis and further purified on DEAE column (12-14).

The molecular weight of the protein was determined as 50,000 by gel filtration and 27,000 by SDS polyacrylamide gel, suggesting that the protein consists of a dimer (15).

The protein incubated with retinal absorbs maximally at 440 nm, indicating the formation of a protonated Schiff-base linkage between retinal and protein (see Fig 1). In fact, on changing the pH from 6.5 to 9.5 in the dark, the 440-pigment is converted into an alkaline form which absorbs maximally at 365 nm, with a pK=8.4 (16) . This is the characteristic absorbance behaviour of protonated (440nm) and unprotonated (365nm) Schiff-bases which form between retinal and amino groups .

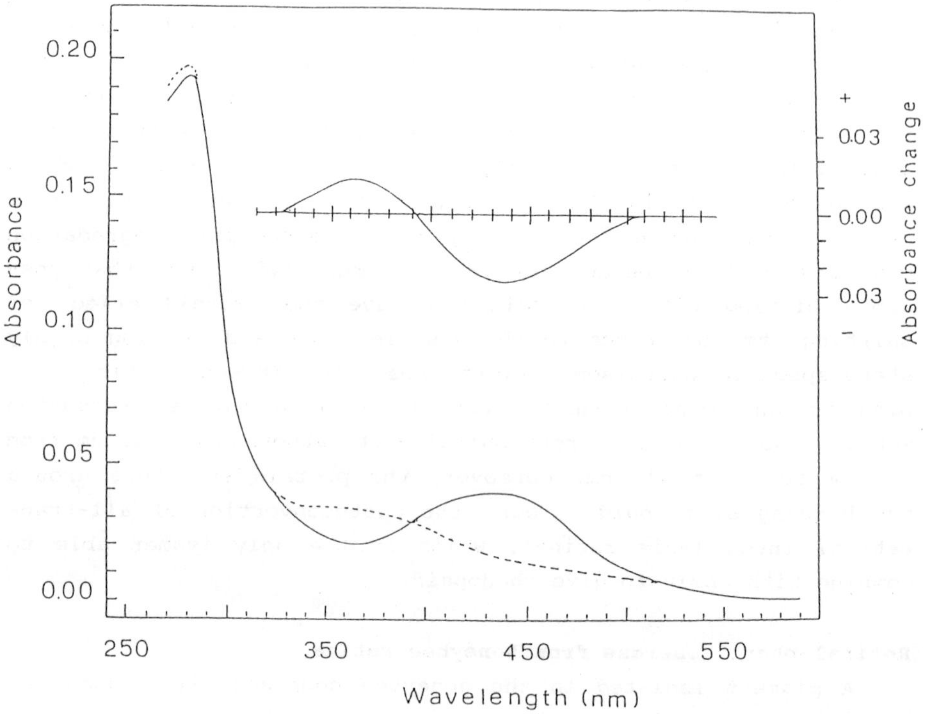

Fig. 1 *Absorbance spectrum of bee photoisomerase before (continuous line) and after 40 min irradiation (dotted line) with light of wavelength 490 nm. The inset shows the absorbance difference between the two curves. Adapted from (19).*

The stoichiometry of the binding between all-trans retinal and protein was studied and the results indicated that one molecule of retinal is able to bind one molecule of protein (dimer) with a dissociation constant of $K=2x10^{-6}$ M (15).

The pigment bleaches upon illumination: Fig.1 shows that the absorbance at 440 nm decreases and that is accompanied by the formation of a photoproduct with absorbance maximum at 370 nm.

The extinction coefficient of 440-pigment was calculated as being approximately 47,000 M^{-1} cm^{-1} (16). The photoproduct formed upon irradiation has an extinction coefficient estimated to be 24,000 M^{-1} cm^{-1}. This value, considerably lower than that of the original pigment, suggests a conversion of the chromophore to a cis-isomer. When analysed by high performance liquid chromatography (HPLC) the photoproduct was in fact identified as 11-cis retinal (see Fig.2).

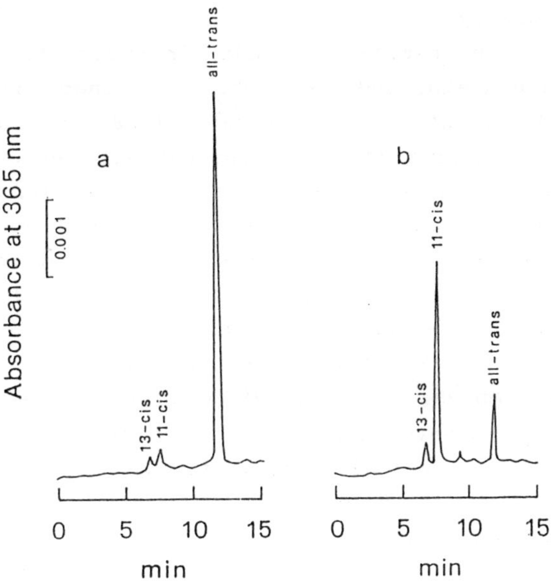

Fig. 2 *HPLC analysis of the chromophore of bee photoisomerase a) before and b) after 40 min irradiation with monochromatic light of wavelength 490 nm. Adapted from (19).*

The pigment also bleaches in the dark in the presence of hydroxylamine, but not in the presence of cyanoborohydride. Bleaching in the presence of the latter only occurs in the light, suggesting that the binding site is buried in the interior structure of the protein. Light would cause a change in conformation and a consequent deprotonation of the Schiff-base linkage which would thereby become accessible to the reducing agent (16).

The protein from honeybee is able to bind retinol as well as retinal. When incubated with retinol, the pigment shows an absorbance maximum at 330 nm which is typical for retinol free in solution, indicating that the binding of the protein with retinol is presumably not covalent. The binding site for retinol is presumably the same as that of retinal as suggested by the complete displacement of radioactive retinol bound to protein by the addition of an excess of retinal (15).

Enzymic activity

When all-trans retinal was added in excess to a solution containing the protein, and the mixture was then continuously irradiated with light of a wavelength (528 nm) absorbed only by retinal bound to protein, all-trans retinal was almost exclusively transformed into 11-cis retinal (see Fig 3).

After about 2 hour of irradiation, although with a wavelength of low probability for absorption by the retinal-protein complex, the 11-cis retinal formed was about 50% of the total. The formation of 13-cis retinal - which was already present as a contaminant - remained under 10% of the total.

Control samples, which contained only free retinal or retinal added to heat-denatured protein, irradiated with the same wavelength, showed no retinal isomerization.

The reaction could be accelerated by using a light with shorter wavelength(500 nm), but many other retinal isomers were obtained as a result of photon absorption by free retinal.

Considering that photoisomerization of free retinal produces a mixture of isomers, with 11-cis retinal being not the favoured product, the result of retinal irradiation in the presence of bee photoisomerase is remarkable. The protein is able to direct the isomerization of retinal made by light almost exclusively toward the 11-cis retinal formation.

When 13-cis or 9-cis retinal replaced all-trans retinal as the starting isomer , the same stereospecific photoisomerization to 11-cis retinal was obtained in the presence of bee-photoisomerase (17).

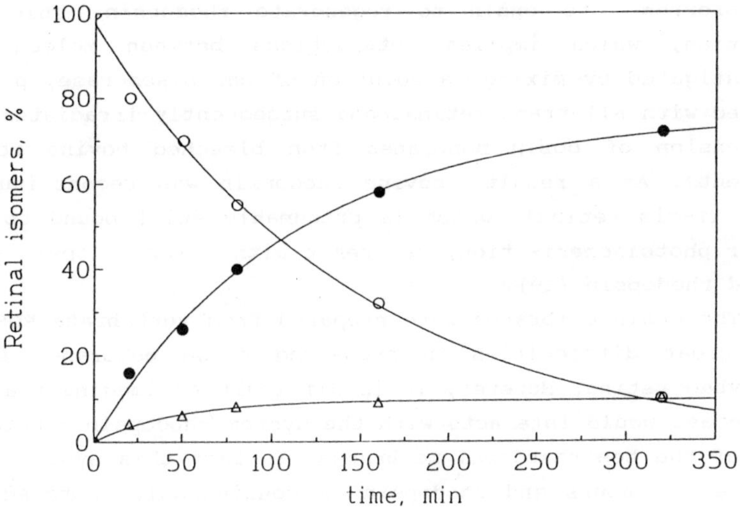

Fig. 3 *Isomerization of all-trans retinal by continuous irradiation with 528 nm light at room temperature in the presence of 5.6 x 10^{-6} M bee photoisomerase in 0.1 M phosphate buffer, pH 7. Initial concentration of all-trans retinal was about three fold molar excess compared to the protein concentration. Retinal isomers are represented by the following symbols: (○) all-trans; (●) 11-cis; (△) 13-cis. Adapted from (14).*

Physiological role

According to Goldsmith (18), photoisomerase is a major photopigment in the bee compound eye accounting for up to 80% of the total photopigments and has been localized in the primary pigment cells by antibody staining and electron microscopy.

Bee photoisomerase could be involved in the fundamental process of visual pigment regeneration. The dark degradation of metarhodopsin was clearly demonstrated for *Calliphora* , but it is also very likely for honeybees, as suggested by the decline of all-trans retinal in honeybee eyes during prolonged dark and orange adaptation (18). All-trans retinal released by the degradation of metarhodopsin could bind to photoisomerase which could transform it into 11-cis retinal in the light of visible range. Most of 11-cis retinal could then be reduced to 11-cis retinol by an alcohol dehydrogenase or alternatively, might be directly transferred to opsin to regenerate rhodopsin. This complex function, which implies interactions between molecules, was investigated by mixing a solution of photoisomerase, previously loaded with all-trans retinal and subsequently irradiated, with a suspension of opsin membranes from bleached bovine rod outer segments. As a result bovine rhodopsin was reconstituted), i.e. 11-cis retinal, which is presumably still bound to protein after photoisomerization, can react with membrane opsin so as to yield rhodopsin (19).

The opsin membranes were prepared from vertebrate ROS due to the great difficulties in preparing these membranes from the honeybee retina. However, it is difficult to imagine that photoisomerase would interacts with the system rhodopsin - metarhodopsin of the bee rhabdomal membranes. In fact this system remains stable for hours and rhodopsin is continuously photoregenerated by the absorption of metarhodopsin. As other kinds of regeneration processes such as dark regeneration, are probably excluded in the bee as in the blowfly (18), photoisomerase could very likely be involved in the slow renewal of rhabdomal membranes which therefore implies the biosynthesis of new molecules of rhodopsin.

Comparison between bee photoisomerase and cephalopod retinochrome

A photosensitive protein, retinochrome, containing all-trans retinal as its prosthetic group was discovered in cephalopods (20). Retinochrome was found in association with membranes of the inner segments of visual cells and was therefore extracted in aqueous solutions containing 2% digitonin. The absorbance maximum of squid retinochrome is at about 495 nm at pH 6.5. When irradiated with orange light it bleaches giving a photoproduct (metaretinochrome) whose absorbance maximum also depends on pH (460 nm at pH 6.5 or 370 nm at pH 9). When this photoproduct was thermally denatured in order to detach the chromophore, and was incubated in the dark with cattle opsin, rhodopsin was formed, which indicates that the chromophore of retinochrome is changed by light from all-trans retinal to 11-cis retinal (21). The formation of squid rhodopsin was later demonstrated using rhabdomal membrane preparations containing opsin and metaretinochrome (22).

There is a striking similarity between retinochrome and bee photoisomerase. This is the property for the two proteins to combine with various geometrical isomers of retinal, such as all-trans, 13-cis and 9-cis, and to catalyse their photoisomerization almost exclusively to 11-cis retinal (17,23). This suggests that there is a very similar active site for both of these enzymes i.e. a particular structure of the protein near the binding site which would favour the photoproduction of 11-cis retinal.

When retinochrome bleached by orange light is further exposed to light in the near ultraviolet spectrum (360 nm), it shows a net regeneration (24); this photoregeneration, which reconverts 11-cis retinal to all-trans retinal, was also observed for the bee photoisomerase. It is therefore probable that 11-cis retinal produced by bee photoisomerase, being insoluble in aqueous medium, will remain attached to the protein, although not necessarily in the same binding site, in order to be transported where it is needed to regenerate rhodopsin.

Although cephalopod retinochrome and bee photoisomerase shares these and other similarities such as a low molecular weight of the apoprotein (24,000 for retinochrome and 27,000 for bee photoisomerase), they differ in their solubility properties. In fact, retinochrome is bound to membranes and needs detergents such as digitonin to be brought in solution, whereas bee photoisomerase is already found water soluble. Therefore rhodopsin regeneration by retinochrome poses some problems since rhodopsin is located in the rhabdomal microvilli of the outer segment of the cephalopod visual cells, whereas retinochrome is mainly stored in the inner segments associated with the membranes of myeloid bodies (25,26).

It has been suggested that some molecules of retinochrome could move from inner to outer segment during light adaptation of squids (25). Nevertheless, rhabdomal microvilli membranes are so far away from myeloid body membranes that a hypothesis for some suitable shuttle of retinal between these structures was made. This hypothesis was suggested by the discovery of a water-soluble retinal binding protein in the squid retina (27-29). In honeybee retina, however, such an intermediary between the isomerizing enzyme and rhodopsin may not be necessary, as the molecule of photoisomerase, being water-soluble, could directly act as a carrier for 11-cis retinal.

A possible scheme for visual pigment regeneration in invertebrates.

Both the fly *Calliphora erythrocephala* and the honeybee *Apis mellifera* seem to be able to form visual pigment only when 11-cis retinal is present having been supplied by a blue/violet-absorbing isomerase (11,18). No other pathways of dark regeneration of rhodopsin from metarhodopsin, nor biosynthesis of metarhodopsin or rhodopsin from all-trans retinal, were observed in those visual systems. A dark biosynthesis of rhodopsin can occur but only when 11-cis retinal is made available by the light-dependent process of isomerization.

Dark regeneration of rhodopsin was never observed in cephalopods. Photoregeneration from metarhodopsin to rhodopsin is thought to be the only fast regeneration pathway , while the slow regeneration process would be supported by the photoisomerase activity of retinochrome. Metaretinochrome, or a soluble intermediary, would in fact be able to provide 11-cis retinal for newly synthesized opsin in order to regenerate rhodopsin during the renewal of photoreceptor membranes (30).

Evidence that a retinochrome-like photoisomerase exists in the retina of a marine gastropod *Conomulex Luhuanus* was reported by Ozaki et al. (31). In this visual system rhodopsin, isolated in the rhabdomal microvillus membranes, has maximal absorbance at 474 nm, retinochrome, located in the photic vesicles of the visual cell body, absorbs maximally at 510 nm. When irradiated with orange light retinochrome was bleached into metaretinochrome changing its chromophore from all-trans to 11-cis retinal.

Other evidence of the existence of a photoisomerase activity was found in the eyes of the arthropod crayfish (32). In this visual system infact, the slow regeneration of rhodopsin after irradiation with orange light (which converts rhodopsin to metarhodopsin) was found to be accelerated by the action of blue light.

The finding of photoisomerase activity in different visual

systems such as those of bees, flies, cephalopods and recently in the lateral eye of *Limulus* leads to the simplified scheme of Fig. 4 in which the regeneration of rhodopsin should occur via two pathways only: photoregeneration and renewal, both of which are light-dependent processes.

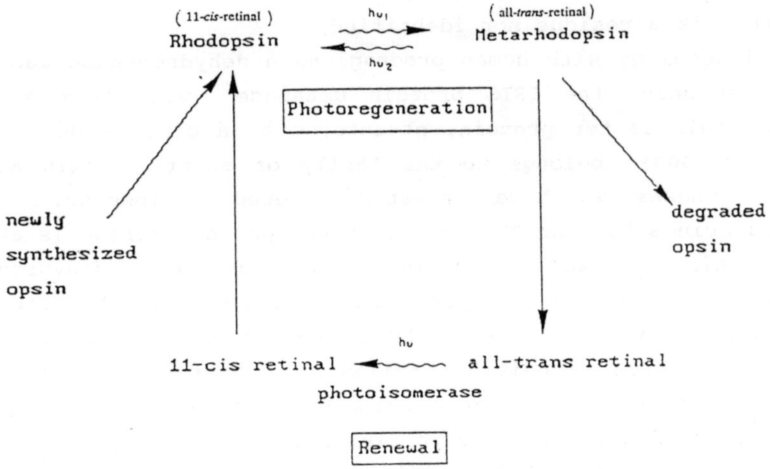

Fig. 4 Visual pigment cycle showing the two light-dependent pathways of rhodopsin regeneration. Photoreactions are represented by wavy arrows.

In planning future experiments on rhodopsin regeneration, it may be useful to examine the possibility of extending the simple cycle of Fig. 4 to other invertebrate visual systems so that the problem of rhodopsin regeneration could result greatly simplified.

Partial protein sequence

Since the N-terminal of the bee-photoisomerase was found to be blocked, the protein was digested with Staphylococcus aureus V8 protease . Four peptides obtained from the digestion were electroeluted after separation on SDS-PAGE (34). After sequencing, three peptides were found strongly contaminated with V 8 fragments and the fourth peptide gave the following sequence of 29 aminoacids:

GLDRAIFIAXDVSKNEQFQESFKXVXDTY

(where x is a residue not identified)
A good homology with human prostaglandin dehydrogenase was found after scanning the ISIS protein database with this sequence (34). This latter protein, structured as a dimer (MW of monomer: 29,000), belongs to the family of short - chain alcohol dehydrogenases, which were first discovered in insects.

This brings to mind the fact that bee photoisomerase is able to bind retinal as well as retinol, so it may have dehydrogenase activity. However, experiments performed on photoisomerase solutions in the presence of NADPH and retinal or NADP and retinol in different conditions failed to detect any dehydrogenase activity. It could be a mere coincidence that such a good homology occurred only for this partial sequence.

The sequence of the binding site was determined by adding radioactive retinal to a solution of photoisomerase and then reducing the Schiff-base linkage between retinal and protein with sodium borohydride. After this reduction, the chromophore retinal remains covalently attached to the protein. Peptide fragments were obtained from cyanogen bromide digestion of the sample. In this way, the sequence of the radioactive peptide fragment was compared to that of the native control where the binding between retinal and protein was hydrolized. The binding site resulted located in the first lysine (third aminoacid) of the following fragment:

TAKIITPXVLKXXH

This sequence is quite different from the binding site sequence of cephalopod retinochrome and various visual pigment rhodopsins (see Table). Only the alanine before the binding site lysine is conserved throughout rhodopsins but not in the retinochrome photoisomerase. The proline at the 10th position that is conserved in rhodopsins and in retinochrome is sostituted by a leucine in the bee-photoisomerase. The great difference between the binding site sequence of the bee-photoisomerase and those of visual pigment rhodopsins is probably related to the difference in photoconversion of retinal, while the difference with the binding site of retinochrome, which has the same catalyst property for converting all-trans to 11-cis retinal, is maybe due to the different structures of the two proteins (bee-photoisomerase is water-soluble while retinochrome needs detergents to be brought in solution).

TABLE

Aminoacid sequences of the retinal binding site of various photopigments

photoisomerase	(honeybee)	T A K I I T P X V L K X X H	
retinochrome	(todarodes)	M S K T G C A L F P L L I	(36)
rhodopsin	(todarodes)	F A K A S A I H N P M I Y	(37)
rhodopsin	(octopus)	F A K A S A I H N P I V Y	(38)
rhodopsin	(drosophila)	F A K S A A C Y N P I V Y	(39)
rhodopsin	(bovine)	F A K T S A V Y N P V I Y	(40)

Work is now in progress on the screening of a cDNA library just prepared from honeybee retina tissue with Lambda GT 11 as vector. The aim of this work is to clone the gene of the bee photoisomerase in order to obtain enough material for determining the protein structure. We think that the knowledge of the structure of photoisomerase is essential to explain the mechanism of stereospecific isomerization which is shared by visual pigment rhodopsin.

Acknowledgements

This work was supported by P.F. Ingegneria Genetica, C.N.R.

References

1 R. Hubbard ,R.C.C. St. George . The rhodopsin system of the squid. *J.Gen.Physiol. 41* (1958) : 501 - 528.

2 P.K.Brown., P.S. Brown . Visual pigments of octopus and cuttlefish. *Nature. 182* (1958) : 1288 - 1290.

3 K. Hamdorf , J. Schwemer . Photoregeneration and the adaptation process in insect photoreceptors. In :A.W. Snyder, R. Menzel , (eds) *Photoreceptors Optics*, Springer, Berlin, 1975 ,pp.263 - 289.

4 K. Hamdorf . The physiology of invertebrate visual pigments. In :H.Autum (ed) *Handbook of Sensory Physiology, vol VII/6A* . Springer, Berlin , 1979,pp.145 - 224.

5 P. Hillman, S. Hochstein and B. Minke. Transduction in invertebrate photoreceptors: role of pigment bistability. *Physiol. Rev. 63* (1983) 668 - 772.

6 R. H. White. The effect of light deprivation upon the ultrastructure of the larval mosquito eye. III. Multivesicular bodies and protein uptake. *J. Exp. Zool. 169* (1968) 261 - 278.

7 A. Perrelet. Protein synthesis in the visual cells of the honeybee drone as studied with electron microscope radioautography. *J. Cell. Biol. 55* (1972) 595 - 605.

8 L. J. Goldman, S. N. Barnes and T. H. Goldsmith. Microspectrophotometry of rhodopsin and metarhodopsin in the moth *Galleria*. *J. Gen. Physiol. 66* (1975) 383 - 404.

9 A. D. Blest. Photoreceptor membrane turnover in arthropods: comparative studies of breakdown processes and their implications. In: T. P. Williams and B. N. Baker (eds) *The effect of constant light on visual processes*. lenum. New York London. (1980) 217 - 245.

10 J. Schwemer. Renewal of visual pigment in photoreceptors of the blowfly. *J. Comp. Physiol. A 154* (1984) : 535 - 547.

11 J. Schwemer. Pathways of visual pigment regeneration in fly photoreceptor cells. *Biophys. Struct. Mech. 9* (1983) : 287 - 298.

12 I.M. Pepe, A. Perrelet and F. Baumann, Isolation by polyacrylamide gel electrophoresis of a light-sensitive vitamin A-protein complex from the retina of the honeybee drone, *Vision Res. ,16* (1976) 905-908.

13 I.M. Pepe ,C. Cugnoli. Isolation and characterization of a water - soluble photopigment from honeybee compound eye. *Vison Res. 20* (1980) : 97 - 102.

14 J. Schwemer ,I.M. Pepe ,R. Paulsen and C. Cugnoli. Light induced trans - cis isomerization of retinal by a protein from honeybee retina. *J. Comp. Physiol. 154* (1984) : 549 - 554.

15 I.M. Pepe, C. Cugnoli, M. Peluso, L. Vergani and A. Boero, Structure of a protein catalyzing the formation of 11-cis retinal in the visual cycle of invertebrate eyes, *Cell Biophys., 10* (1987) 15-22.

16 I.M. Pepe ,J. Schwemer ,R. Paulsen. Characteristics of retinal - binding proteins from the honeybee retina. *Vision Res. 22* (1982) : 775 - 781.

17 C. Cugnoli, R. Mantovani, R. Fioravanti and I. Pepe, 11-cis retinal formation in the light catalyzed by a retinal-binding protein from the honeybee retina, *FEBS Lett., 257* (1989) 63-67

18 W.C.Smith and T.H. Goldsmith, The role of retinal photoisomerase in the visual cycle of the honeybee, *J. Gen. Physiol., 97* (1991) 143-165.

19 I.M.Pepe, C.Cugnoli and J. Schwemer, Rhodopsin reconstitution in bleached rod outer segment membranes in the presence of a retinal-binding protein from the honeybee, *FEBS Lett., 268* (1990) 177-179.

20 T.Hara and R. Hara, New photosensitive pigment found in the retina of squid, *Nature, 206* (1965) 1331-1334.

21 T. Hara and R. Hara, Rhodopsin and retinochrome in the squid retina, *Nature 214* (1967) 573-575.

22 T. Seki, R. Hara and T. Hara. Reconstitution of squid rhodopsin and cattle rhodopsin by use of metaretinochrome in their respective membranes. *Exp. Eye Res. 34* (1982) 609-621.

23 T. Hara and R. Hara, Isomerization of retinal catalysed by retinochrome in the light, *Nature 242* (1973) 39-43 .

24 T. Hara and R. Hara, Regeneration of squid retinochrome, *Nature 219* (1968) 450-454.

25 T. Hara and R. Hara, Distribution of rhodopsin and retino chrome in the squid retina, *J. Gen. Physiol. , 67* (1976) 791-805.

26 K. Osaki, R. Hara and T. Hara, Dependency of absorption characteristics of retinochrome on pH and salts, *Expl. Eye Res. , 34* (1982) 499-508.

27 K. Osaki, A. Terakita, R. Hara and T. Hara, Isolation and characterization of a retinal binding protein from the squid retina, *Vision Res. ,27* (1987) 1057-1070.

28 T. Hara, Visual photoreception of the squid : photopigment regeneration In : T. Hara (ed) *Molecular Physiology of Retinal Proteins* Yamada Science Foundation, Yamada, 1988, pp.305-310.

29 A. Terakita, R. Hara and T. Hara. Retinal-binding protein as a shuttle for retinal in the rhodopsin-retinochrome system of the squid visual cells. *Vision Res. 29* (1989) 639-652.

30 R. Hara and T. Hara. Squid m-retinochrome. *Vision Res. 24* (1984) 1629-1640.

31 K. Ozaki, A. Terakita, R. Hara and T. Hara. Rhodopsin and retinochrome in the retina of a marine gastropod, *Conomulex Luhnanus Vision Res. 26* (1986) 691-705.

32 T. W. Cronin and T. H. Goldsmith. Dark regeneration of rho dopsin in crayfish photoreceptors. *J. gen. Physiol. 84* (1984) 63-81.

33 W.C.Smith, M.A.Freiedman and T.H.Goldsmith. Retinoids in the lateral eye of *Limulus*: Evidence for a retinal photoisomer ase. *Visual Neurosci. 8,* (1992) 329-336.

34 I.M.Pepe, C.Cugnoli, J.N.Keen and J.B.C.Findlay. Retinal photoisomerase from honeybee compound eye : physiological

role. In: *Structures and functions of retinal proteins* J.L.Rigaud ed., J.Libbey Eurotext, London, vol.221 (1992) 275-278.
35 Akrigg, D., Bleasby, A.J., Dix,N.I.M., Findlay, J.B.C., North, A.C.T., Parry-Smith, D., Wootton, J.C., Blundell, T.L., Gardner, S.P., Hayes, F., Islam, S., Sternberg, M.J.E., Thornton, J.M., Tickle, I.J. and Murray-Rust, P. A protein sequence/structure database. *Nature 335,* (1992) 745 - 746.
36 Hara-Nishimura,I., Matsumoto,T., Mori,H., Nishimura,M. Hara, R. and Hara, T) *FEBS Lett. 271,* (1990) 106-110.
37 Hara-Nishimura,I., Kondo, M., Nishimura,M., Hara,R., and Hara,T. *FEBS Letter 317,* (1993) 5-11.
38 Ovchinnikov,Yu.A., Abdulaev, N.G. and Tsuda,M. *FEBS Lett. 232,* (1988) 69-72.
39 Zuker,C.S., Cowman,A.F. and Rubin,G.M. *Cell 40,* (1985) 851-858.
40 Hargrave,P.A., McDowell,J.H., Curtis,D.R. and Argos,P. *Biophys.Struct.Mech. 9,* (1983) 235-144.

EXTRAOCULAR PHOTOSENSITIVITY IN INVERTEBRATES: A LOOK INTO FUNCTIONAL MECHANISMS AND BIOPHYSICAL PROCESSES (*)

CARLO MUSIO

Istituto di Cibernetica, C.N.R.,
Via Toiano 6, I-80072 Arco Felice (Napoli), Italy

ABSTRACT

This paper provides a brief survey on a particular kind of light-sensitivity shared by organisms belonging to the lower animal *phyla*. This unusual sensory capability, named extraocular photoreception, does not occur in ocular and in eye-like or retinic structures. On the contrary, it is carried out by single or clustered cells scattered throughout the whole body or it is strictly confined to some neuronal cells or neural pathways. In invertebrates, extraocular photoreception is involved in the regulation of several behavioural expressions such as photoperiodism, locomotion and timing process of circadian rhythms. It represents a light-detection pathway parallel to the structured visual system. A complete explanation of the mechanisms underlying this type of photosensitiveness is particularly meaningful to elucidate the evolutionary lineage of photoreception in invertebrates, matching these processes, by a functional point of view, with the extraretinal functions still present in some lower vertebrates.

1. INTRODUCTION

The functional evolution of the photoreceptors and, consequently, of the mechanisms of light detection developed from diffuse non-specialized cellular structures to localized and specialized ones. These processes evolved from an ancestral cell membrane specialized for photosensitivity, to the localization of photosensitivity in restricted zones of the cell membrane (through the inclusion of photopigment as integral membrane proteins) and, finally, to an increasing of photosensitive surface by means of the formation of fine processes or lamellated membranes [13]. This third evolutionary step of the photoreceptorial system culminated in the organization of specialized structures such as the ciliary (in vertebrates) and microvillar (in invertebrates, with restricted exceptions) membrane photoreceptor [13, 19, 56].

The evolutionary development of the structure-function relationship produced a growing degree of complexity of photoreceptors and eyes in invertebrates as well as in vertebrates. Evidently, complex photoreceptor arrays correspond to complex photophysiology and of consequence, to complex animal behaviour [5].

Nevertheless, both invertebrates and vertebrates maintain another way of light detection, that parallels the photoreception conveyed by the eyes [30, 40, 41, 68-70].

This kind of photosensitivity is termed *extraocular photoreception* (EOP), or, at times, extraretinal photoreception.

(*) *to Angela, without doubt.*

In EOP the photic information is mediated through photosensitive cells, either solitary or clustered, scattered over the body surface or located in neural areas. The EOP can be classified as below:

(i) *Dermal Photosensitivity*, in which the cells (belonging to non-nervous tissues) do not present morphological features identifiable as a generally accepted photoreceptor.

(ii) *Diffuse Photosensitivity*, in which the scattered and/or isolated cells are structured as known photoreceptive cells.

(iii) *Neural Photoreception* and *Neuronal Photoreception*, in which the photosensitive field is localized in neural tissue and in single or clustered nervous cells, respectively.

In invertebrates, EOP, like photoreception conveyed by eyes, is involved in behavioural phenomena such as phototropism, phototaxis, photoperiodism, locomotor activity and reproductive processes. EOP system is also associated to timing processes, the most important of which is the entrainment of circadian rhythms [10].

Generally, the EOP system acts together with retinal structures and is functionally connected with them. Until now, there is no evidence of the coexistence of both dermal and neural EOP in the same animal (except for the photosensitivity in *Ciona intestinalis,* [18]). On the other hand, EOP is the only type of light responsiveness in the lowest *phyla*, which lack a developed visual system.

Unfortunately, there are no exhaustive reports in the current literature of the above aspects, due to experimental difficulties, encountered in any given animal. In fact, single behavioural or physiological or biophysical approaches have been carried out on a variety of animals (fly, scorpion, squid, crayfish, snail, marine worm, sea squirt), but a multiaspect study comprehensive of all possible EOP mechanisms is lacking.

However, even if far from being a complete elucidation, the findings available in invertebrates clarify some important questions about the evolution of the photoreceptor systems as well as about the comparative relationships between EOP and retinal photoreception.

The field is quite active on both the experimental and the theoretical fronts. In fact, also refferring to the line of evolution of photoreception outlined by Yoshida [70] (who propose a development from a direct responsiveness of cells, via EOP, respectively through the diffuse system to a localized structure and through the neural to neuronal system), we have in some cases difficulties to interprete the coexistence of EOP and retinal mechanisms in a given animal.

In most cases EOP cannot be considered a meaningless evolutional residue, because it plays a crucial role in many functions essential for the animal's survival; *e.g.*, in insects, EOP is related to the influence of light on brain neuroendocrine activity [63]. In this respect, an interesting case is provided by the increase of EOP in advanced groups of insects in which visual pathways are entirely deserted in favour of extraretinal photoreceptors [63].

Furthermore, the existence of EOP system also in vertebrates confirms that EOP is able to perform and/or control several functions (timing processes related to circadian rhythms or to sexual cycles, hormonal changes...) with or without a link to the visual system.

To confirm this remark, more selective data are needed, above all those regarding the identification of photoreceptor(s) and the photosensitive molecule(s) responsible for certain behaviours. This should be the only way to achieve a better understanding of the relationships between EOP and behaviour. Related to this, several papers are yet available and a lot of works are in progress about the similarities between biochemical and biophysical aspects of EOP and retinal phototransduction both in invertebrates and vertebrates.

2. BEHAVIOURAL RESPONSES

Invertebrates exhibit distinct and precise behavioural responses that are mainly associated with the dermal and/or diffuse EOP system. Behavioural reactions can be considered to belong to the reflex or to the kinetic type. In some cases, responses elicited by EOP interact with other behavioural activities, of which the best studied is the entrainment of circadian rhythm.

2.1. *Reflex responses*

There are many types of reflex responses but the most prevailing is the shadow reaction. The shadow reaction is elicited by a light-on -off stimulus, directed over a portion of the animal body, and it is detected as a withdrawal or an expansion under the illumination of the animal body. Several reflex responses trigger phototactic reaction (see below) and need the integration and the coordination of nervous element for their effectiveness.

A useful approach to study the shadow reflex is to search for the electrical correlates originating from dermal receptors. In this way, electrical activities have been studied in the bivalve mollusc *Lima scabra* [43]. After excision of the eyes, that produces large spikes at the cessation of illumination, small spikes which in all probability are activities of the small nerve fibers originated from the dermal photoreceptors occur during an off-response. An enhancement of this activity becomes evident when a small black ribbon is moved across the illuminated area: dermal units have shorter latency than the large spikes that originate from the ocular system.

Many invertebrates show behavioural responses to visible light even without localized photoreceptors: this occurs mainly in Cnidaria, which show dermatoptic responses that are governed also by multimodal mechanisms. A typical example of photosensitivity without detectable photoreceptors is represented by the cnidarian *Hydra*, of which some aspects regarding the photomotobehaviour will be briefly outlined in the section 3.1 of this paper and more detailed by Taddei-Ferretti *et al.* in this volume [62].

The sea anemone *Metridium senile* is a particular case in which the muscle tissue is directly photosensitive [46]. In fact, parietal muscle of the endoderm contracts within 2-10 s, after illumination by blue light at an intensity of 8.6×10^8 $s^{-1} \times cm^{-2}$ quanta.

Also the photic response of another sea anemone *Calamactis praelongus* was studied electrophysiologically [34]. This study succeeded in distinguishing two types of response mechanism: one is a local due to non nervous contractions of circular and parietal muscles, leading to the column bending, the other is an extensive due to neurally coordinated responses, including contraction of retractor muscles that themselves are insensitive to light. Thus, light is either perceived by sensory cells that are connected to the muscles responsible of the contractions in the former case, or by the muscles themselves in the latter one.

Sea anemones represent also an intriguing case in which the photic reflex response is affected by the presence of symbionts. Previous behavioural analysis noted different response patterns of elongation-contraction: dark-adapted *Anthopleura elegantissima* that harbor endosymbiotic zooxanthellae expand within 5 to 10 min, after exposure to light illumination [53].

Recently, Sawyer *et al.* [57] have examined and quantified for the first time the photoresponse of *Anthopleura elegantissima* dissecting the neurophysiological activities of the cellular subsystems underlying the response, the ectodermal slow system 1 (SS1), the endodermal slow system 2 (SS2) and the conducting nerve net (TCNN). These authors hypothesized that the photoreception occurs in the endoderm because SS2 and TCNN are involved and SS1 is not. Presumably, the photoreception takes place within sensory cells in a local nerve net, with the information being transmitted to the muscles and with the muscles themselves light-sensitive.

The neural correlates of the photobehaviour in *A. elegantissima* resemble those of *C. praelongus* above reported. In both animals the effect of irradiance on the frequency of activity in the conducting systems may depend on the behavioural state of the anemone. Intrinsic activity is used to maintain body shape and external stimuli are coded by changes in activity in the relevant conducting system.

2.2. *Kinetic reactions*

In most instances dermal sensitivity appears to be associated with a phototactic response. Such behavioural responses enable the animal to withdraw to a region of greater or lesser light intensity and appear to be related to the animal's survival.

Typical direct tactic photoresponses are showed by Protozoa and other microorganisms [30, 31]. However, examples regarding this *phylum* will be not considered in this review, as these organisms represent a case in which a single cell composes a whole organism as well as act at the same time as effector and receptor. For these reasons Protozoa are not strictly an example of EOP and we shall consider only cases occurring in multicellular organisms.

The sea squirt, *Ciona intestinalis*, is of evolutionary interest because it is a transition site between an invertebrate and vertebrate. The adult *Ciona* is phototropic and orients itself in the direction of the light, accompanied by the opening of the siphons. The animal also contracts and elongates upon light excitation even when the siphons are removed, and, according to Millot [40], possesses a dermal light sense. It also has a neural photoreceptor system associated with the opening and closing of its siphon [18].

Phototactic responses have also been found in insects [63]. For example, the alfaalfa weevil, *Hypera postica*, responds to illumination in the red region of the electromagnetic spectrum by orienting toward the light source. This reaction is related to the animal feeding on plants [63].

2.3. *Interactions between EOP and other behavioural activities*

Invertebrate EOP system appears to be strictly linked to the entrainment of circadian rhythms. While such systems can work in conjunction with a visual system, they can function also without it.

The daily rhythm of behavioural activity in *Aplysia* does not require the presence of eyes though these play a role in modulating the activity rhythm [32]. Block & Smith [11] identified cerebral photoreceptors that underlie entrainment of the cyrcadian rhythms of locomotion and nervous activity. In isolated ganglia's preparations of *Aplysia*, they detected responses to the light termiation in most posterior pedal nerves and pleuro-abdominal connectives. The changes of pedal nerve activity following the light termination were determined by photoreceptors located in the cerebral ganglion. Extraocular photoreceptors can also entrain ocular oscillators and the circadian rhythm of the *Aplysia* overt behavioural activity is in part controlled by both extraocular oscillators and photoreceptors. Furthermore, in some instances the EOP system mediates the response entirely, such as the entrainment of the circadian oscillator in the 6th abdominal ganglion of *Aplysia* [32].

In the horseshoe crab *Limulus*, nerve impulses evoked by light have been recorded from afferent fibers from the telson, even though neither photoreceptors nor their terminations in the central nervous system have yet been identified. This suggests that circadian extraocular photoreceptors exist in the animal tail and that they concur, together with the lateral eye the median ocelli and the ventral photoreceptors, in the timing of the visual circadian rhythms [54].

Truman [63] reviewed the data associated with insect EOP systems and circadian rhythms. Photoperiodic sensitivity is important in adults of some species. Entrainment of stridulatory rhythms occurs in the grasshopper and entrainment of flight rhythms occurs in the silkmoth. In pupae of some species termination of diapause will occur when exposed to a particular photoperiodic regime. In certain species the receptor is an EOP receptor.

In many marine animals, migration is also a behavioural response that is associated with EOP. Migrating rhythms have been closely studied in the squid *Todarodes sagittatus* and in the turbellarian *Convoluta roscoffensis* [69]. *Convoluta* lives in the sand at night during high tide, and emerges onto the surface at low tide during the day time. If the animals are brought to the laboratory under constant light, vertical migration continues for up to 7 days in synchrony with the tides. The rhythm persists only in constant light and not in constant darkness.

Finally, EOP system is related to reproductive development. In the garden slug *Limax maximus* the extraocular receptor measures the duration of increasing daylight. This perception of the increasing day length results in the secretion of a "maturation hormone" by the brain and in turn initiates reproductive development [69].

In the butterfly *Papilio xanthu*, light stimulation induces various movements of genitalia such as penis withdrawal [4]. This stimulation activates two pairs of genitalia EOP sites that are probably involved in regulatig copulation behaviour [3].

3. Functional Mechanisms

This section will focus the neurophysiological analysis of the functional mechanisms responsible of the responses that are mediated by elements involved in the dermal EOP as well as by the activity of elements identified directly as light-sensitive in neural and neuronal EOP systems.

3.1. *Dermal EOP*

As stated before, dermal EOP is a photosensitivity carried out by unidentified responsible elements. To this regard, speaking about the search of functional mechanisms would seem quite contradictory. Thus, in this section, we will indicate some cases in which the photobehaviour, lacking known photoreceptive cells, needs the integration of nervous system elements. Related to this, dermal EOP (apart the historical definition) could be considered either a precursor or an evolutive starting point to the neural/neuronal EOP via the diffuse photosensitivity.

In the sea urchin *Diadema antillarum*, the spines respond consistently by jerking to an increase or a decrease in light intensity. In the echinoids, the superficial nervous system is formed by nerves that emerge through one of the pore pair at the ambulacral margin. Thus the gradations in sensitivity appear to correspond with the distribution of the superficial nervous system. In this way, the dermal light sensitivity is a result of the superficial nerve fibers being photosensitive [70].

Several species of *Hydra* respond to light by contraction and by tactic movements, the correlates of which can be easily recorded electrophysiologically.

Rushforth *et al.* [55] showed in *Hydra pirardi* an inverse relationship between the latency of contraction responses and the intensity of light, the responses being unaffected by removal of tentacles. The complex mechanisms involved in the *Hydra* light-affected behaviour have been elucidated by Passano and McCullough. In a classic series of papers [49-52], they named the electric correlates of the coordinated contraction as "contraction pulse train" composed by a variable number of "contraction pulses", which are abruptly paused by illumination and are generated by pacemaker. These pacemakers are able to integrate various sensory inputs and to trigger coordinate responses in myofibrillar layer as well. Such pacemakers seem to be localized in the hypostome and are directly inhibited by light. These authors found another pacemaker that elicits "rhythmic potentials" and seems to be located in the pedal zone of the animal. Furthermore, these pacemakers are inhibited by sudden illumination.

In the last twentyfive years, the relationship between photic stimulation and the modulation of the periodic behaviour in *Hydra* have been widely studied by Taddei-Ferretti and coworkers; and the main results and future hints are reviewed in a paper of this volume [62].

3.2. *Neural EOP*

This kind of photosensitivity takes place within various non-neuronal part of the nervous system. It could be referred also to dermal responses that have been recognized through the functional activity of nervous elements. The location of the photosensitive loci involved is often hard to identify and the assumption of a nervous source for photoreception should arise from the nerve activity recording or from the lack of whole neural structures. In this regard, a major problem is encountered when we try to analyze a neural component of EOP. In fact, there is always, with rare exceptions, a degree of uncertainty if the activity arising from the structure under study is really directly photoinduced rather than triggered by the activity of associated structures that are the primary site of phototransduction. Furthermore, the putative photosensitivity recorded from nerves can be referred to axonal projections coming from unidentified neuronal elements (central or peripheric). In this case, it should be more appropriate speaking of a neuronal EOP instance (see the example of the butterfly *Papilio xanthu*, in the next section).

As we saw before, the off-response in *Aplysia* can be recorded from all major branches of the posterior pedal nerves and further studies of nerve transections have inferred that the photoreceptors may lie in the cerebral ganglion [11, 32].

The caudal photosensitivity of crayfish is a classical example of neural EOP, although the photosensitivity is not restricted to the caudal region [67]. There are fibers in the abdominal nerve cord that respond either directly or indirectly to light stimulation by producing paired spikes or EPSPs of 3-4 mV. In the next section, the nature and the identification of the neurons responsible of such photosensitivity will be discussed.

In the mollusc *Spisula solidissima*, the neural photosensitivity has been recorded from the pallial nerve near the visceral ganglion [29]. Spontaneous discharges of the proximal end of the nerve were inhibited by illumination on the nerve slightly distal to the visceral ganglion. Cessation of the illumination caused an increase in the discharge pattern and withdrawal response of the siphon. Kennedy proposed an explanation (based on differential effects of colored lights) of the phenomena by assuming a difference in the time courses between excitatory (slow) and inhibitory (fast) processes. At the base of these phenomena there should be a dual mechanism differing in time constants, function, and responsible photopigments; one was fast, inhibitory and blue sensitive and the other, slow, excitatory and red sensitive [29].

The best case in which the structural correlates for neural photosensitivity have been postulated is that of the mollusc *Mercenaria mercenaria* [66]. In the siphonal nerve of this animal single axon activities were recorded also under cessation of illumination. The photosensitive area was located within a circle area of 85 μm in diameter on the inner siphonal wall where the distal ends of the axons were distributed. As no photoreceptor cells belonging to the recorded axons were found, the site of phototransduction must have been the axonal branches themselves.

To end this section, among the examples reported by Yoshida [70] on instances of neural photosensitivity in scorpions it is worth mentionig the five metasomatic segments of *Heterometrus fulvipes* and the telsonic nerve of *Urodacus*.

3.3. *Neuronal EOP*

In his fundamental review Yoshida stated that "Since it is possible to study activities of individual neurons, the works can be more precise than those on any other types of extraocular systems" [70]. The papers currently available in literature do not disclaim this elementary but true sentence, on the contrary they enhance its meaning.

In the central nervous system of diverse invertebrates, many neurons have been identified as photosensitive confirming the usefulness of the electrophysiological approach in ascribing typical characteristics of neuronal EOP to the system under analysis. Far from being comprehensive, this section will attempt to summarize old and most recent findings on this kind of EOP.

Cnidarian offer a good example of neuronal EOP through the central neurons of the hydromedusa *Polyorchis*. An introductory paper showed the physiological evidence that motoneurons of the swimming pulse system were directly photosensitive and able to receive synaptic input from other conducting systems [1]. In these neurons action potentials that have been evoked by EPSP's were observed during the light-off behavioural responses. These results support a role of these neurons in the control of the swimming behaviour as well as of a possible correlate of the shadow response. However, ten years later, Arkett and Spencer [6] demonstrated that the photosensitive *locus* is not represented by swimming motoneurons but by an identified neuronal system ("O" system), located in the outer nerve ring and presynaptic to the swimming motoneurons, that send its cellular projections (dendritic and/or axonal) to each ocellus. The "O" system showed properties common to rhabdomeric photoreceptors of invertebrates, such as graded responses to light intensity (hyperpolarization during decrease and depolarization during increase) and electrical coupling between neighboring cells [7]. These results demonstrated the graded nature of the shadow reflex and the integration of photic information towards the conducting neural nets by the electrical coupling of the neuronal photoreceptors.

The giant neurons of the visceral ganglion in *Aplysia* have been most extensively studied. The first evidence on *A. fasciata* and *A. depilans* ganglia were reported by Arvanitaki and Chalazonitis in a seminal paper [8]. These authors found a double mechanism of spiking during on and off light stimulation indicating that the former is an excitatory event and the latter is a primary inhibitory one, each associated with a heme protein and a carotenoid pigment respectively. The findings on European species were confirmed on an American species, *A. californica* [2, 12]. Furthermore, a complete biophysical analysis was performed on these cells leading the unexpected result of a resemblance between these neurons and atypical ciliated photoreceptor cells of invertebrates that function like vertebrate ciliated photoreceptors (see the section 4). However, the functional meaning of such marked photosensitivity in these neurons still remain unknown.

The presence of two types of reaction (depolarizing and hyperpolarizing) to illumination has been found also in a group of photosensitive neurons of the subesophageal ganglia of the snail *Helix pomatia* [47, 48], even though the depolarizing response is prevailing. The ionic mechanisms involved and the mediation of the light effect by cyclic nucleotides will be treated in the section 4.2.

The ganglionic complex of the marine pulmonate *Onchidium verruculatum* represent a classical preparation demonstrating neuronal EOP [27]. The photosensitive neurons are clustered in the pleuroparietal and in abdominal ganglia. They have the same structure of other non-photosensitive cells and do not show any typical morphology of photoreceptor cells such as cilia or microvilli. They react to light with both depolarizing or hyperpolarizing photoresponse resulting from closing or opening the light-sensitive channels. These activities resemble those of vertebrate and invertebrate retinal (or ocular) photoreceptors, respectively. in this regard, such structurally non-specialized neuronal photoreceptors could represent an ancestral photoreceptor cell for both ciliary and microvillar types. Despite the great amount of data concerning the biophysical mechanisms underlying the photoresponse and visual transduction (see the section 4), the functional role of these photosensitive neurons remains unclear. It has been suggested that these cells could regulate the transmission of sensory information, mostly of tactile type, that trigger the reflex underlying the mantle-levating movement. Recently, a study on colour discrimination in a neuron of this type, which confirmed the functional role stated above, also proposed a further involvement of such cells in regulating other neural or behavioural activity patterns [45].

In the octopus *Eledone cirrhosa*, EOP systems have been identified in the epistellar body within the mantle cavity on the stellate ganglion. Using isolated preparations of the above structures, the degree of light penetration has been measured. The amplitude of the receptor generator potential increased with increased light intensity. Increased response amplitude occurred upon light flashes of icreased duration and the decay of the photoresponse was slower as light duration increased. These results showed that sufficient light can penetrate through the body wall and the light activation processes of these extraocular photoreceptors [16].

Detection of light by neuronal EOP has been identified in the abdominal ganglia of various crustaceans. Despite the fact that in the majority of invertebrates the function of EOP identified in the brain has yet to be interpreted, in crustacea the functional role of the abdominal EOP systems has been well established.

In *Procambarus clarkii* the caudal photoreceptors, located in the 6th abdominal ganglion, are one of the first so-called *identified neurons* to be described [67]. Quantitative aspects of the light response showed the nature of the response that can be elicited by light fluctuations up to 1 KHz. The onset of bright light causes a prolonged, high-frequency discharge. The basis of the photic discharge is a relatively slow membrane depolarization, but until now the ionic mechanisms are not yet known. The role of these photosensitive neurons is linked to the photonegative nature of the animal. The functional role is extended to the triggering of the tail reflex and to the walking behaviour of the animal. Responses triggered by illumination of the caudal ganglion involve both walking, in reverse direction, and rhythmical flexion/extension of the abdomen. In a recent study, leg movements and the abdominal motor pattern were evoked by repetitive stimulation of a single caudal photoreceptor at frequencies normally achieved during the cell's response to light [58]. These results confirmed without doubt the link between caudal photoreceptor response and the abdominal motor pattern and strengthened the link between the caudal photoreceptor response and backward walking. Thus, the command system is normally activated by the caudal photoreceptors to produce the abdominal motor response.

Furtherly, the activity of brain EOP systems have been examined in insects by means of pharmacological analysis. In the locust *Locusta migratoria*, paired vasopressin-like immunoreactive neurons (VPLI) of the suboesophageal ganglion receive synaptic excitation from extraocular photoreceptors, of which the precise location still remains unknown, via a cholinergic descendent intereurons (DI) [9]. The excitatory drive to VPLI neurons occurs only in darkness and is absent in the light, this photosensitivity persists also after ablation of compound eyes, optic lobes and ocelli. The EOP system is sensitive to light of 497±7 nm (SD) (blue-green) and has an absorption spectrum of a rhodopsin-like pigment [9]. The transmitter of this EOP system is histamine: in fact, application of histamine mimics the EOP-mediated inhibition of the DI neurons [33]. This result indicates that EOP or extraretinal systems, like the visual system, of insects share histamine as the primary photoreceptor transmitter.

Until now, we have reported examples of neuronal EOP systems in which the photoreceptors are represented by central neurons. Another type of neuronal EOP cells are sensory neurons located in non-nervous periphery that project axonal branches, via nerves, to the CNS.

By this way, a typical case of peripheral EOP of the neuronal type is provided by the genital photoreceptors (despite their particular dermal location) of the butterfly *Papilio xanthu*. In this animal two pairs of photoreceptive sites (four per individual) have been identified in the pigmented cuticle of the genitalia. The photoresponse of the genital photoreceptors can be recorded from the branches of two abdominal nerves (N5 and N6) that contain the axons of the extraocular photoreceptors. Each nerve shows a sustained train of single spikes to the ganglion under light flash stimulation. The amplitude of these spikes is larger than that of other recorded spikes in the same erves, suggesting the presence of a single thick axon that comes from the photoreceptive site [3, 4]. The identified four extraocular photoreceptors are not involved in resolving images (even though they appear as separate eyespots) but they simply function as detector of the light covering of the body surface. A possible role of these extraocular photoreceptors in the regulation of copulating behaviours has been yet reported in the section 2.3. Furthermore, ultrastructural evidences have supported the previous electrophysiological analysis confirming definitively the first example of peripheral EOP in insects and proposing a phaosome-containing sensory neuron as the cellular correlate of the EOP [42].

The structural organization of the butterfly genital EOP is similar to that of another photosensory neuron acting as an extraocular photoreceptor, the ventral nerve photoreceptor (VNP) of the horseshoe crab *Limulus polyphemus*.

Limulus has two external ocular systems, the lateral and the median, that underlie the phototactic behavioural response. In addition, at early stage of animal life extraocular photoneurons (VNP cells) are located over the ventral surface and during the development many of these cells become separated from the brain and displaced along the ventral nerve. For this reason, the VNP cells have been classified by Yoshida [70] as a type of "internally scattered photoreceptive cells" associated with the CNS and not merely restricted to the epidermal surface.

Millecchia *et al.* [37] reported Hartline's personal findings that provided the first recording of a light-induced signals in the *Limulus* brain, although the author ascribed wrongly those signals to "olfactory nerves". Afterward, Millecchia and coworkers in their well-established papers [14, 38-39] provided the first description of the functional anatomy and the basic physiology of the *Limulus* VNP.

The VNP cells are mostly single and are separated from each other, they have a diameter of about 50 μm and a length of 150-200 μm with a highly infolded plasmatic membrane resulting in an array of strongly packed microvilli, the membrane capacitance is between 10^{-9} and 10^{-8} F. Further investigations (for a review see ref. 44) showed that the VNP is divided into two segments, the photoreceptive segment (rhabdomeral) R lobe and the non-photoreceptive segment (arhabdomeral) A lobe. The phototransductive protein machinery and photoexcitatory processes as well as the adaptation mechanisms take part within the R lobe, also the light-sensitive channels are located in this portion of the cell. A VNP cells has about 10^6 microvilli and 10^6 light-sensitive ion channels, each microvillus contains between 1000 and 4000 rhodopsin molecules and a light-sensitive channel density of 1 per microvillus.

Upon illumination VNP cells respond with a light-induced current that has an inward direction at negative, and outward direction at positive membrane potentials; the reversal potential is at +15 mV suggesting that mainly sodium ions flow into the cell through the light-activated channels, although the Na^+ Nerst potential (+95 mV) indicates a further permeability for K^+ and other ions as well. The biophysics of the visual transduction of VNP, related also to new findings about the cells structure, will be detailed in the section 4.2.

In spite of a great amount of works elucidating the phototransductive mechanisms, the functional role of VNP in the *Limulus* overall behaviour remain still ignored.

The VNP could be involved in the downward or upward movement of the tail at light-on or -off respectively [65]. Furthermore according to Hartline, combining both electrophysiological and transection techniques on eye nerves Snodderly [59] recorded impulses from the optic tract upon illumination of the ventral eyes showing that the activity of VNP can affect neurons that are to be activated by the lateral eye.

These results fit well with the data previously reported in the section 2.3 in which the combined action of tail EOP, lateral eye and VNP, among others, could concur to the timing of visual circadian rhythms [54].

4. Biophysical Processes

The search of the biophysical mechanisms and processes underlying EOP is a crucial point in this field. The behavioural analysis and other experimental approaches, as we have seen, not always guarantees any sharp indications in favour of a decisive relationship between light-detection and effector (either a behaviour or a cell activity) by the EOP system under analysis.

In recent years, the development of new techniques in the field of biochemistry and cellular biology, combined with the incredible growth of advanced electrophysiological techniques, has provided useful tools to investigate the biophysical nature of the elements involved in the photoreceptive (*e.g.*, photopigment) and phototransductive mechanisms (*e.g.*, light-sensitive ion channel, Ca^{++} role) and pathways (*e.g.*, second messenger chain) involved in EOP.

4.1. *Photosensible pigments*

The identification and the characterization of photopigment involved in the EOP system is quite difficult, particularly for what it concerns the dermal and the diffuse photosensitivity.

Early studies based their speculation on spectral sensitivity measurements evaluated as the variations of known behaviour parameters or cell activities upon different wavelength light stimulation. In most cases, the maximum sensitivity was found between 460 and 500 nm wavelengths. Responses also occur in the 550-900 nm range.

The data currently available indicate that usually the photoreceptor pigment is a carotenoid. In deep-sea cephalopods rhodopsin has also been extracted (see below). In several cases a haeme or porphyrin pigment molecule has been implicated and identified. However, too few pigments have been extracted from receptor sites that can be precisely identified with EOP systems.

The sea anemone *Anthopleura* shows a flexion toward the rim of the oral disk and a retraction from extended position upon exposure to wavelengths of 250-400 nm and 340-600 nm, respectively [15]. Despite several unsuccessful attempts to identify the photopigment, *Hydra* represents also an interesting case in which the periodic behaviour can be modulated by different chromatic conditions of stimulation and of the natural illumination background [62].

The peak sensitivity in the action spectra for most insects is around 450 nm. The alfaalfa weevil *Hypera postica* has a behavioural sensitivity peak in the red region of the spectrum from 650 to 900 nm [36]. Two species of scorpions *Heterometrus fulvipes* and

H. gravimanus have a reddish or yellowish-brown tail and the spectral sensitivity has a peak around 586 nm and another peak at 440 nm [20].

The photosensitive neurons of *Onchidium* and *Aplysia* are orange pigmented and the spectral sensitivity indicates the presence of a haemoprotein and a carotenoid. In cephalopods a photopigment was spectroscopically and biochemically identified, as the same as rhodopsin in the eye, in epistellar body and parolfactory vesicles respectively in *Eledone* and in *Loligo*. In the squid *Todarodes* the first quantitative study on molluscs found that retinochrome exists together with rhodopsin, in parolfactory vesicles, with peaks at 480 and 490 nm respectively [25].

Recently, a new approach has been developed in order to assign photosensitive characteristics by means of immunocytochemical and immunihistochemistry techniques.

In the blowfly *Calliphora* the S-Antigen (arrestin)-immunoreaction, a marker for retinal and extraretinal photoreceptors, provided the evidence that EOP may be located in various parts of the brain [17]. In the cephalopod *Ruditapes* immunohistochemical experiments using anti-squid-rhodopsin serum in conjunction with FITC or streptavidin-biotin revealed specific binding of the anti serum to the microvillar layer of the outer and inner epithelial cells of the distal portion of the siphon [28]. Both these studies provide a structural basis for further physiological experiments toward a definitive identification of the putative extraocular photoreceptors.

4.2. *Light-sensitive channels and phototransduction chain*

Thanks to the traditional and, above all, to the new advanced electrophysiological techniques, the study of the functional properties of the photoresponse is orienting towards the "single cell approach". This kind of approach is fruitful when a given cell has already been identified as a photoreceptor as well as to discover new instances of photosensitivity.

The best studied EOP models by means of this approach are those cells identified as neuronal photoreceptors, or those cells belonging to the central nervous system with the soma located outside the brain in peripheral sensory regions.

Since the middle of 80's, Gotow and coworkers have extensively studied the ionic mechanisms involved in the photoresponse of the neuronal photoreceptors of *Onchidium* [22-24]. Identified abdominal neurons (A-P-1) respond to light with a decrease of K^+ conductance, due to the closing of light-sensitive channels, associated with a membrane depolarization. This finding resembles, except for the response polarity, the light-induced hyperpolarizing potential in vertebrate photoreceptors that is also induced by a conductance decrease [22]. Instead, differences are evident if the A-P-1 ionic behaviour is compared to the same photoresponse in ocular and extraocular photoreceptors of other invertebrates. In fact, it is known that most invertebrate ocular photoreceptors respond to light with a membrane depolarization produced by an increase of membrane conductance to Na^+ and K^+ ions following the opening of light-sensitive channels. Similarly, an increase of conductance occurs also in extraocular photoreceptors such as the photosensitive neurons identified in the abdominal ganglion of *Aplysia* - even though in these neurons the receptor potential is hyperpolarizing [2, 12], analogously to what happen in invertebrate ciliary ocular photoreceptors [21].

The photoresponse of *Onchidium* A-P-1 resembles that of vertebrate photoreceptors also in the role of internal messenger. Injection of cyclic guanosine monophosphate (cGMP) in the dark produces an outward current, associated with an increase of conductance, that is suppressed by illumination (suggesting an hydrolisis of cGMP by light) [23]. Thus, the cGMP mediates *Onchidium* A-P-1 phototransduction, because the photoresponse is the suppression of the cGMP-induced current. Furthermore the photocurrent is amplified by pressure-injection of inositol 1,4,5-triphosphate (IP_3), indicating also a role of this messenger in the visual cascade. In this way, the light-sensitive channels of the extraocular photoreceptors seem to be regulate by cGMP in the dark and by IP_3 in the light. [23]

These findings have been confirmed in a paper reporting one of the first recording of a light-sensitive ion channel belonging to a neuronal photoreceptor [24]. By means of cell-attached patch clamp configuration, a light-sensitive K^+ channel was recorded. This channel is active (opened) during dark and is closed by light. In the inside-out configuration, a channel that appeared to be the same as the light-sensitive channel was activated (opened) by application of cGMP [24].

The photosensitive neurons in the left parietal ganglion of the snail *Helix* provide another well-studied model of ionic mechanisms underlying the light detection [47].

In these cells the light produces a slow inward current associated with a decrement of slope conductance. This light-induced current is due to the suppression of K^+ conductance and the addition of internal concentration of cGMP mimics the effect of light. In fact, the I-V curves of the light-induced and the cGMP-induced currents follow a similar course and have a common reversal potential [48]. In this case the difference respect to the *Onchidium* photosensitive neurons is evident, because cGMP acts to produce an outward current that is suppressed by light. This evidence indicates that in *Onchidium* neurons the light activates a phosphodiesterase that, as in vertebrate photoreceptors, reduces cGMP.

Apart from its evolutionary value in the development of the photoreceptive function, the *Limulus* ventral nerve photoreceptor (VNP) is certainly the well-established invertebrate model (among those currently used) for investigating the light-induced biophysical processes. Since the pioneering works of Millecchia and coworkers [14, 37-39], a huge amount of studies carried out by several authors along the years revealed the exceptional suitability of the *Limulus* VNP to perform quantitative electrophysiology and biochemical procedures as well.

For the sake of convenience and far from to be comprehensive, a very brief summary of the *Limulus* VNP main biophysical characteristics will be given, leaving the reader to some excellent reviews (from which the data below reported are gathered) for more detailed results and ample references [44, 60].

The function of *Limulus* VNP appears very effective thanks to the synergic action of light- and voltage-activated conductances. The latter conductances have been measured in number of four, while three different light-activated conductances concur to shape three different components to the receptor macroscopic current (yet described in the section 3.3.). These components recover with different rates and have different reversal potentials and behavioural kinetics as well, indicating different excitation mechanisms.

In this respect, current available data propose three transduction pathways leading different terminal transmitter: 1) an unknown transmitter produced by the Phospholipase C (PLC*)→ IP_3→ Ca^{++} enzymatic cascade; 2) cGMP; 3) cyclic adenosine monophosphate (cAMP). The starting points of these chains, after the triggering by a photon absorption, could be different G proteins activated by the same metarhodopsin molecule, even though the presence of other types of metarhodopsin cannot be discarded.

Also calcium ions play a crucial role in the *Limulus* VNP phototransductiveprocess. In a first instance, calcium was supposed as terminal transmitter produced by the IP_3 enzymatic cascade while a strong injection of Ca^{++} in VNP causes a conductance increase. Further evidences set aside this charge and they speak in favour of a Ca^{++} role as messenger for the adaptation process. Although free Ca^{++} ions do not pass through the light-sensitive channels, they are important in the activation of these channels establishing and regulating feedbacks of the enzymatic cascades underlying the production of transmitters and the intracellular calcium level as well.

5. Concluding Remarks

The results summarized in the previous sections (above all that on biophysical processes) depict a certain complexity, indicating several differences between EOP cells of species belonging to the same *phylum*, similarities between EOP cells and ocular photoreceptors of different species, as well as some correspondences between EOP cells of some species of invertebrates and vertebrate photoreceptors.

We know that the sharp functional differences between invertebrate and vertebrate photoreceptors, deriving from their microvillar and ciliar structure respectively, is not always true. Furthermore, we know that also invertebrates show ciliar photoreceptors that in some cases share with EOP photosensitive cells similar ionic mechanisms [21].

For these reasons, the experimental data on the biophysical aspects of photoresponse in EOP, in particular those regarding the physiology of light-sensitive channels, could be indicative of the difficult work that must be done to search for the ionic mechanisms involved in the phototrasduction chain. In any event, it appears clear that the application of patch-clamp technique associated with microspectroscopy, immunoistochemistry and molecular biology should represent the experimental future of this field.

Many questions are still open about the EOP mechanisms. The crucial point is why EOP, with its functional and structural limitations, as well as its constraints in spatial and temporal resolving power, has evolved as an important sensory mechanism for behaviour in animals in which a well-developed visual system exists and functions effectively.

In this regard, the explanation that EOP is an ancestral way of light-detection, from which ocular structures and higher visual capabilities have been developed, is not sufficient. At the same time, we cannot say that EOP is a phylogenetic residue carried as an evolutional ballast by higher species.

Studies of EOP in vertebrates confirm this assumption. The presence of EOP is essential to regulate several behaviours on which the animal's survival depends: the entrainment of the biological clock controlling locomotor behaviour, thermoregulatory behaviour, compass orientation, the control of seasonal production and release of hormones, and many others [35]. Finally, findings of light penetration in mammalian brain (including humans), where the pineal gland contains photoreceptors with sensitivity peaks in the 525-565 nm range, should be quoted [61, 64].

At this point, the best explanation is provided by Wolken and Mogus with the following suggestive hypothesis: "EOP acts as a *backup mechanism* so that certain biological functions would continue in the absence of any visual system" [68].

The same authors indicate the correct methodological way to pursue a better understanding of the structure and function of EOP: i) localization of EOP system(s), ii) identification of the photoreceptor and its structure, iii) spectral sensitivity of the receptor, iv) electrophysiological correlates, v) behavioural action spectrum and finally, vi) identification and characterization of the photopigment.

Unfortunately, these hints point to sensitive limitations, due to the impossibility to realize experimentally all the above items on the same animal. At present, hoping to reach a full description of EOP mechanisms in a given animal, the only weapons at our disposal are the physiological comparison of photosensory systems among different species and the correct use of conceptual efforts applied to the phylogenetic evolution.

ACKNOWLEDGEMENTS - I wish to express my gratitude to *Dr. Cloe Taddei-Ferretti* for providing me with the chance to proceed this enterprise and for her fruitful suggestions and stimulating discussions. A friendly tribute to *Dr. Enrico Nasi* (Boston University) for invaluable revising and comments on the manuscript. This work was partially supported by the Istituto Italiano per gli Studi Filosofici, Napoli, I.

REFERENCES

[1] Anderson P.A.V. & G.O. Mackie (1977) Electrically coupled, photosensitive neurons control swimming in a jellyfish. *Science* 197:186-188.

[2] Andresen M.C. & A.M. Brown (1979) Photoresponses of a extraretinal photoreceptor in *Aplysia. J. Physiol* 287:26-282.

[3] Arikawa K. & K. Aoki (1982) Response characteristics and occurrence of extraocular photoreceptors of lepidopteran genitalia. *J. Comp. Physiol. A* 148:483-489.

[4] Arikawa K. & K. Aoki (1984) Inhibitory effect of genital photoreceptors on the activity of abdominal motoneurones in *Papilio xuthus. J. Insect Physiol.* 30:853-860.

[5] Arkett S.A. (1990) Hydromedusan photophysiology: an evolutionary perspective. In: *Evolution of the First Nervous Systems* (P.A.V. Anderson, ed.), New York: Plenum, pp. 373-388.

[6] Arkett S.A. & A.N. Spencer (1986a) Neuronal mechanism of a hydromedusan shadow reflex. I. Identified reflex components and sequence of events. *J. Comp. Physiol.* 159:201-213.

[7] Arkett S.A. & A.N. Spencer (1986b) Neuronal mechanism of a hydromedusan shadow reflex. II. Graded responses of reflex components, possible mechanism of photic integration, and functional significance. *J. Comp. Physiol.* 159:215-225.

[8] Arvanitaki A. & N. Chalazonitis (1961) Excitatory and inhibitory processes initiated by light and infra-red radiations in single identifiable nerve cells (giant ganglion cells of *Aplysia*). In: *Nervous Inhibition* (E. Florey, ed.), Oxford: Pergamon, pp. 194-231.

[9] Baines R.A. & J.P. Bacon (1994) Pharmacological analysis of the cholinergic input to the locust VPLI neuron from an extraocular photoreceptor system. *J. Neurophysiol.* 72:2864-2874.

[10] Bennett M.F. (1979) Extraocular light receptors and circadian rhythms. In: *Handbook of Sensory Physiology, Vol. VII/6A, Comparative Physiology and Evolution of Vision in Invertebrates* (H. Autrum, ed.), New York: Springer, pp. 641-663.

[11] Block G. & J.T. Smith (1973) Cerebral photoreceptors in *Aplysia. Comp. Biochem. Physiol.* 46A:115-121.

[12] Brown A.M. & H.M. Brown (1973) Light responses of a giant *Aplysia* neuron. *J. Gen. Physiol.* 62:239-254.

[13] Burr A.H. (1984) Evolution of eyes and photoreceptor organelles in the lower phyla. In: *Photoreceptors and Vision in Invertebrates* (M.A. Ali, ed.), New York: Plenum, pp. 131-178.

[14] Clark A.W., R. Millecchia & A. Mauro (1969) The ventral photoreceptor cells of *Limulus*. I. The microanatomy. *J. Gen. Physiol.* 54:289-309.

[15] Clark E.D. & D.J. Kimeldorf (1970) Tentacle responses of the sea anemone *Anthopleura xanthogrammica* to ultraviolet and visible light. *Nature* 227:856-857.

[16] Cobb C.S. & R. Williamson (1996) Extraocular photoreceptors in the octopus, *Eledone cirrhosa*. *S.E.B. Annual Meeting*, Lancaster (UK), abstr. A10.27.

[17] Cymbrowski B. & H.-W. Korf (1995) Immunocytochemical demonstration of S-antigen (arrestin) in the brain of the blowfly *Calliphora vicina*. *Cell Tissue Res.* 279:109-114.

[18] Dilly P.N. & J.J. Wolken (1973) Studies on the receptors in *Ciona intestinalis*. IV. The ocellus in the adult. *Micron* 4:11-29.

[19] Eakin R.M. (1982) Continuity and diversity in photoreceptors. In: *Visual Cells in Evolution* (J.A. Westfall, ed.), New York: Raven, pp. 91-105.

[20] Geethabali X. & K.P. Rao (1973) A metasomatic neural photoreceptor in the scorpion. *J. Exp. Biol.* 58:189-196.

[21] Gomez M. & E. Nasi (1994) The light-sensitive conductance of hyperpolarizing invertebrate photoreceptors: a patch-clamp study. *J. Gen Physiol.* 103:939-956.

[22] Gotow T. (1989) Photoresponse of an extraocular photoreceptor associated with a decrease in membrane conductance in an opisthobranch mollusc. *Brain Res.* 479:120-129.

[23] Gotow T. & T. Nishi (1991) Roles of cyclic GMP and inositol triphosphate in phototransduction of the molluscan extraocular photoreceptor. *Brain Res.* 557:121-128.

[24] Gotow T., T. Nishi & H. Kijima (1994) Single K^+ channels closed by light and opened by cyclic GMP in molluscan extra-ocular photoreceptor cells. *Brain Res.* 662:268-272.

[25] Hara T. & R. Hara (1980) Retinochrome and rhodopsin in the extraocular photoreceptor of the squid, *Todarodes*. *J. Gen. Physiol.* 75:1-19.

[26] Hartline H.K. & C.H. Graham (1932) Nerve impulses from single receptors in the eye. *J. Cell. Comp. Physiol.* 1:277-295.

[27] Hisano N., H. Tateda & M. Kuwabara (1972) Photosensitive neurones in the marine pulmonate mollusc *Onchidium verruculatum*. *J. Exp. Biol.* 57:651-660.

[28] Karakisawa H., S. Tamotsu A. Terakita & K. Ohtsu (1994) Identification of putative photoreceptor cells in the siphon of a clam, *Ruditapes philippinarum*. *Zool. Sci.* 11:667-674.

[29] Kennedy D. (1960) Neural photoreception in a lamellibranch mollusc. *J. Gen. Physiol.* 44:277-299.

[30] Klein R.M. (1992) Effects of green light on biological systems. *Biol. Rev.* 67:199-284.

[31] Lenci F., F. Ghetti, G. Colombetti, D.-P. Häder & P.-S. Song, eds (1991) *Biophysics of Photoreceptors and Photomovements in Microorganisms*. New York, London: Plenum.

[32] Lickey M.E., G.D. Block, D. Hudson & J.T. Smith (1976) Circadian oscillators and photoreceptors in the gastropod *Aplysia*. *Photochem. Photobiol.* 23:253-273.

[33] Lundquist C.T., R.A. Baines & J.P. Bacon (1996) Extraocular photoreceptors in insects. *S.E.B. Annual Meeting*, Lancaster (UK), abstr. A10.11.

[34] Marks P.S. (1976) Nervous control of light responses in the sea anemone *Calamactis praelongus*. *J. Exp. Biol.* 65:85-96.

[35] Menaker M., ed. (1976) Extraretinal photoreception. *Photochem. Photobiol.* 23:213-306.

[36] Meyer J.R. (1977) Head capsule transmission of long-wavelength light in the *Curculionidae*. *Science* 196:524-525.

[37] Millecchia R., J. Bradbury & A. Mauro (1966) Simple photoreceptors in *Limulus polyphemus*. *Science* 154:1199-1201.

[38] Millecchia R. & A. Mauro (1969a) The ventral photoreceptor cells of *Limulus*. II. The basic photoresponse. *J. Gen. Physiol.* 54:310-330.

[39] Millecchia R. & A. Mauro (1969b) The ventral photoreceptor cells of *Limulus*. III. A voltage clamp study. *J. Gen. Physiol.* 54:331-351.

[40] Millott N. (1968) The dermal light sense. In: *Invertebrate photoreceptors* (J.D. Carthy & G.E. Newell, eds), New York, London: Academic, pp. 1-36.

[41] Millott N. (1978) *Extra-ocular Photosensitivity*. Burham: Meadowfield Press.

[42] Miyako Y., K. Arikawa & E. Eguchi (1993) Ultrastructure of the extraocular photoreceptor in the genitalia of a butterfly, *Papilio xuthus*. *J. Comp. Neurol.* 327:458-468.

[43] Mpitsos G.J. (1973) Physiology of vision in the mollusk *Lima scabra*. *J. Neurophysiol.* 37:371-383.

[44] Nagy K. (1991) Biophysical processes in invertebrate photoreceptors: recent progress and a critical overview based on *Limulus* photoreceptors. *Quart. Rev. Biophys.* 24:165-226.

[45] Nishi T. & T. Gotow (1992) A neural mechanism for processing colour information in molluscan extra-ocular photoreceptors. *J. Exp. Biol.* 168:77-91.

[46] North W.J. (1957) Sensitivity to light in the sea anemone *Metridium senile* (L.). II. Studies on reaction time variability and effects of change in light intensity and temperature. *J. Gen. Physiol.* 40:715-733.

[47] Pasic M. & G. Kartelija (1995) The reaction of *Helix* photosensitive neurons to light and cyclic GMP. *Neuroscience* 69:557-565.

[48] Pasic M., D. Ristanovic, D. Zecevic & G. Kartelija (1977) Effects of light on identified *Helix pomatia* neurons. *Comp. Biochem. Physiol.* 58A:81-85.

[49] Passano L.M. & C.B. McCullough (1962) The light response and the rhythmic potentials in *Hydra*. *Proc. Natl. Acad. Sci. USA* 48:1376-1382.

[50] Passano L.M. & C.B. McCullough, (1963) Pacemaker hierarchies controlling the behaviour of hydras. *Nature* 199: 1174-1175.

[51] Passano L.M. & C.B. McCullough (1964) Co-ordinating systems and behaviour in *Hydra*. I. Pacemaker system of the periodic contractions. *J. Exp. Biol.* 41:643-664.

[52] Passano L.M. & C.B. McCullough (1965) Co-ordinating systems and behaviour in *Hydra*. II. The rhythmic potential system. *J. Exp. Biol.* 42:205-31.

[53] Pearse V.B. (1974) Modification of sea anemone behavior by symbiotic zooxanthellae: expansion and contraction. *Biol. Bull.* 147:641-651.

[54] Renninger G.H. & S.C. Chamberlain (1993) Modulation of photoreceptor function in *Limulus polyphemus* by a central circadian clock. In: *Sensory Systems of Arthropods* (K. Wiese *et. al.*, eds.), Basel: Birkhäuser, pp. 307-316.

[55] Rushforth N.B., A.L. Burnett & R. Maynard (1963) Behavior of *Hydra*: contraction responses of *Hydra pirardi* to mechanical and light stimuli. *Science* 139:760-761.

[56] Salvini-Plawen L.v. (1982) On the polyphyletic origin of photoreceptors. In: *Visual Cells in Evolution* (J.A. Westfall, ed.), New York: Raven, pp. 137-154.

[57] Sawyer S.J., H.B. Dowse & M. Shick (1994) Neurophysiological correlates of the behavioral response to light in the sea anemone *Anthopleura elegantissima*. *Biol. Bull.* 186: 195-201.

[58] Simon T.W. & D.H. Edwards (1990) Light-evoked walking in crayfish: behavioural and neuronal responses triggered by the caudal photoreceptor. *J. Comp. Physiol.* 166:745-755.

[59] Snodderly Jr. D.M (1971) Processing of visual inputs by brain of *Limulus*. *J. Neurophysiol.* 34:588-611.

[60] Stieve H. & K. Nagy, Phototransduction in *Limulus* photoreceptors and enigmas of the transduction mechanism in photoreceptor cells of invertebrates, *in this volume*.

[61] Svaasand L.O. & R. Ellingsen (1983) Optical properties of human brain. *Photochem. Photobiol.* 38:293-299.

[62] Taddei-Ferretti C., C. Musio, S. Santillo & A. Cotugno, *Hydra* photosensitivity, *in this volume*

[63] Truman J.W. (1976) Extraretinal photoreception in Insects. *Photochem. Photobiol.* 23:215-225.

[64] Van Brunt E.E., M.D. Shepard, J.R. Wall, W.F. Ganong & M.T. Clegg (1963) Penetration of light into the brain of mammals. *Ann. N. Y. Acad. Sci.* 117:217-227.

[65] Wassermann G.S. (1973) Unconditioned response to light in *Limulus*: mediation by lateral, median and ventral eye loci. *Vision Res.* 13:95-105.

[66] Wiederhold M.L., E.F. MacNichol Jr. & A.L. Bell (1973) Photoreceptor spike responses in the hardshell *Mercenaria mercenaria. J. Gen. Physiol.* 61:24-55.

[67] Wilkens L.A. (1988) The crayfish caudal photoreceptor: advances and questions after the first half century. *Comp. Biochem. Physiol.* 91C:61-68.

[68] Wolken J.J. & M.A. Mogus (1979) Extra-Ocular photosensitivity. *Photochem. Photobiol.* 29:186-196.

[69] Wolken J.J. & M.A. Mogus (1981) Extraocular photoreception. In: *Photochemical and Photobiological Reviews, Vol. 6* (K.C. Smith, ed.), New York, London: Plenum, pp. 181-199.

[70] Yoshida M. (1979) Extraocular photoreception. In: *Handbook of Sensory Physiology, Vol. VII/6A, Comparative Physiology and Evolution of Vision in Invertebrates* (H. Autrum, ed.), New York: Springer, pp. 582-640.

Discrimination of natural colours and receptor spectral sensitivity functions

Misha Vorobyev

Institute of Evolutionary Physiology and Biochemistry, Thorez pr. 44, 194223 St.Petersburg, Russia / Institute für Neurobiologie, Freie Universität Berlin, Königin-Luise-Str. 28-30, 14195 Berlin, Germany.

Introduction

To detect and identify objects the visual system has to discriminate them from each other and from the background on the basis of different spectral reflectances. Discrimination performance depends on the shape, wavelength position of the receptor spectral sensitivity functions (SSFs) and on the photoreceptors' signal to noise ratio.

Two questions are considered 1. How does discrimination depend on the wavelength position of the SSFs? 2. Can the proper shaping of the SSFs improve discrimination?

To answer these questions a quantitative measure of the quality of discrimination must be introduced. I used two measures:

1. The number of the distinguishable colours of the given class of objects (Govardovskii & Vorobyev,1989),
2. Percentage of indistinguishable objects (Vorobyev et al 1994).

Two classes of surface spectral reflectances are considered:

1. All theoretically possible spectral reflectances(Govardovskii & Vorobyev 1989),
2. The reflectances of biologically significant objects - plants (1056 flowers, 230 leaves and 80 fruits) (Menzel & Shmida,1993, Chittka et al, 1994)

Theory

In the case of trichromatic colour vision system each colour can be represented as a point in the three-dimensional colour space, where the quantum catches of short-wavelength (S), middle-wavelength (M) and long-wavelength (L) receptors are placed along the co-ordinate axis. Because the visual pigments possess broad-band overlapping spectra (Fig 1), no light can produce pure green or blue sensation. I am going to consider the colours of real objects - non-selfemitting surfaces. The colours corresponding to all theoretically possible non-selfemitting surfaces occupy only a part of the colour space - an object-colour solid (fig 2).

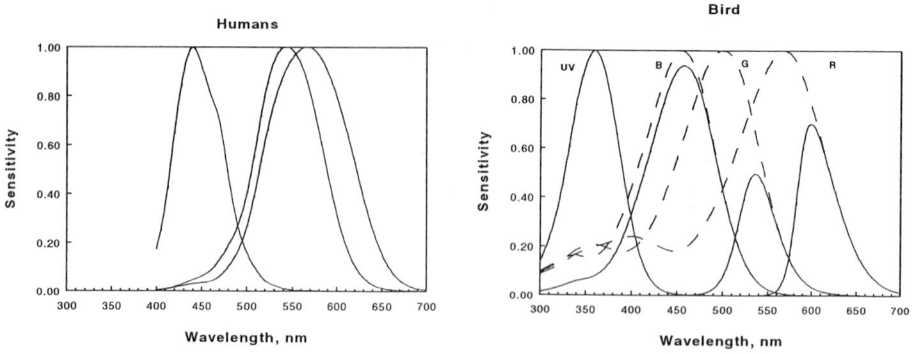

Fig. 1

Spectral sensitivity functions of human beings and bird *Leiothrix lutea* (Maier & Bowmaker, 1993). Solid curves - the real spectral sensitivity functions, the dashed curves - the SSFs of a hypothetical bird without coloured oil droplets.

The colour solid was introduced by the German scientist Luther (1927) and by the Russian scientist Nyberg (1928). The boundaries of sections of colour solid by the planes of equal brightness are known as MacAdam limits(MacAdam, 1935).

The colour of the surface with reflectance $\rho(\lambda)$ illuminated by the light with spectral power distribution $P(\lambda)$ is described by the tristimulus value $X_i = \int \rho(\lambda) * P(\lambda) * S_i(\lambda) d\lambda$, where $S_i(\lambda)$ are the spectral sensitivity functions of 3 colour types of receptors. The boundaries of the colour solid are formed by optimal colour stimuli. These stimuli have spectral reflectance curves of a very simple type: $\rho(\lambda)$ is either zero or unity, and in moving through the visible spectrum, there are generally not more than two transitions between these values. Optimal colour stimuli are imaginary stimuli in the sense that no actual objects surfaces have reflectance curves with abrupt transitions of this kind. The theory of optimal colour stimuli was developed by Shrödinger (1920) and Rösh (1929).

The object-colour solid so defined is, in general, an asymmetric "torpedo"-shaped body, with apices at the origin and at the white point with the tristimulus values $X_i = W_i$, corresponding to the object-colour stimulus involving the perfect reflecting surface with $\rho(\lambda) = 1$ at all wavelengths. The shape and volume of the object-colour solid depend on the shape and wavelength position of the SSFs. The less the overlap of SSFs the larger the volume

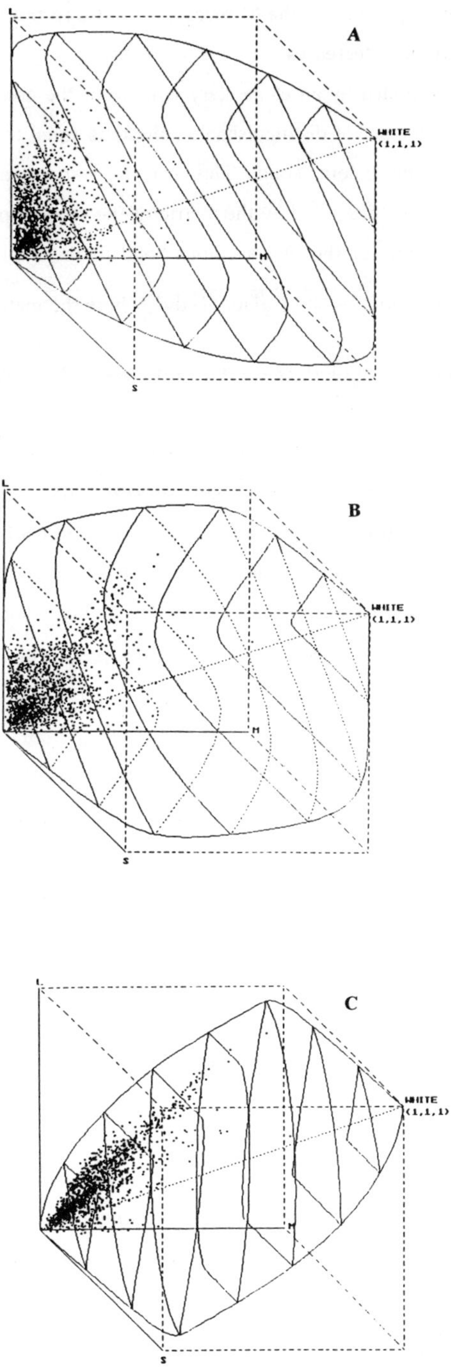

Fig. 2

Object-colour solids, depending on the wavelength positions of the SSFs.

A - λ max of S receptor - 340 nm, of M - 380, of L - 540 nm.

B - λ max of S receptor - 340 nm, of M - 430, of L - 540 nm (typical set of photoreceptors of Hymenoptera).

C - λ max of S receptor - 340 nm, of M - 510, of L - 540 nm.

Cross-sections by the planes of equal brightness are shown. White point is normalised to Wi=1.

The dots correspond to the colours of plants.

of the object colour solid: when the wavelength position of the M receptor is close to the S or L receptors the colour body flattens and its volume decreases.

Each point in the colour space is surrounded by an elementary volume of the colours indistinguishable from a given one. The total number of distinguishable colours is equal to the number of this elementary volume filling the colour solid. This can be expressed in mathematical terms. Colour discrimination can be described by the metric of the colour space (Helmholtz, 1896, Shrödinger,1920). The volume of the object-colour solid in the metrical space is equal to the number of distinguishable colours - $N= \int_V \sqrt{Det(G)}\, dv$, where the metrical tensor - G depends on the position of the point in the colour space (Govardovskii & Vorobyev, 1989).

Discrimination performance is inevitably limited by the photoreceptor noise caused by quantum fluctuations of absorbed light. All other noise sources decrease the colour discriminability. Therefore, the quantum fluctuation limited metric allows estimation of the upper limit of the number of distinguishable colours. The amplitude of noise caused by quantum fluctuations is proportional to the square root of the number of effectively absorbed quanta $\sqrt{Xi}$, yielding the metric Gik=Ki/Xi if i=k, 0 if k#i, where Ki is the proportionality factor. In this case the number of distinguishable colours depends only on the level of illumination and on the SSFs.

The metric can also be derived from the behavioural data. Backhaus (1991) has developed the model of colour discrimination of the honeybee. This model takes into account that it is rather difficult to train a bee to distinguish stimuli which differ only in brightness. Therefore the colour space can be described by a two dimensional diagram (Backhaus, 1991, Chittka, 1992). The number of distinguishable colours can be evaluated as an area of the projection of colour solid on the opponency plane (Vorobyev et al., 1994)

The points corresponding to the colours of real objects form a cloud which lies within the object-colour body. The number of distinguishable colours in this case can be estimated as the volume (area) of the cloud.

The colours are indistinguishable from each other if the distance between them is lower than the distance defined by the discrimination threshold. To estimate the number of indistinguishable objects, for each object point, the number of objects indistinguishable from the given one must be counted, and this number must be averaged. Obviously, the number of indistinguishable colours depends on the mutual position of the points corresponding to different objects. Therefore the number of indistinguishable objects can not be derived from the number of colours. If the total number of objects is very high and the position of the points in the colour space are independent from each other the percentage of indistinguishable objects is

$$P=\int_V f^2(X_i)/\sqrt{Det(G)}\ dv$$

, where f(x) is the probability distribution of the object colours in the colour space.

Colour discrimination and the wavelength positions of SSFs

How many colours can animals discriminate? Are the wavelength positions of the SSFs in some animals placed optimally? How does the optimal position of receptors depend on the set of coloured objects and on the criterion of optimality? Figure 3 shows the number of distinguishable colours of all theoretically possible non-selfemitting surfaces, colours of 1377 plant objects, and percentage of indistinguishable objects (plants) depending on the wavelength positions of the SSFs. In each of the three variations, the wavelengths of two receptors were clamped at the positions corresponding to the maxima of receptors of bees; the third receptor was shifted. As one would expect from the properties of object-colour solid, the colour discrimination improves when the S receptor is shifted to the short wavelength, L - to the long-wavelength and M is placed the middle.

Bees and other Hymenoptera seems to have a set of photoreceptors which is very close to optimal: the position of M receptor corresponds to the maximum on the curve of distinguishable colours and to the minimum of indistinguishable objects, the shift of S and L receptors can only slightly improve the colour discrimination. However, not all animals have such a good set of photoreceptors. The SSFs of M and L receptor in primates are very close to each other (Fig. 1), therefore humans can discriminate 4 times less colours than bees. It seems that the humans visual system is the worst of all trichromatic animals.

The optimal wavelength position of SSFs does not depend on the set colours used in the analysis. The number of all theoretically possible distinguishable colours and the number of plant colours have nearly the same dependence on the receptor wavelength position (compare the panels of the Fig 3). It is always favourable to reduce the overlap between the SSFs.

Do the reflectance spectra of natural objects give the additional information in comparison with colours of all non-selfemiting surfaces? To investigate this question in more detail we performed the variations of three types of receptors independently. For each position of the S receptor the number of distinguishable colours and indistinguishable objects were calculated

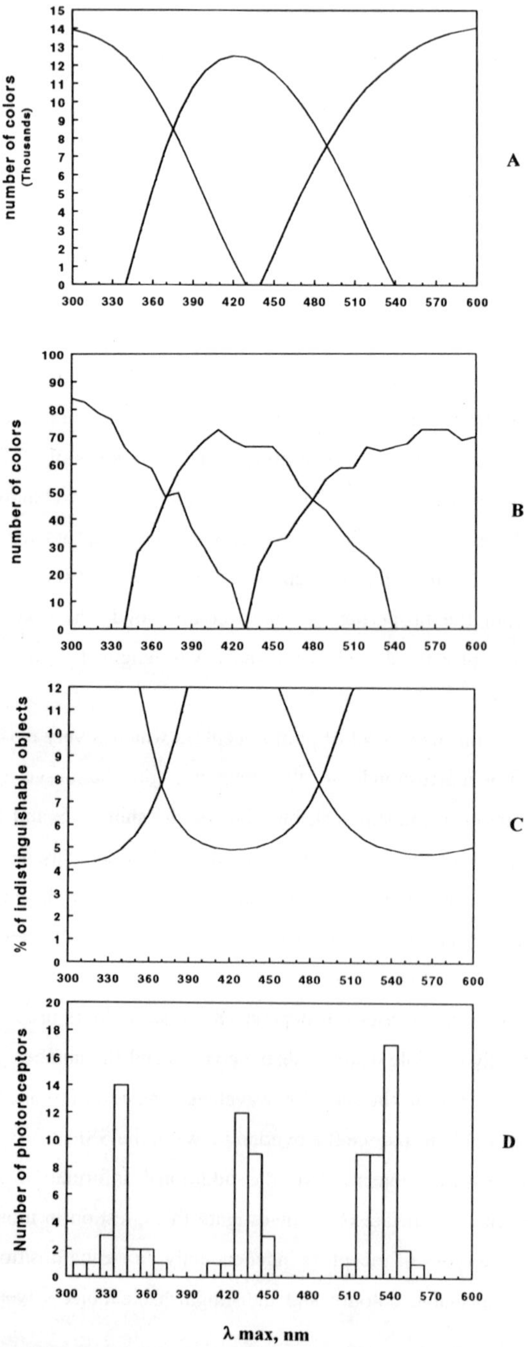

Fig. 3

The number of distinguishable colours and indistinguishable objects depending on the wavelength positions of photoreceptors. In each of three variations the wavelength positions of two receptors were clamped at the wavelength positions where they most frequently occur in Hymenoptera. (λmax =340 nm, 430 nm, 540 nm).

A - number of all theoretically possible surface colours, distinguishable at the illumination level corresponding to 10000 photons per receptive field during the summation time.

B - number of distinguishable colours of plants. The metric was derived from the behavioral data.

C - the percentage of indistinguishable objects (plants). The discrimination threshold was derived from the behavioural data.

D - the absolute number of photoreceptors recorded at every given λ max for 40 different species of Hymenoptera (Peitch et al, 1992).

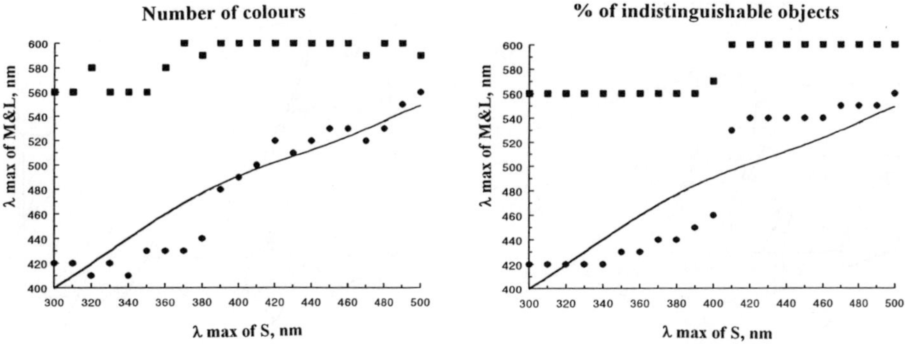

Fig. 4

The optimal positions of the L and M receptors as a function of the wavelength position of S receptor. Rounds show the position of M, squares - of L receptors, when the number of distinguishable colours of plants and the percentage of indistinguishable objects (plants) were used as a criterion of the quality of the colour vision system. The solid line shows the position of M receptor, optimal for the discrimination of all theoretically possible colours.

as a function of the position of the L and M receptors. For each of these functions of the wavelength positions of L and M receptors the optimal position of the L and M receptor was found. Optimum position of the M receptor for the flower colours does not depend on the position of the S receptor when S is varied from 300 nm to 370 and its position corresponds to 410 - 440 nm(Fig 4). Thus, optimum position of the blue receptor does not depend on position of UV receptor. This is not valid when all theoretically possible colours are considered. The same is even more obvious when the number of indistinguishable colours is chosen as a criteria of optimality. In addition, the optimal position of the M receptor does not change when S receptor varies from 410 to 460 and corresponds to 540 nm . Thus, the optimum position of the M - green receptor does not depend on the position of the S - blue receptor. This conclusion can not be derived if all theoretically possible colours are taken into account.

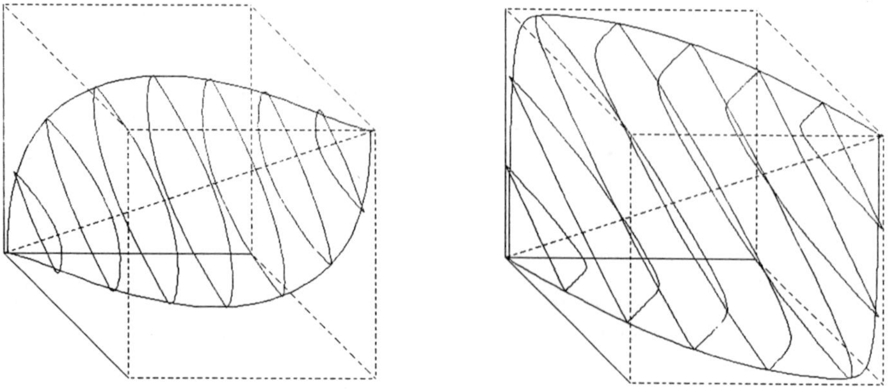

Fig. 5

The object colour solid of land turtle *Testudo horsfieldi*. Left panel without oil droplets, right panel - with oil droplets. The turtle was supposed to possess P562, P506, and P465 combined with red, orange, and colourless droplets, resp.

The number of distinguishable colours and the shape of SSFs

Is an increase in the number of distinguishable colours and decrease in the number of indistinguishable objects one of the goals of the evolution of colour vision systems? Are there any examples of evolutionary adaptations which improve colour discrimination?

Retina of birds and reptiles contain brightly coloured oil droplets. The droplets act as cut-off filters which suppress the short-wave branch of the visual pigment spectral sensitivity curves and thus change the shape of SSFs, producing the narrow-band, spectrally tuned colour receptors (Fig 1) . This leads to the increase in the volume of the object-colour solid (Fig 5). On the other hand, a certain amount of light is absorbed in the coloured droplets, thus the quantum catches decrease, and the photoreceptor noise increases. Computations were performed for trichromatic and tetrachromatic cone systems apparently present in the retinas of turtles and birds (Govardovskii, Vorobyev 1989). Introducing the coloured oil droplets into the retina leads to an increase in the number of distinguishable colours (2 for the trichromatic vision of turtle and 2.7 for the tetrachromatic vision of bird). It can be suggested that the proper shaping of the SSFs which leads to the overall gain in colour discrimination is the main selective advantage which accounts for the development of coloured oil droplets during evolution.

Conclusions

Improvement of the overall colour discrimination can explain the wavelength positions and shaping of the receptor SSFs in some animals. The unfavourable positions of the SSFs in many animals and in human beings may be explained in various ways: 1. The spectral positions are tuned to a very specific class of object. 2. There has not been enough time for evolution to tune the wavelength positions of SSFs precisely 3. The actual wavelength positions of SSFs are tuned to other visual tasks (Nagle & Osorio, 1993).

Acknowledgements

The presented results were obtained in collaboration with Prof. Victor I. Goovardovskii (St. Petersburg), Dr. Lars Chittka and Prof. Randolf Menzel (Berlin). I thank these people for friendly co-operation, and I thank Lars Chittka and Randolf Menzel for hospitality during my stay in Berlin. I am especially grateful to Victor Govardovskii - the author of many ideas of this paper - for bringing my attention to the problem. I thank Philomena Maher for correcting the English.

The study was supported by Alexander on Humboldt Foundation.

REFERENCES

1. Backhaus, W. Vision Res. 31, 1381-1397 (1991)
2. Chittka, L. J. Comp. Physiol. A. 170, 553-545 (1992)
3. Chittka, L., et. al. Vision Res. 34,1489-1508 (1994)
4. Govardovskii, V.I. & Vorobyev, M.V. Sensory systems, 3, 150-159. (1989) (In Russian).

5. Helmholtz, H. von. Handbuch der Physiologischen Optik (2nd ed.) Vos, Humburg, 1896.

6. Luther, R. Tecn. Phys., 8, 997-1012 (1927)

7. MacAdam, D.L. J. Opt. Soc. Am. 25, 361 (1935)

8. Maier, E.J. & Bowmaker, J.K. J. Comp. Physiol. A 172, 295-301 (1993)

9. Menzel, R. & Shmida, A., Biological Reviews, 68, 81-120 (1993)

10. Nagle, M.G., Osorio, D. Proc. R. Soc. London B, 252, 209-213 (1993)

11. Nyberg, N. Zt. Sch. Phys. 52, 406 (1928).

12. Peitsch, D., Feitz, A., Hertel, H., de Souza, J., Ventura, D.F., & Menzel R. J. Comp. Physiol. A. 170, 23-40 (1992).

13. Rösch, S. Fortschr. Mineral., Krist., Petrogr. 13,143 (1929)

14. Schrödinger, E. Ann. Physik. 62, 603 (1920a)

15. Schrödinger, E. Ann. Physik. 63, 481 (1920b)

16. Vorobyev M., Chittka L., & Menzel R. In: Proceeding of the conference "Ecology of vision", Oristano, Sardinia. (1994)

17. Wyszecky, G. & Stiles, W.S. Color Science: Concepts and Methods, Quantitative Data and Formulae (2nd ed.) ,Willey, N.Y., 1982.

HYDRA PHOTOSENSITIVITY

CLOE TADDEI-FERRETTI, CARLO MUSIO,
SILVIA SANTILLO & ANTONIO COTUGNO
Istituto di Cibernetica, CNR, 80072 Arco Felice (NA), Italy

Introduction

An interesting problem in photoreception research is offered by the case of the Cnidarian *Hydra*: it does not bear identified photosensitive structures, even not as scattered photosensitive single cells; in spite of this, the animal is characterized by a well defined photosensitivity, that results in short-term or long-term lasting modifications of its periodic behaviour. It is a case of extraocular dermal or neuronal photosensitivity [15, 38].

The behaviour of the animal consists in a periodic alternation of shortenings and elongations, which are carried out respectively by the myofibrils of the epitheliomuscular cells of the ectoderm (arranged longitudinally with respect to the long axis of animal) and by the myofibrils of the same type of cells of the endoderm (circularly arranged, *i.e.* orthogonally with respect to the previous ones) [7, 23].

Such behaviour has a vital value, because it is related to osmoregulation [1], digestion and locomotion.

Various types of bioelectric activity are linked to such behavioural expression.

A periodic big slow wave (BSW), *i.e.* a DC variation of the whole animal biopotential, has the same duration as that of the behavioural period. At one end of the animal it has a trend that is opposed to that at the other end; in such a way a potential difference between the two ends varies periodically with the behaviour expression [29].

At definite phases of such slow potential variations, two types of fast bioelectric variations arise [7-9, 16-19, 22-23, 25, 29]: contractions pulses (CP's) with an intensity of a few tenths of mV, arranged in trains of contraction pulses (CPT's), with a duration of about 30 s, triggered mainly at the hypostomal end, transmitted longitudinally, with a small spread in the direction of the body circumference, and linked to the ectodermal activity; and series of rhythmic potentials (RP's), with an intensity of a few tens of mV and a duration of the series of a few minutes, triggered mainly at the basal end, and transmitted circularly, slowly spreading in the longitudinal direction and linked to the endodermal activity. The CPTs arise at the falling phase of the BSW of the proximal end, while the RPs arise at the rising phase, and *vice-versa* with respect to the BSW of the distal end.

The nervous elements, made of bipolar and multipolar ganglion neurons [34] with polarized and unpolarized synapses [37], constitute a net lying between the ectoderm and endoderm layers. The net is topologically quasi-isotropic. However, the distribution of the ganglion neurons of the network is lower along the colum and higher at the two ends [2], and especially at the basal end [4, 34]; thus, the nervous net can be anatomically considered as two subnetworks. Furthermore, a thin nerve ring is present in the subhypostomal region in the species *oligactis* [6, 14]; an aggregation of few neurons at the base of each tentacle in the species *littoralis* [12-13, 36]; an agglomeration of sensory neurons in the hypostome and a collar of neurons at the base in the species *vulgaris* [6].

The mechanism responsible of the *Hydra* periodical behaviour should be due to the functional interaction between the two subnetworks, each of which alternately inhibits, and is inhibited by, the other one; the activity of each of them should result as a post-inhibitory rebound after the received inhibition [24].

Scattered sensory neurons have mechano- and chemo-receptive role, but there is not yet evidence that they have also photoreceptive functions [11].

Materials and Methods

The biopotentials were picked up from specimens of *Hydra vulgaris* and *magnipapillata*, by using a polyethylene suction Ag/AgCl electrode.

A description of the culturing medium and other details of the biopotential recording and of the photic stimulation can be found elsewhere [28, 33].

Review of experimental data

The main experimental results on the *Hydra* photoresponse, obtained by other researcher and by us, are here reported [3, 16-22, 25-33, 35].

The duration of the behavioural period (inter-CPT interval) in undisturbed conditions depends on various parameters of the background illumination, such as intensity and wavelength. In particular, the duration varies inversely with the wavelength in the range 450-700 nm.

The behavioural sequence expression can be modulated by light pulse stimuli, that shift the phase of the behavioural sequence at which the stimulus is applied, bringing it to a new phase.

A white light pulse affects the BSW shape: if given during the RP series, it momentarily decreases the slope of both BSWs, while if given during a CPT it produces an inversion of the trend of both BSWs giving rise to the opposite BSW trend at which the RPs arise.

The phase-response curve expressing the sensitivity to a single photic stimulus has a shape which depends on various parameters of the stimulus, which affect the amount and the direction (advance or delay of the subsequent CPT) of the response; such parameters are: i) duration; ii) relative intensity; iii) application phase; iv) polarity; v) wavelength. In particular, the amount of the response is maximum at the beginning of the CPT. During a CPT, the response to a white light pulse is an advance, during the

RP series it is a delay, while the reverse happens for a darkness pulse; also a 550 nm light pulse produces an advance if applied at the beginning of a CPT, while a 450 nm pulse at the same phase produces a delay.

It has been ascertained that:

(a) the periodic behaviour is entrainable by an external *Zeitgeber* by means of repetitive light pulse stimulation having a period different from the behavioural one;

(b) it is possible to suppress one part of the behavioural activity by means of repetitive light pulse stimulation of suitable frequency, which continuously resets the system at a fixed phase different from the suppressed one.

Both the possibility to entrain the behavioural sequence and to suppress part of it are a consequence of the existence of a phase response curve, which indicates a variability of photosensitivity along the behavioural phases.

If - when the length of the period of the behavioural sequence in undisturbed conditions has been varied by means of a variation of the steady background illumination wavelength - one applies a pulse stimulus, then (if one expresses the amount of the response as a fraction of the total period amount) the response should not vary with respect to the response that one obtains with such stimulus (all parameters of which being unvaried) at the previous background illumination wavelength. On the contrary, the response to a light pulse stimulus of fixed parameters varies with the background illumination wavelength variation.

Experiments carried out by a combination of two light pulses of different wavelength, being invaried the background illumination wavelength, have shown:

i) with two pulse stimuli of the same duration and intensity and different wavelength applied at the same time, the response values are intermediate with respect to those obtained with a stimulus of the higher or lower wavelength alone having the same duration and intensity of that of the combined stimulation;

ii) with two pulse stimuli of the same duration and intensity and different wavelength applied in close succession, the response values are intermediate with respect to those obtained with a single pulse of the higher or lower wavelength alone having the same intensity and double duration of that of each of the two previous stimuli; in addition, the foreseen effect as due to the second half of the combined pulse prevails;

iii) the above prevailing effect of the second pulse of a combined light pulse stimulation is decreased by the duration increase of the first of the two stimuli applied in close succession.

Discussion

All the above experiments clearly indicate quantitative light effects on the periodic behaviour of *Hydra*, in spite of the absence of recognizable photosensitive structures, suggesting the presence of photosensible molecules.

The results speak also in favour of the complexity of the mechanism responsible of the photoresponse, in spite of the simplicity of organization of such diblastic animal.

The fact that, for a light pulse stimulus of fixed application phases (along the shortening-elongation behavioural cycle), intensity, duration and wavelength, the response (shift of the next CPT) varies with the wavelength of the background illumination although the amount of the response be expressed in normalized units (as a fraction of the total behavioural period length, which varies with the wavelength of the background illumination), means that the modulation of the behavioural period duration due to the background wavelength variation does not occur isotropically along all the phases.

It is worth to mention also an asymmetry in the photoresponse for what it concerns CPs and RPs: while a light pulse applied during a CPT immediately stops the CP series, which is replaced by the RP series (and only later the CP series occurs, although it constitutes an advanced CPT), a light pulse applied during an RP series inhibits such series only momentarily, but such series is not replaced by the CP one [16, 18, 22, 25].

The interaction between the wavelength effects of the background and of the pulse stimuli could intervene either at phototransductive molecular level, or at further stages culminating in the photoresponse.

The fact that the response to short and long wavelengths differs not only in its amount but also in its direction could be interpreted either by the existence of two photosensitive pigments, or by two reciprocally reversible wavelength-dependent states of a single photoconvertible pigment. Up to now, it is only known that the level of astacin, assumed by food and stored in the epitheliomuscular cells, decreases with the illumination intensity level increase [5]. We were not able to clearly identify the hypothesized pigment(s) by means of microspectrophotometric methods; this could suggest that the pigment(s) is (are) uniformely diffused either in the cytoplasm, or on the cell membrane of the already unknown photoreceptive cells. These cells could be either sensory neurons, or ganglion neurons, or epitheliomuscular cells of the ectoderm or of the endoderm. Up to now, only chemo- or mechano-receptive functions have been attributed to the *Hydra* sensory neurons [11].

While biopotentials from dissociated ectoderm and endoderm cells of *Hydra* have been already obtained intracellularly [10], we are planning patch-clamp investigations in order to identify the cell type involved in the phototransduction.

Acknowledgements

CTF is indebted to L.M. Passano who introduced her to this research field. This work has been partially granted by the CNR Special Project "Non Linear Dynamics and the Role of Noise in Biological Systems".

References

1. Benos, D.J. & R.D. Prusch, (1973), Osmoregulation in *Hydra*: column contraction as a function of external osmolality, *Comp. Biochem. Physiol.*, 44A, 1397-1400.
2. Bode, H., S. Berking, C.N. David, A. Gierer, H. Schaller & E. Trenkner, (1973), Quantitative analysis of cell types during growth and morphogenesis in *Hydra*, *Wilhelm Roux's Arch. Entwicklungsmech. Org.*, 171:269-285.
3. Borner, M. & P. Tardent, (1971), Der Einfluss von Licht auf die Spontanaktivität von *Hydra attenuata*, Pall., *Rev. Suisse Zool.*, 78:697-704.
4. Epp, L. & P. Tardent, (1978), The distribution of nerve cells in *Hydra attenuata*, Pall., *Wilhelm Roux's Arch. Entwicklungsmech. Org.*, 185:185-193.
5. Frigg, M., (1970), Vorkommen und Bedeutung der Carotinoide bei *Hydra*, *Z. Vergl. Physiol.*, 69:186-224.
6. Grimmelikhuijzen, C.J.P., (1985), Antisera to the sequence Arg-Phe-amide visualize neuronal centralization in hydroid polyps, *Cell Tissue Res.*, 241:171-182.
7. Josephson, R.K., (1967), Conduction and contraction in the column of *Hydra*, *J. Exp. Biol.*, 47:179-190.
8. Josephson, R.K. & M. Macklin, (1967), Transepithelial potentials in *Hydra*, *Science*, 156:1629-1631.
9. Kass-Simon, G., (1972), Longitudinal conduction of contraction burst pulses from hypostomal excitation loci in *Hydra attenuata*, *J. Comp. Physiol.*, 80:29-49.
10. Kass-Simon, G., (1977), Spontaneous and evoked potentials from dissociated epithelial cells of *Hydra*, *Nature*, 265:75-77.
11. Kass-Simon, G., & L.A. Hufnagel, (1991), Functional anatomy of *Hydra*'s tentacular ganglion cell, *Proc. 4th International Workshop on Hydroid Development*, Schloss Reisenburg, Günzburg, FRG, p. 57.
12. Kinnamon, J.C. & J.A. Westfall, (1981), A three-dimensional serial reconstruction of neuronal distribution in the hypostome of *Hydra*, *J. Morphol.*, 168:321-329.
13. Kinnamon, J.C. & J.A. Westfall, (1982), Types of neurons and synaptic connections of hypostome-tentacle junctions in *Hydra*, *J. Morphol.*, 173:119-128.
14. Koizumi, O, M. Itazawa, H. Mizumoto, S. Minobe, L.C. Javois, C.J.P. Grimmelikhuijzen & H.R. Bode, (1992), Nerve ring of the hypostome in *Hydra*. I. Its structure, development, and maintenance, *J. Comp. Neurol.*, 326:7-21.
15. Musio, C., (1997), Extraocular photoreception in invertebrates: a look into biophysical processes and functional mechanisms. In: *Biophysics of Photoreception: Molecular and Phototransductive Events* (C. Taddei-Ferretti, ed.), Singapore, New Jersey, London, Hong Kong: World Scientific, in press.
16. Passano, L.M & C.B. McCullough, (1962), The light response and the rhythmic potentials in *Hydra*, *Proc. Natl. Acad. Sci. USA*, 48:1376-1382.
17. Passano, L.M & C.B. McCullough, (1963), Pacemaker hierarchies controlling the behaviour of hydras, *Nature*, 199:1174-1175.

18. Passano, L.M. & C.B. McCullough, (1964), Co-ordinating systems and behaviour in *Hydra*. I. Pacemaker system of the periodic contractions, *J. Exp. Biol.*, 41:643-664.

19. Passano, L.M. & C.B. McCullough, (1965), Co-ordinating systems and behaviour in *Hydra*. II. The rhythmic potential system, *J. Exp. Biol.*, 42:205-231.

20. Rushforth, N.B., (1971), Behavioral and electrophysiological studies of *Hydra*, *Biol. Bull. Woods Hole Mass.*, 140:255-273.

21. Rushforth, N.B., (1973), Behavioral modifications in Coelenterates. In: *Invertebrate Learning*, vol. 1 (W.C. Corning, J.A. Dyal & A.O.D. Willows, eds.), 123-169, New York, London: Plenum.

22. Rushforth, N.B., A.L. Burnett & R. Maynard, (1963), Behavior of *Hydra*: contraction responses of *Hydra pirardi* to mechanical and light stimuli, *Science*, 139:760-761.

23. Shibley, G.A., (1969), Gastrodermal contractions correlated with rhythmic potentials and prelocomotor bursts in *Hydra*, *Am. Zool.*, 9:586.

24. Taddei-Ferretti, C. & S. Chillemi, (1987), Modulation of *Hydra attenuata* rhythmic activity. V. A revised interpretation, *Biol. Cybern.*, 56:225-235.

25. Taddei-Ferretti, C. & L. Cordella, (1975), Modulation of *Hydra attenuata* rhythmic activity: Photic stimulation, *Arch. Ital. Biol.*, 113:107-121.

26. Taddei-Ferretti, C. & L. Cordella, (1976), Modulation of *Hydra attenuata* rhythmic activity. II. Phase response curve, *J. Exp. Biol.*, 65:737-751.

27. Taddei-Ferretti, C., & A. Cotugno, (1991), Interaction between the effects of light pulses of different chromatic content in *Hydra attenuata*, *Hydrobiologia*, 216/217:589-593.

28. Taddei-Ferretti, C, S. Chillemi & A. Cotugno, (1987), Modulation of *Hydra attenuata* rhythmic activity. IV. The mechanism responsible for rhythmic activity, *Exp. Biol.*, 46:133-140.

29. Taddei-Ferretti, C., L. Cordella & S. Chillemi, (1976a), Analysis of *Hydra* contraction behaviour. In: *Coelenterate Ecology and Behavior* (G.O. Mackie, ed.), 685-694, New York, London: Plenum.

30. Taddei-Ferretti, C., C. Musio & A. Cotugno, (1991a), Different weights of the influence of different light pulse position inside a stimulation in *Hydra*. In: *Cybernetics and Biophysics Italian Conference, SIF Conference Proceedings*. 31 (C. Frediani ed.), 123-128, Bologna (Italy): Compositori.

31. Taddei-Ferretti, C., S. Chillemi, V. Di Maio & C. Musio, (1994), Suppression of *Hydra* periodic shortening activity by repetitive phase delay, *Prog. Biochem. Biotechnol.*, 2:147-154.

32. Taddei-Ferretti, C., V. Di Maio, S. Ferraro & A. Cotugno, (1988), *Hydra* photoresponses to different wavelengths. In: *Light in Biology and Medicine*. 1 (R.H. Douglas, J. Moan & F. Dall'Acqua, eds.), 411-416, New York, London: Plenum.

33. Taddei-Ferretti, C., V. Di Maio, C. Musio & A. Cotugno, (1992), Modulation of *Hydra attenuata* rhythmic activity. VI. Combined effects of background and pulse light wavelength, *J. Photochem. Photobiol. B.*, 15:307-315.

34. Tardent, P. & C. Weber, (1976), A qualitative and quantitative inventory of nervous cells in *Hydra attenuata* Pall.. In: *Coelenterate Ecology and Behavior* (G.O. Mackie, ed.), 501-512, New York, London: Plenum.

35. Tardent, P., E. Frei & M. Borner, (1976), The reaction of *Hydra attenuata* Pall. to various photic stimuli. In: *Coelenterate Ecology and Behavior* (G.O. Mackie, ed.), 671-683, New York, London: Plenum.

36. Westfall, J.A., D.R. Argast & J.C. Kinnamon, (1983), Numbers, distribution, and types of neurons in the pedal disk of *Hydra* based on a serial reconstruction from transmission electron micrographs, *J. Morphol.*, 178:95-103.

37. Westfall, J.A., S. Yamataka & P.D. Enos, (1971), Ultrastructural evidence of polarized synapses in the nerve net of *Hydra, J. Cell Biol.*, 51:318-323.

38. Yoshida, M., (1979), Extraocular photoreception. In: *Handbook of Sensory Physiology. Vol. VII/6A. Comparative Physiology and Evolution of Vision in Invertebrates* (H. Autrum, ed.), 582-640, New York: Springer.

Costs and benefits of increasing the dimensionality of colour vision system.

Misha Vorobyev

Institute of Evolutionary Physiology and Biochemistry, Thorez pr. 44, 194223 St.Petersburg, Russia / Institut für Neurobiologie, Freie Universität Berlin, Königin-Luise-Str. 28-30, 14195 Berlin, Germany.

Introduction

Three types of colour vision systems have been described in the animal kingdom - dichromatic, trichromatic and tetrachromatic. One might ask why the number of colour receptor types does not exceed 4, and why many animals have only 2 spectral receptor types?

Two factors, obviously, limit the number of colour receptor types. 1. The higher the dimensionality of the colour vision system, the more sophisticated processing of the colour signal required. 2. The spatial resolution of n-dimensional colour vision system is theoretically limited by the area of n cones. In human beings, the eyes' optics allow a better resolution than this theoretical limit. It, presumably, explains the phenomenon of Brewster colours (Williams & Collier, 1983) - a fine pattern of black and white lines sometimes appears to be covered with a splotchy pattern of desaturated colours.

Is the quality of colour perception itself always improved as the number of spectral receptor types increases? Two different tasks of the colour vision system can be considered.

1. Discrimination of objects on the basis of colour.
2. Identification of objects on the basis of their colour in conditions of changing illumination (colour constancy).

Colour discrimination

To compare the ability of visual systems of different dimensionalities (e. g. dichromacy vs. trichromacy) to discriminate colours, we computed the number of all theoretically possible distinguishable colours at various levels of illumination. Each colour can be represented as a point in n-dimensional colour space, where n is the number of colour receptor types. The

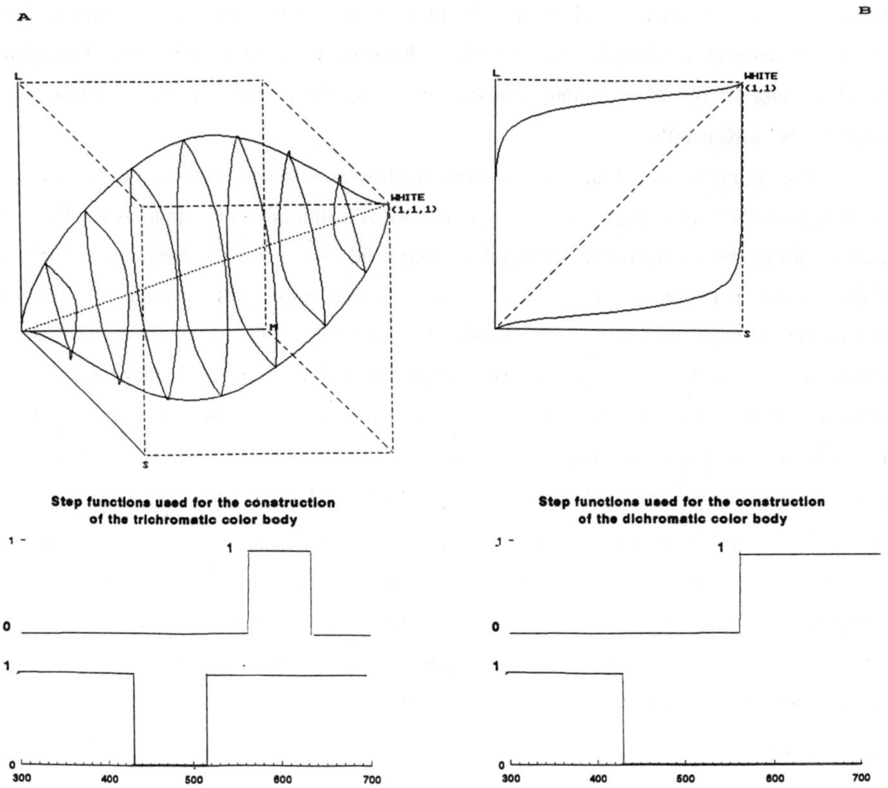

Fig. 1

Object-colour solid of a trichromatic visual system (left), and of a dichromatic visual system (right). For the trichromatic visual system the cross-sections by the planes of equal brightness are shown. The boundaries of colour solid are formed by optimal colour stimuli. The reflectance curves of optimal colour stimuli - the step functions - are shown.

colours corresponding to all theoretically possible reflecting surfaces occupy only a part of the colour space - an object-colour solid (Nyberg, 1928, Luther 1927) (Fig 1). The colour of the surface with reflectance $\rho(\lambda)$ illuminated by light with spectral power distribution $P(\lambda)$ is described by n-stimulus value $Xi=\int\rho(\lambda)*P(\lambda)*Si(\lambda)\, d\lambda$, where $Si(\lambda)$ are the spectral sensitivity functions of n colour types of receptors. The boundaries of the colour solid are formed by optimal colour stimuli. The $\rho(\lambda)$ of optimal colour stimuli is either zero or unity, and in moving through the visible spectrum, there are generally not more than n-1 transitions between these values (1 for a dichromatic and 2 for a trichromatic visual system) (Fig 1). The object-colour solid is an asymmetric "torpedo"-shaped body, with apices at the origin and at the white point, corresponding to the object-colour stimulus involving the perfect reflecting surface with $\rho(\lambda)=1$ at all wavelengths.

Each point in the colour space is surrounded by an elementary volume of the colours indistinguishable from a given one. The total number of distinguishable colours is equal to the number of this elementary volumes filling the colour solid. At low luminance levels, discrimination performance is limited by the photoreceptor noise caused by quantum fluctuations of the absorbed light. The number of absorbed quanta has a Poisson distribution which can be approximated by a Gaussian distribution with dispersion equal to the averaged number of absorbed quanta -Xi. The actual stimuli, corresponding to the colour with the mean co-ordinates - Xi with a given probability Q, lie within an n-dimensional ellipsoid with the centre at Xi. The volume of this ellipsoid is $U=A(Q,n)\cdot \Pi Xi^{1/2}$, where the proportionality factor A can be calculated from the χ^2 distribution with n degrees of freedom. If the discrimination threshold were adjusted to the noise, the indistinguishable stimuli would lie within this ellipsoid. The colour stimuli of the surface with colour Xi lie outside the ellipsoid with a probability P=1-Q, therefore P is equal to the probability of false alarm: identical surfaces are perceived as different. The total number of all theoretically distinguishable colours is $N = \int_V 1/U dv$.

The number of the quanta reflected from the white surface and absorbed by the i-types of cones within the receptive field during the summation time is equal to Li. Substitution Yi=Xi/Li would move the white point to the point Yi=1. N is maximal when Li=I/n, where I is the total number of absorbed photons within the receptive field during the summation time. Thus $N= I^{n/2} /(A(P,n) \sqrt{n})\int_V 1/\Pi\sqrt{Yi}\, dYi$.

The number of distinguishable colours is proportional to the light intensity raised to the power n/2. This means that at low light levels monochromatic vision (n=1) ensures better discrimination, whereas with increasing intensity multidimensional systems become more efficient (Vorobyev & Govardovskii 1993) (Fig 2). At the level of P=0.1 the Old World monkeys' trichromatic visual system would be better than the dichromatic one at a light intensity higher than 1000 photons, and better than monochromatic at the level of 100 photons. The light intensity at which the dichromatic system allows better colour discrimination than the trichromatic increases when P decreases (Fig. 3). For the given level of discrimination

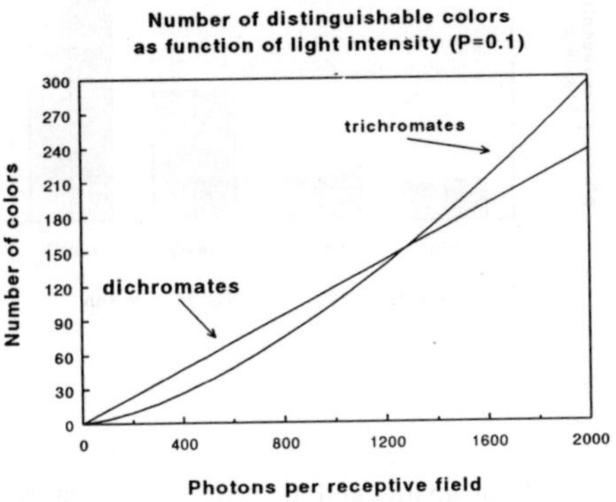

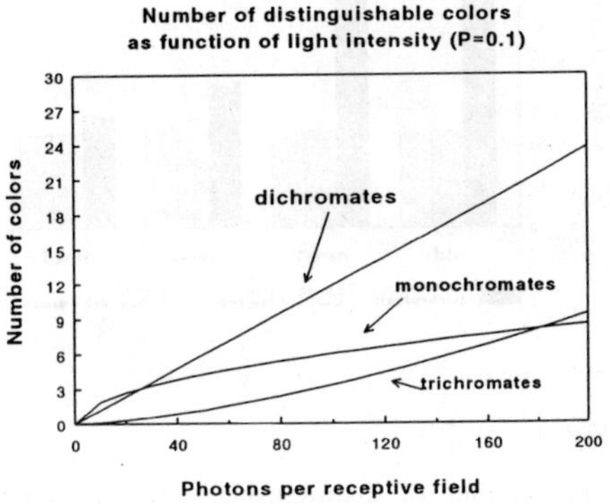

Fig. 2

The number of distinguishable colours as a function of light intensity for the human trichromatic blue-red dichromatic and monochromatic visual systems at the moderate levels of illumination (left) and at the low levels of illumination (right). Probability of false alarm - P=0.1.

Fig. 3

Light intensity of switch from di- to trichromacy as a function of the probability of false alarm.

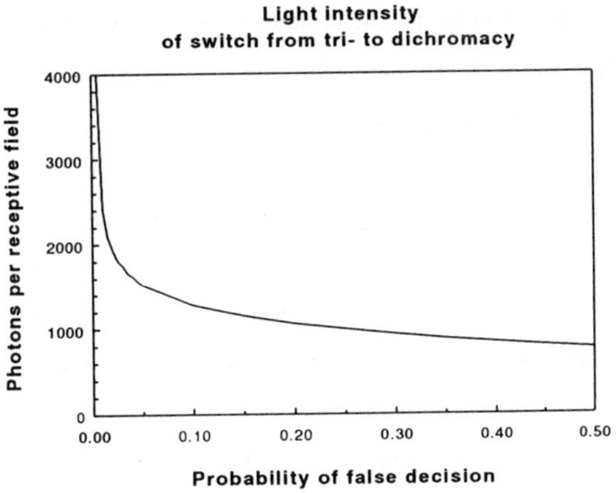

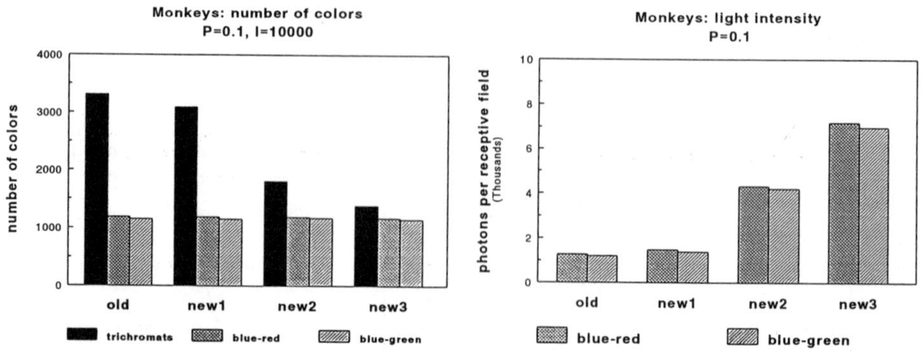

Fig. 4

The number of distinguishable surface colours for trichromatic and dichromatic primates (right), (the light intensity I=10000 photons per receptive field during the summation time, probability of false alarm - P=0.1), and the light intensity at which the switch from dichromacy to trichromacy becomes profitable (left). The Old World monkey were assumed to have the human set of colour receptors. Three types of trichromatic visual systems of New World monkey *Callithrix* were considered. New1 - P424, P545, P565; New2 - P424, P545, P558; New3 - P424, P558, P565. Each of these trichromatic systems was compared with corresponding blue-red and blue-green dichromatic systems.

threshold the higher the dimensionality of the colour vision system, the higher the frequency of false alarm. Thus, increasing the frequency of false alarms is one of the costs of increasing the dimensionality of colour vision system.

The colour vision system New World monkey offers an interesting example of the possible evolutionary switch from dichromacy to trichromacy or from trichromacy to dichromacy. The males of these animals are dichromates, while females are often trichromates. The visual pigments of the monkey *Callithrix* have one type of short-wavelength pigment (blue) and 3 types of long-wavelength (green and red) pigments. The combination of one type of long-wavelength pigment with the blue one produces three types of dichromatic vision (males) and the combination of blue pigment with two long-wavelength pigments produces 3 types of trichromatic visual systems (females) (Travis et al, 1988). We calculated the number of colours which these monkeys can discriminate (Vorbyev & Govardovskii 1993) (Fig. 4) In this case, the light intensity at which the switch from dichromacy to trichromacy becomes profitable (Fig 4) corresponds to the light intensity below which the visual system of males is better than that of females.

Generally speaking, dichromacy would allow for better colour discrimination in conditions of low illumination. This may explain the widespread dichromacy in mammals whose retina might have been adapted to nocturnality for a long time (Walls,1942).

Colour constancy.

Several animal species whose visual ranges extend from UV to red, e.g. birds, have 4 colour receptor types, whereas animals whose visual range is close to that of humans are often trichromatic (1). I will analyse two types of visual systems: with and without the UV sensitive receptor, assuming that the corresponding visual ranges are 300-700 nm (UBGR for UV-blue-green-red) and 400-700nm (BGR for blue-green-red).

To compare the ability of visual systems of different dimensionalities (e.g. trichromacy vs. tetrachromacy) to perceive colours constantly, a criterion of quality of colour constancy must be introduced. To assure colour constancy the visual system has to extract information about the reflectance spectra (SRs) of the objects from their colours and use this information to predict the appearance of the objects in conditions of changing illumination. The stability of surface colour appearance can be equated with recovery of SRs by the visual system (Brill, 1978, Malloney & Wandel, 1986); the SRs describe the surface colour and do not change when the illumination is changed. Therefore the quality of colour constancy can be expressed in terms of the best theoretically possible accuracy of reconstruction of the SRs from the information available to the visual system - photoreceptor quantum catches.

The error of approximation reaches its lower limit when the illumination spectra is recovered by the visual system absolutely accurately. Therefore, in order to estimate the best possible accuracy of approximation of SRs, we considered the reconstruction of SRs in conditions of known illumination. We reconstructed the SRs directly from the quantum catches using the linear algorithm which provides an average minimal error of approximation (Vorobyev et al, 1994). Approximation of SR - $\rho a(\lambda) = f_0(\lambda) + \sum_{i=1}^{i=n} (X_i + \delta_i) * f_i(\lambda)$, where n is the

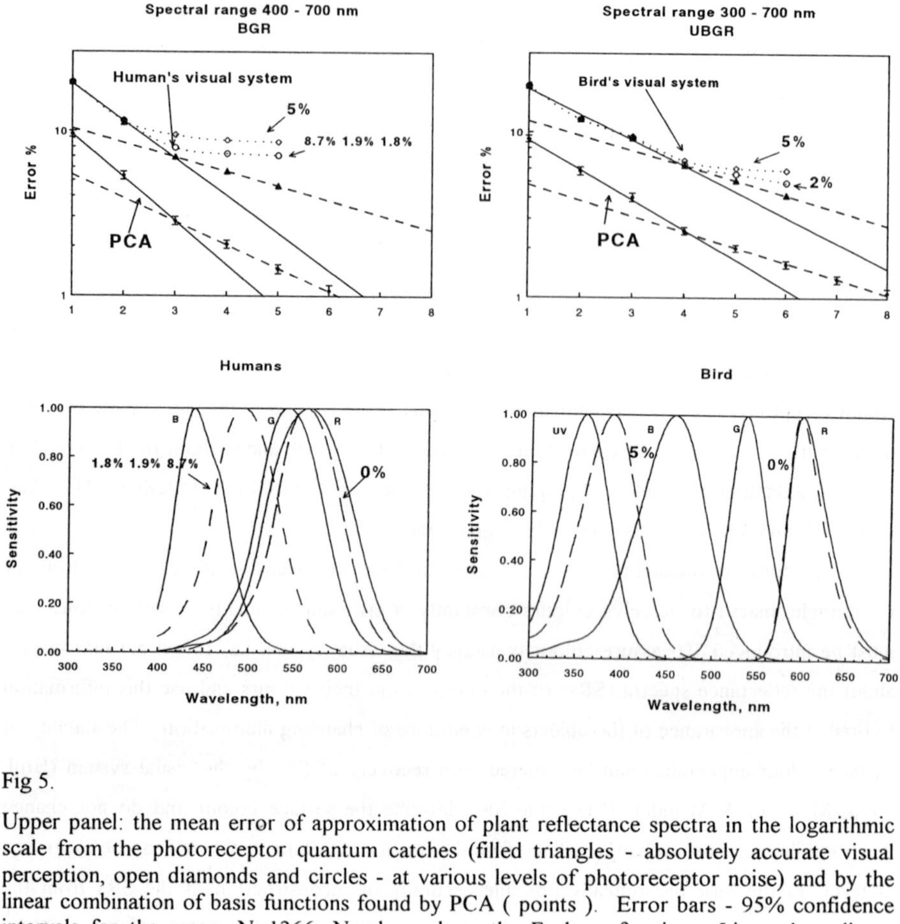

Fig 5.

Upper panel: the mean error of approximation of plant reflectance spectra in the logarithmic scale from the photoreceptor quantum catches (filled triangles - absolutely accurate visual perception, open diamonds and circles - at various levels of photoreceptor noise) and by the linear combination of basis functions found by PCA (points). Error bars - 95% confidence intervals for the mean, N=1366. Numbers show the Fechner fractions. Lines show linear regression.

Lower panel: spectral sensitivity functions used in the analysis. Solid curves - the real spectral sensitivity functions of "ultra violet" UV, "blue" - B, "green" G and "red" - R receptors, the dashed curves - the additional receptors for the hypothetical optimal tetrachromatic systems of humans and pentachromatic systems of *Leiothrix* depending on the photoreceptor noise. The numbers show the Fechner fractions. The optimal set of photoreceptors for the hypothetical systems of the lower dimensionalities did not depend on the photoreceptor noise level. The following sets of photoreceptors were found to be optimal: R - for the monochromatic, B-R for the dichromatic systems of humans and bird; B-G-R for the hypothetical trichromatic system of bird.

number of colour receptor types, photoreceptor quantum catches $X_i = \int_{\lambda min}^{\lambda max} \rho(\lambda)*P(\lambda)*S_i(\lambda)d\lambda$, $P(\lambda)$- D65 standard illumination.

To specify the statistical properties of SRs of biologically relevant materials we performed a principal component analysis of SRs of 1366 plant objects (fruits, leaves and flowers) (Menzel & Shmida,1993, Chittka et. al.,1994) both in the BGR and in the UBGR visual ranges. The slope of the logarithm of the error-of-approximation curve has a sharp change at 3 basic functions in the BGR-range and at 4 in the UBGR-range (Fig 5, points with error bars), which corresponds to the number of colour receptor types of many animals whose visual systems are sensitive within these spectral ranges.

In order to assess the quality of visual systems for colour constancy, we used the trichromatic colour receptor set of humans and the tetrachromatic of an animal whose visual system has been investigated in detail - the Passeriform bird *Leiothrix lutea* (Maier, Bowmaker, 1993). We then removed single receptor types from these systems or added one or two receptor types (Fig. 5). The wavelength positions of additional receptors were adjusted so that they provided the best possible accuracy of approximation of the SRs (Fig. 5, dashed curves). The strategy used to remove a receptor type from a system was to find the least important one for colour constancy. The performance of these artificially generated systems was compared with that of the real visual systems (Fig 5). Despite the fact that the actual spectral positions of receptors for the trichromatic system in the BGR range and for the tetrachromatic in the UBGR range were not placed optimally, increasing the number of colour receptor types yielded only marginal improvement in the accuracy of approximation. On the other hand, the loss of only the least important colour receptor fundamentally decreased the accuracy (Fig 5, filled triangles). Thus, constructing a tetrachromatic system on the basis of the human visual system and a pentachromatic on the basis of the visual system of *Leiothrix* may not be profitable.

Photoreceptor noise inevitably limits the accuracy of visual perception. If the inaccuracy for all receptors were equal to 5%, the tetrachromatic visual system of *Leiothrix* would be much better than the trichromatic system, and the pentachromatic system would provide nearly the same accuracy of approximation as the tetrachromatic. At the same level of photoreceptor noise the accuracy of reconstruction of the SRs from the information available to the human trichromatic system would be only marginally higher than that for the dichromatic system (Fig 5, open diamonds). From this one can conclude that the trichromatic visual system with noisy colour inputs may not be profitable compared to a dichromatic one.

According to the psychophysical determination of visual thresholds the Fechner fractions for "red" and "green" channels are 1.8% and 1.9%, whereas for the "blue" channel it is 8.7% (Wyszecki & Stiles, 1982) which exceeds 5%. Because of the significant overlap of spectral sensitivity functions of human "red" and "green" receptors, their quantum catches are highly correlated and thus the accuracy is more strongly influened by noise in these spectral channels than in the "blue" channel. Our calculations show that the trichromatic human visual system's potential ability for colour constant perception, with noisy "blue" input and less noisy "red" and "green" inputs, is higher than a dichromatic, and nearly the same as a hypothetical tetrachromatic system (Fig 5, open circles).

Thus, increasing the number of colour receptor types without increasing the width of the visual range would not be profitable either for humans or birds. In the case of accurate visual perception the loss of one type photereceptor would impair colour constancy, however the humans' trichromatic visual system would not be profitable in comparison with trichromatic if the photoreceptor noise levels were higher than 5%.

Acknowledgements

The presented results were obtained in collaboration with Prof. Victor I. Goovardovskii (St. Petersburg), Dr. Lars Chittka and Prof. Randolf Menzel (Berlin). I thank these people for friendly co-operation, and I thank Lars Chittka and Randolf Menzel for hospitality during my stay in Berlin. I am especially grateful to Victor Govardovskii - the author of many ideas of this paper - for bringing my attention to the problem. I thank Philomena Maher for correcting the English.

The study was supported by Alexander on Humboldt Foundation.

REFERENCES

1. Brill, M. J. J. Theor. Biol. 71, 473-478 (1978)

2. Buchsbaum, G. J. Franklin Inst. 310, 1-26 (1980)

3. Chittka, L., et. al. Vision Res. 34, 1489-1508 (1994)

4. Govardovskii V.I. & Vorobyev M.V. Sensory systems, 3, 150-159 (1989) (In Russian).

5. Luther, R. Tecn. Phys., 8, 997-1012 (1927)

6. Maier E.J. & Bowmaker J.K. J. Comp. Physiol. A 172, 295-301 (1993)

7. Maloney, L. & Wandel B. J. Opt. Soc. Am. A, 3,29-33(1986)

8. Menzel R. & Shmida A., Biological Reviews, 68, 81-120(1993)

9. Neumeyer, C. In: Evolution of the Eye and Visual System. 284-305, ed. J.R Cronly-Dillon & R.L. Gregory, RC press, Boca Raton,1991

10. Nyberg, N. Zt. Sch. Phys. 52, 406 (1928).

11. Travis, D.S., Bowmaker, J.K. & Mollon J.D. Vision Res. 28, 481-490 (1988).

12. Vorobyev, M. & Govardovskii V., Proceeding of the 21th Göttingen Neurobiology Conference, 396 (1993)

13. Vorobyev, M., Menzel, R. & Chittka L. Science, subbmited.

14. Walls, G.L. The Vertebrate Eye and ist Adaptive Radiation. Granbrook Institute of Science., Bloomfield Hills, Michigan, 1942

15. Williams, D.R. & Collier, R.J. Science, 221,385 (1983)

16. Wyszecky, G. & Stiles, W.S. Color Science: Concepts and Methods, Quantitative Data and Formulae (2nd ed.) ,Willey, N.Y., 1982.

A PHYSIOLOGICAL MODEL DESCRIBING HONEYBEE PHOTORECEPTOR RESPONSES TO INCREMENTAL AND DECREMENTAL LIGHT STIMULI

Uli Gerster & Werner Backhaus

Institut für Neurobiologie, c/o Institut für Biophysik, Freie Universität Berlin, Thielallee 63, 14195 Berlin, FRG

1. Introduction

The aim of our physiological simulations of the transduction process in photoreceptors of the honeybee (*Apis mellifera*) worker is to develop physiologically adequate models of the photoreceptors for extending the theory of color vision and color choice behavior [1] by temporal properties. Physiologically adequate means in this context that only those physiological properties of the photoreceptors are taken into account which essentially contribute to the time-courses of the membrane potential in response to light stimulation.

2. Physiology of phototransduction

A typical response of a dark or light adapted photoreceptor to an increase of light intensity consists of a first transient peak potential and a following plateau phase reached after about 50ms [2], (see examples fig. 3a, b). Within a few seconds, the photoreceptor reduces its sensitivity (adaptation) and the plateau potential declines to a steady-state value. The change of the membrane potential is due to a change of the membrane permeability to Na^+ and K^+ which is induced by a second messenger cascade (permeability changing cascade, PCC). The drop from peak to plateau potential and the adaptation are due to an increase of intracellular Ca^{2+} [drone: 3, 4, 5; *Limulus*: 6]. The sources for intracellular Ca^{2+} are intracellular stores [7] and a light induced influx from the extracellular space [8]. Although Ca^{2+} has multiple effects and feedback sites in the transduction cascade, it turns out in our simulations of membrane potential traces that it is not necessary to include a detailed mechanistic description at this stage of modeling.

3. Results

3.1 The physiological model

Based on the physiological findings mentioned above, a physiological model (fig. 1) of phototransduction is developed for the honeybee worker. In the model, illumination leads via the PCC to an increase of the Na^+/K^+-permeability P of the photoreceptor membrane. The kinetics is described by an exponential function. $P = P_{max} \cdot (1-\exp(-t/\tau_1))^n$, $P_{max} = g_1 \cdot I$, which describes the kinetics of events mediated by chemical reaction cascades. P_{max} is proportional to the light intensity I with a gain factor g_1. τ_1 is the time constant. It turns out in the simulations that n=2 gives the best fits to the measured data (see below). The intracellular Ca^{2+}-concentration is also described by two (with respect to the two different sources) exponential functions of this type with different time constants:

$$[Ca^{2+}]_{i,fast} = [Ca^{2+}]_{i,fast,max} \cdot (1-\exp(-t/\tau_{fast}))^2 ,$$

$$[Ca^{2+}]_{i,slow} = [Ca^{2+}]_{i,slow,max} \cdot (1-\exp(-t/\tau_{slow}))^2 .$$

The fast component is responsible for the drop from peak to plateau potential and the slow component is responsible for the decline from plateau to steady-state potential. The maximum values of $[Ca^{2+}]_i$ for the fast and the slow component are again proportional to the light intensity with specific gain factors:

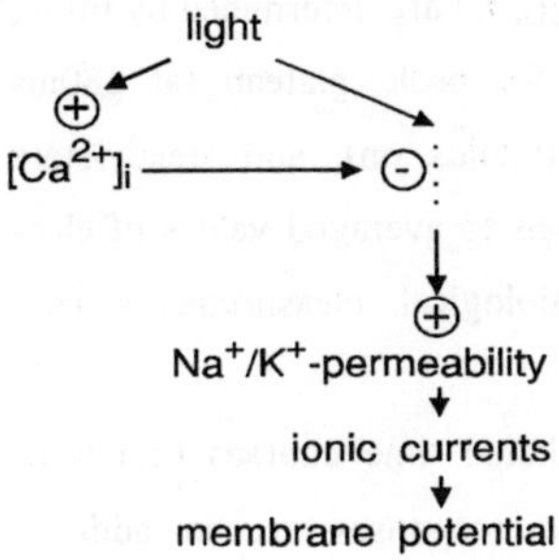

Fig. 1 Physiological photoreceptor model.

$$[Ca^{2+}]_{i,fast,max} = g_2 \cdot I, \qquad [Ca^{2+}]_{i,slow,max} = g_3 \cdot I.$$

Without specifying the acting site of Ca^{2+} in the PCC, the adaptation due to Ca^{2+} is simply described by reducing P_{max} according to a Ca^{2+}-concentration dependent factor:

$$P_{max,adap} = P_{max} \cdot \left(1 - \frac{[Ca^{2+}]_i}{K + [Ca^{2+}]_i}\right),$$

where $[Ca^{2+}]_i$ is the total intracellular concentration of Ca^{2+}, K is the concentration of Ca^{2+} at which $P_{max,adap}$ is half of P_{max}. From the resulting change of the Na^+/K^+-permeability, the ionic currents are calculated according to the Nernst-Planck equation [9]:

$$j = -Pd \cdot \left(\frac{dc}{dx} + zc\frac{F}{RT}\frac{dE}{dx}\right)$$

(j: mass current density, d: membrane thickness, c: ion concentration, z: ion valence, F: Faraday constant, R: general gas constant, T: temperature, E: membrane potential, x: position variable). The membrane potential is:

$$E(t) = \frac{zF}{c_m}\int_{t_0}^{t} j \cdot dt ,$$

with the specific membrane capacity $c_m = 1\mu F/cm^2$.

3.2 Simulations of electrophysiological measurements

With the presented physiological model, electrophysiological measurements [2] in photoreceptor cells of honeybee workers are simulated and the unknown parameters (n, τ_1, τ_{fast}, τ_{slow}, g_1, g_2, g_3, K) are determined by fitting values for peak, plateau (at 180ms after stimulus on), and steady-state potentials to averaged values of electrophysiological measurements (fig. 2).

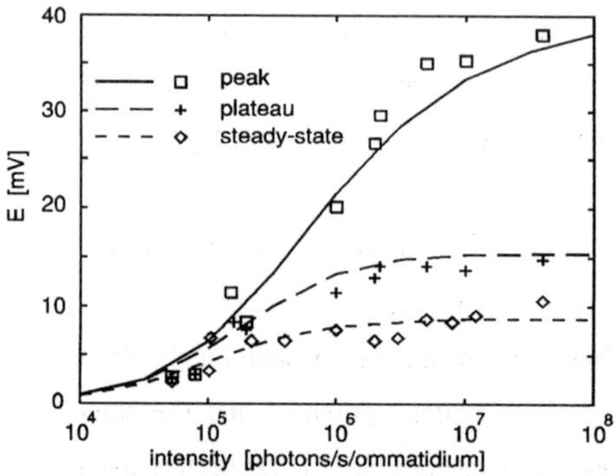

Fig. 2 Comparison between predicted (lines) and averaged values of measured ([2], symbols, number of measurements between 6 and 22) peak, plateau, and steady-state potentials. Parameters used: n=2, τ_1 =6.5ms, τ_{fast} =20ms, τ_{slow} =10s, g_1 =$6 \cdot 10^{-13}$m, g_2=$1.3 \cdot 10^{-2}$ Ms, g_3=$1.7 \cdot 10^{-2}$ Ms, K=35μM.

Predicted time courses of photoreceptor responses are in addition compared to single cell recordings (fig. 3). Since single cell recordings deviate from the averages, for which the model is developed, the gain factors g_1, g_2, and g_3 have to be modified accordingly. All the other parameters remain unchanged. In addition to the responses of dark adapted photoreceptors, the model describes also the responses of light adapted photoreceptors (fig. 3b). Although nothing was assumed in the model about afterpolarization effects, the simulation describes also the typical afterhyperpolarization correctly. With the same model, which was developed for photoreceptor responses to incremental stimuli, it is also possible to describe responses to decremental stimuli

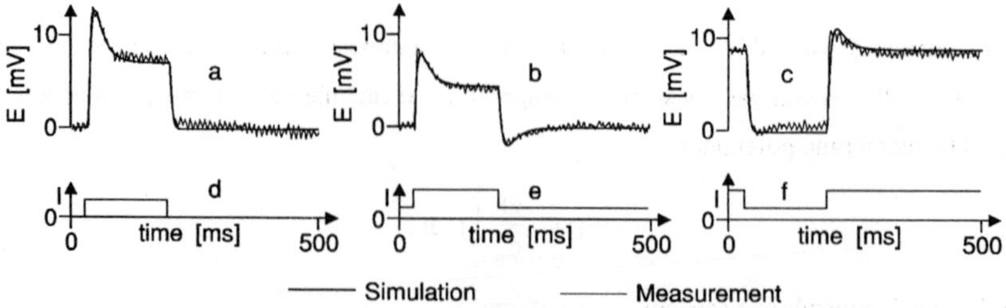

Fig. 3 Typical measured single cell recordings (Peitsch, 1992) and simulated photoreceptor responses of dark adapted (a) and light adapted (b: incremental, c: decremental light stimuli) photoreceptors. d,e,f: stimulus monitor.

(see fig. 3f, stimulus monitor) using the same time constants (τ_1, τ_{fast}, τ_{slow}) as for incremental contrasts (fig. 3c). Again, only g_1, g_2, and g_3 have to be modified in order to account for the deviations of the individual photoreceptor cell recording from the average.

4. Conclusion

The presented physiological model accurately describes averaged peak, plateau and steady-state potentials of dark adapted photoreceptors although it uses some simplifying assumptions. It describes almost the entire physiologically relevant light intensity range with one set of parameters. Although nothing was specifically assumed about responses of light adapted photoreceptors to incremental and decremental light stimuli, the typical shapes of the time courses (peak, plateau and afterhyperpolarization) are nevertheless properly described by the model. In order to describe time-courses of individual cell recordings, the deviations of individual single cell recordings from the average can be compensated by modifying only the gain factors g_1, g_2, and g_3 in the model. These results show that the presented physiological model, although it uses some simplifying assumptions, is already sufficient to describe membrane potential traces due to light stimuli of almost the entire physiological intensity range.

Acknowledgement

We would like to thank Dagmar Peitsch for kindly providing us with data files from single cell recordings of her Ph.D. work.

References

[1] W. Backhaus (1993). Color vision and color choice behavior of the honeybee. In: Recent Progress in Neurobiology of the Honeybee. *Apidologie*, special issue, 24 (6), 309-331.

[2] D. Peitsch (1992). Contrast responses, signal to noise ratios and spectral sensitivities in photoreceptor cells of hymenopterans. Ph.D.-Thesis, Free University Berlin, FRG.

[3] C.R. Bader, F. Baumann & D. Bertrand (1976). Role of intracellular calcium and sodium in light adaptation in the retina of the honey bee drone (*Apis mellifera, L.*). *J. Gen. Physiol.*, 67, 475-491.

[4] S. Levy (1979). Mesure du Ca intracellulaire libre dans les photorecepteurs du faux-bourdon (*Apis mellifera*), au moyen de microelectrodes selectives au Ca: effets de la photostimulation. M.S. Thesis, University of Geneva, Switzerland.

[5] M. Raggenbass (1983). Effects of extracellular calcium and of light adaptation on the response to dim light in honey bee drone photoreceptors. *J. Physiol.*, 344, 525-548.

[6] S. Levy & A. Fein (1985). Relationship between light sensitivity and intracellular free Ca concentration in *Limulus* ventral photoreceptors. *J. Gen. Physiol.*, 85, 805-841.

[7] A. Ziegler & B. Walz (1990). Evidence for light-induced release of Ca^{2+} from intracellular stores in bee photoreceptors. *Neurosci. Lett.*, 111, 87-91.

[8] A. Ziegler & B. Walz (1989). Analysis of extracellular calcium and volume changes in the compound eye of the honeybee drone, *Apis mellifera*. *J. Comp. Physiol. A*, 165, 697-709.

[9] M. Planck (1890). Ueber die Potentialdifferenz zwischen zwei verdünnten Lösungen binärer Electrolyte. *Ann. Phys. Chem.*, 40, 561-576.

APPLICATION OF LASER CONFOCAL MICROSCOPY FOR Ca^{2+} MEASUREMENTS IN *LIMULUS* VENTRAL PHOTORECEPTORS.

Kyrill Ukhanov and Richard Payne, *Dept. of Zoology, University of Maryland, College Park, MD 20742, USA*

Introduction

Ca^{2+} involvement in phototransduction in invertebrates is one of the most intriguing and arguable issues. A number of studies suggests participation of Ca^{2+} in excitation and adaptation of invertebrate photoreceptors [1,2]. It was well established by the previous studies that Ca^{2+} is released in *Limulus* ventral photoreceptors as a final act of rhodopsin-Gprotein-PLC-IP3 chain of reactions [3,4]. Moreover direct injections of Ca^{2+} were proved to induce in the photoreceptors a current having the same reversal conductance as of the light-induced one [2]. Nevertheless, previous attempts have failed to show that light-induced Ca-release precedes the electrical excitation of the cells [5-7].

Recently we studied light-induced Ca-release in *Limulus* photoreceptors using several fluorescent dyes and a conventional micro-fluorimeter that collected fluorescence from the entire rhabdomeral (R)-lobe of the cell. We concluded that the time-course of the rising edge of the Ca signal was probably distorted by spatial averaging of localized changes of Ca^{2+} [8]. In order to minimize spatial averaging, we have now employed a laser confocal microscopy to measure Ca-signal within different areas of *Limulus* ventral photoreceptors.

Materials and Methods

Limulus ventral nerves were mounted in a chamber on the stage of Zeiss Axiovert-135 inverted microscope. Single photoreceptors were pressure injected with 1 mM fluo-3, Ca Green-5N or indo-1 dissolved in 100 mM potassium aspartate, 10 mM HEPES, pH 7.0. For quantitative measurements we used ratiometric procedure with indo-1 as a high affinity dye to measure low Ca^{2+} concentrations and a mixture of CaGreen 5N with ANTS [8] to measure Ca concentrations in the range over a hundred micromolar. Only those cells capable of producing quantal events as a criterion for their viability were considered for experiments. All measurements were performed on a Zeiss LSM 410 laser confocal scanning microscope using the 488 nm line of an Ar-Kr laser or 364 nm line of an Ar ion laser. The membrane voltage of the photoreceptors and a light monitor were also stored on a computer.

The photoreceptor and the dye were simultaneously stimulated by a stationary laser spot, focused onto the most sensitive region of the R-lobe. The volume within which the Ca signal was measured did not exceed 10 μm^3 (10^{-14}l).

Results

Under the conditions of laser light stimulation, the Ca signal rose simultaneously with the beginning of the electrical response as detected with fluo-3 or CaGreen 5N (Fig. 1). The Ca signal peaked within 50 ms, rising nearly as fast as membrane depolarization (Fig. 1A). Positioning of the laser spot further away from the most sensitive area of the R-lobe resulted in decrease of the rate of rise though Ca^{2+} concentration still peaked at 160 μM (Fig. 1B). Even in the latter case the estimated rate of rise of Ca^{2+} concentration at the earliest time of the release could be as high as 1 mM/sec.

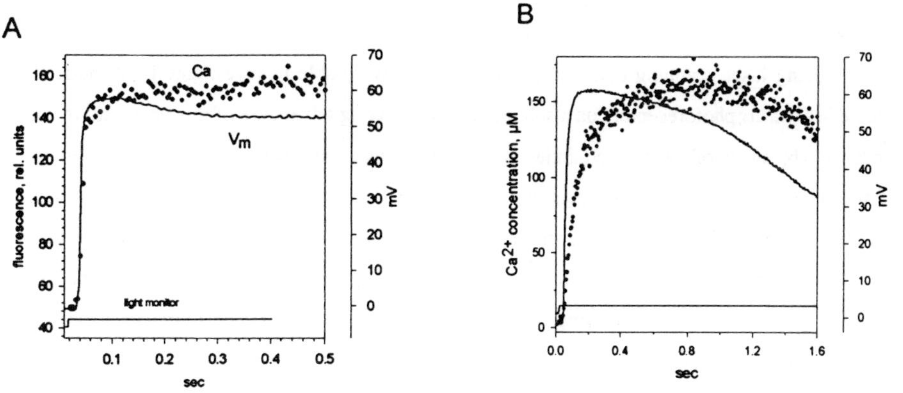

Fig. 1. Ca-release measured from the single confocal spot in the R-lobe of *Limulus* ventral photoreceptors. (A) Ca signal was monitored by using fluo-3. (B) Cell injected with CaGreen 5N/ANTS mixture to determine Ca^{2+} concentration.

Using a line scanning mode combined with loading the photoreceptors with indo-1, we observed that elevation of Ca^{2+} concentration spreads throughout the R-lobe, beginning at its outer rim (Fig. 2). Once initiated Ca-release rapidly moves across an entire R-lobe within 300 msec. Another interesting feature was that Ca-release induced only a faint elevation of Ca^{2+} concentration in the A-lobe, implying effectively management by intracellular Ca-buffers within the R-lobe.

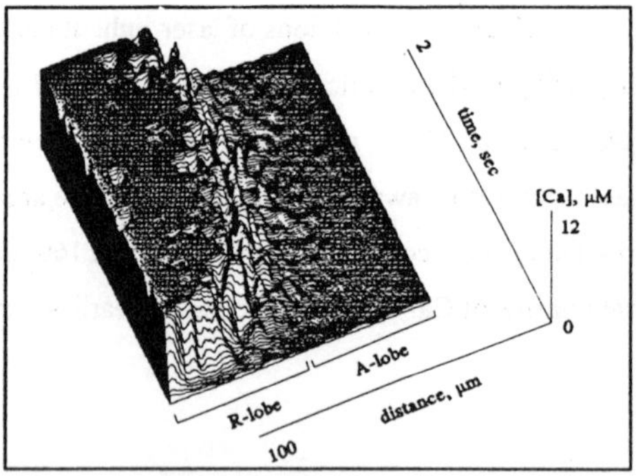

Fig. 2 Spatio-temporal distribution of Ca^{2+} concentration along the line scanned across *Limulus* ventral photoreceptor. Graph comprises of 512 individual lines at 4 ms intervals. The length of the scanning line is 100 μm. The left border of the graph corresponds to the outer edge of the R-lobe.

Discussion

Using laser scanning confocal microscopy we were able to investigate light-induced Ca-release in *Limulus* photoreceptors on a subcellular level. Taking advantage of this powerful technique we have to be aware of some uncertainties introduced by using highly focused local light stimulation of the photoreceptors. This becomes especially important when comparing our data with those previously reported. In all previous studies, the photoreceptors were stimulated with the diffused light. When the photoreceptor is stimulated with a light spot its response has smaller latent period and rises for a longer time than following diffuse light of the same physiological intensity [9]. In other words under this conditions nonlinear components of the light response may be greatly enhanced. Another important consideration is that actual detection of Ca^{2+} occurs only in the small confocal volume while most sensitive areas (i.e. microvilli) might be located beyond this volume in the lightpath of the measuring light.

One of the most impressive manifestations of the light-induced Ca release is its spatial heterogeneity. As was tested by confocal microscopy as well, endoplasmic reticulum (ER) is contiguous within the whole ventral photoreceptors [10]. Therefore it could appear that there are some specific microdomains most active in releasing and removing Ca^{2+} ions. We might assume that the regions of ER with the highest density of IP3-receptors probably faces microvilli where all the phototransduction machinery is located. This is in line with what is predicted for the *Drosophila* phototransduction [11]. After the termination of a light stimulus or in the course of a prolonged light illumination the portion of the ER within the R-lobe could participate in the spread of the Ca release. Our data supports an earlier finding that Ca^{2+} elevation in the A-lobe of *Limulus* ventral photoreceptors are rather delayed and insignificant compared to that in the R-lobe under the conditions of light stimulation [12].

The question of whether there are some specific microdomains of the ER responsible for the initial Ca release which might trigger opening of the ion channels may be successfully tackled using laser scanning confocal microscopy.

References

1. Lisman, J.E., and Brown, J.E. (1972). The effects of intracellular iontophoretic injection of calcium and sodium ions on the light-response of *Limulus* ventral photoreceptors. J. Gen. Physiol. *59*, 701-719.

2. Payne, R., Corson, D.W., and Fein, A. (1986a). Pressure injection of calcium both excites and adapts *Limulus* ventral photoreceptors. J. Gen. Physiol. *88*, 107-126.

3. Payne, R., Corson, D.W., Fein, A., and Berridge, M. (1986b). Excitation and adaptation fo *Limulus* ventral pphotoreceptorsby inositol 1,4,5-trisphosphate result from a rise in intracellular calcium. J. Gen. Physiol. *88*, 127-142.

4. Payne, R., Walz, B., Levy, S., and Fein, A. (1988). The localization of calcium release by inositol trisphosphate in *Limulus* ventral photoreceptors and its control by negative feedback. Phil. Trans. Roy. Soc. (B) *320*, 359-379.

5. Payne, R., and Flores, T.M. (1992). The latency of the response of *Limulus* photoreceptors to inositol trisphosphate lacks the calcium-sensitivity of that to light. J. Comp. Physiol. (A) *170*, 311-316.

6. Stieve, H., and Benner, S. (1992). The light-induced rise in cytosolic calcium starts later than the receptor current of the *Limulus* ventral photoreceptor. Vision Res. *32*, 403-416.

7. Walz, B., Zimmerman, B., and Seidl, S. (1994). Intracellular Ca^{2+} concentration and latency of light-induced Ca^{2+} changes in photoreceptors of the honeybee drone. J. Comp. Physiol. (A) *174*, 421-431.

8. Ukhanov, K.Y., Flores, T.M., Hsiao, H-S., Mohapatra, P., Pitts, C.H., and Payne, R. (1995). Measurement of cytosolic Ca2+ concentration in *Limulus* ventral photoreceptors using fluorescent dyes. J.Gen. Physiol., *105*, (in press).

9. Payne, R., and Fein, A. (1986) The initial response of *Limulus* ventral photoreceptors to bright flashes. Released calcium as a synergist to excitation. J. Gen. Physiol., *87*, 243-269.

10. Feng, J.F., Carson, J.H., Morgan, F., Walz, B., and Fein, A. (1994) Three-dimensional organization of endoplasmic reticulum in the ventral photoreceptors of *Limulus*. J. Comp. Neurology, *341*, 172-183.

11. Hardie, R.C., and Minke, B. (1993). Novel Ca^{2+} channels underlying transduction in *Drosophila* photoreceptors: implications for inositide-mediated Ca^{2+} mobilzation. Trends Neurosci. *16*, 371-376.

12. Levy, S., and Fein, A. (1985). Relationship between light-sensitivity and intracellular free calcium in *Limulus* ventral photoreceptors. J. Gen. Physiol. *85*, 805-841.

PHOTORECEPTION AND PHOTOTRANSDUCTION IN VERTEBRATES

VISUAL TRANSDUCTION BY RETINAL RODS

L. F. GIULIO
Dipartimento di Morfofisiologia Veterinaria, Università di Torino, Viale Mattioli 25, 10125 Torino, Italy

The purpose of future studies on the fine structure of the light-trapping device of photoreceptors, the so-called visual cells outer segment, is, as Worthington says, mainly to "provide a more detailed picture so that correlation between structure and function can became a reality" [11]. The cooperative use of very different probing techniques has provided in the last years a valuable sum of informations on the way the various molecular types may be assembled in the photoreceptor membranes. From this point of view, the vertebrate rod outer segment (ROS) is the most extensively studied, and best understood object; the main purpose of this short lecture will be to try to give a simplified (perhaps oversimplified) picture of the dynamic aspects of the molecular architecture of the photoreceptor membrane of the ROS.

Chemical composition studies of ROS membranes indicate that about 40% of the dry weight of the outer segment is lipid and 60% protein, of which over 80% is Rhodopsin (Rho). This signifies, *inter alia*, that for every molecule of the pigment there are about 60-90 lipid molecules (57-63 phospholipids per Rho [Milianich & Schwartz, unpublushed data quoted by 6]). There are five principal kinds of phospholipids: phosphatidylcoline (40-44%), phosphatidylserine (13-14%), phosphatidylethanolamine (38-39%), sphingomyelin (3%), and phosphatidylinositol (2%). The cholesterol is present in a very small percentage (2-3%).

The lipid components of cell membranes are usually found as a "liquid crystalline" (mesomorphic condition), and are characterized by a periodically ordered long-range organization with a somewhat disordered short-range organization.

The fluidity of the cell membrane structure can be investigated with thermotropic transition tests and in the presence of highly unsaturated lipids fluidity it is high; it decreases as the percentage of more saturated lipids increases, *i.e.* when lipids contain unsaturated bonds, phase transition occurs at lower temperatures.

In the sequence of events which leads to visual excitation, geometrical configuration of the chromophore of photosensitive pigment for the absorption of a single quantum of light (cis-trans isomerism) seems most likely to be the first in order of time. As Wald said "the only thing that light does in any visual system we know is to isomerize retinene" [10]. Events occurring afterward, in fact, come into the category of thermal ("dark") reactions.

The isomerization of the chromophore from 11-cis to all-trans eliminates the steric correspondence between the chromophore itself and the protein fraction of the photosensitive pigment, joined together by a Schiff base linkage between the terminal aldehyde group of retinal (according to the Commission on the Nomenclature of Biological Chemistry, *J. Amer. Soc.*, 1960, 82, 5575) and the aminogroup of lysine 296 in helix VII of the opsin. In this way, the close "lock-and-key" relationship between retinal and protein ends and the process begins which will lead, at least in the unstable meta-rhodopsin of Vertebrates, to hydrolysis in opsin and all-trans retinal (process of bleaching).

The steric configuration of the 11-cis retinal is of the greatest importance. In fact, it allows, by the sufficiently intimate approach between the isoprenoid side chain and the chromophore ring (β-ionone) on one side, and protein fraction in helix VII of the opsin on the other, the intervention of Van der Waals attraction forces and some degree of charge displacement [4]. The effect of the absorption of one quantum, and therefore of the following isomerization, is at least twofold: the electric tension between chromophore and protein disappears, and the parts of the protein which were "covered" by the chromophore are now revealed. This second process exposes two to three sulphydryl groups per molecule, and a proton-binding group with a pK of about 6.6 (in the molecule of the photosensitive pigment the Schiff base linkage is protonated, *i.e.* has attached an H^+, and the opsin, moreover, assumes a negative charge which, in intimate association with the retinal isoprenoid side chain, draws the positive charge up into the conjugated system; all of which enhances the resonance forms and shifts the absorption spectrum towards the red).

The frog ROS, in unbleached condition, exhibits a positive uniaxial birefringence. The change of birefringence from dark to light adaptation could be, at first sight, a simple consequence of the photolitic process: the Rho, in fact, would affect more strongly the transverse refractive index (*i.e.* the ordinary ray) and the shift toward the shorter of the last photoproducts could cause a Δn (positive uniaxal birefringence) increase. The total birefringence of the ROS decreases with increasing temperature. Thermotropic phase transitions (*i.e.* phase changes induced by heat) was observed in both dark and light adapted states, at transition temperature, T_t, from 42 to 46 °C. This changes in the physiological state of the membranes precedes the disorientation of the pigment molecules. From +52-53 °C onwards, Rho molecules appear to be completely disoriented as though in solution.

To explain the lack of photodichroism in axially illuminated photoreceptors one could suppose that the visual pigment molecules possess some rotational freedom in multilayered structures of photorecepor cells.

The "closely packed" pigment molecules could present a dinamic disorder at their absorption oscillator direction level, disorder which is presumably due to continous brownian motion.

The rotational relaxation time was calculated by Cone [1], using the wellknown equation :

$$\rho_o = 4\pi\eta r^3/kT$$

One could estimate a rotational relaxation time (ρ_o) for Rho in fresh dissected retinas of 4-50 μs at +10 °C (ca. 20 μs at +20 °C). Between the dynamic aspects of photoreceptor membranes, we have the translation diffusion coefficient of the Rho. This event appears to be connected with the average time τ between collisions with another Rho molecule, measured as:

$$\tau = s^2/4D$$

The time between collisions (τ) is then 3.9 μs at +20 °C; this signifies a collision frequency in the range of 10^5-10^6 s^{-1}. A possible consequence of multiple collisions by thermal motion might be, according to De Vries [2], the "activation of the photochemical substance".

In conclusion, Rho rotates around an axis, parallel to the long axis of the outer segment; it also is endowed with a translational motion, the translational velocity being of 0.3 ms^{-1}.

Another aspect, and the first fast photoelectric effect within the photoreceptor cells, is the photovoltage of very short latency "early receptor potential" (ERP), biphasic response with the initial phase consistent in a corneal positive deflection (R1) of latency <0.5 μs at +25 °C, the spectral sensitivity of which is congruent with that of Rho. This phase can be isolated from the following corneal negative deflection (R2), by cooling the tissue to about 0 °C, and is still detectable at -35 °C. *Inter alia*, warming frog retinas to +64 °C abolishes irreversibly R1 and R2 (and Rho appears completely disoriented, as if in solution). The amplitude of ERP depends also on the grade of alignment in the space of current microgenerators identifiable with the molecules of Rho.

Retinal photoreceptors generate, in the dark, discrete electrical events, indistinguishable from those evoked by light. The random spontaneous events are strongly temperature-dependent, and have been attributed to thermal isomerization of the chromophore of Rho. But thermal generation of dark events requires activation energies in the range between 23 and 27 kcal mol^{-1} for photoisomerization of the chromophore. The photoreceptor noise results from the thermal isomerization of a relatively unstable form of Rho, one in which the Schiff base linkage between the chromophore and the protein is unprotonated [5, 7].

According to this hypothesis, a two-step ground-state process generates photoreceptor dark noise. Step one is the deprotonation of the 11-cis Schiff base linkage, and step two is the thermal isomerization of the 11-cis reatinal to all-trans form.

It has been demonstrated that one retinal rod can be stimulated by one quantum acting upon it. The question then arises as to whether one single quantum can not also be sufficient for seeing. Again the facts of temporal and spatial summation could hardly be explained if one quantum were sufficient for vision, the smallest number of quanta required, according to Hecht *et al.* [3], is 5-7 photons. Van der Velden [8, 9] found 2 photons. Since the few photons in question are distributed over several hundred receptor cells, the probability that one ROS captures two photons is nearly null. Absorption of a single photon, in conclusion, must thus produce a significant signal in one receptor, and each such signal must contribute to vision.

Light absorption by the visual pigment Rho triggers, through G-protein coupling, a cascade of events in the outer segment of the rod cell of the Vertebrate retina that results in membrane hyperpolarization and nerve excitation. The response of a retinal rod cell to a weak flash of light is mediated by a receptor/GTP binding protein (Rhodopsin Transducin) signal transduction system, and terminates within a second.

In the Vertebrate light response, inactive heterotrimeric Transducin ($G_{t\alpha\beta\gamma}$-GDP) binds the photoactivated Rho (Meta-Rhodopsin II) and releases GDP. GTP binds firmly to the α-subunit of the receptor-bound Transducin causing $G_{t\alpha}$-GTP to separate from both the receptor and the βγ heterodimer. The $G_{t\alpha}$-GTP complex activates its effector, cGMP-specific phosphodiesterase ($\alpha\beta\gamma_2$cGMP-PDE) by binding to its inhibitory γ-subunit. Activated αβcGMP-PDE lowers the concentration of cGMP, which closes cGMP-gated channels and hyperpolarizes the retinal rod cell, thereby generating the nerve impulse. $G_{t\alpha}$-GDP releases PDEγ leading to a reformation of both the inactive cGMP-PDE$\alpha\beta\gamma_2$ complex and the inactive $G_{t\alpha\beta\gamma}$-GDP heterotrimer which is a primer for the reactivation by metarhodopsin II.

In summary, photobleaching of rod photoreceptor activates the visual cascade system leading to a decrease in cyclic GMP and the closure of GMP-gated channels in the ROS plasma membrane. Finally, retinal rods respond to light with a membrane hyperpolarization.

We are able to conclude with the words of George Wald: "I think that an outer segment represents the highest degree of isolation of function to be found in the whole of physiology. Its only business is excitation" [10].

References

[1] Cone, R.A. (1972) Rotational diffusion of rhodopsin in the visual receptor membrane. *Nature* 236:39-43.

[2] De Vries, H. (1956) Physical aspects of the sense organs. *Prog. Biophys. & Biophys. Chem.* 6:207-264.

[3] Hecht, S., S. Shlaer & M.H. Pirenne (1942) Energy, quanta and vision. *J. Gen. Physiol.* 25:819-840.

[4] Hubbard, R.J. (1958) The thermal stability of rhodopsin and opsin. *J. Gen. Physiol.* 42:259-280.

[5] Hubbard, R.J, D. Bownds & T. Yoshizawa (1965) The chemistry of visual photoreception, *Cold Spring Harbor Symp. Quant. Biol.* 30: 301-315.

[6] Schwartz, S, G. Santillan, G.K. Blasie & R. Dratz (1979) *VII Int. Congr. on Photobiology*, Roma, Symposia Abstr., p. 302.

[7] Stell, W.K. & J.-H. Kock (1984) Structure, development and visual acuity in the goldfish retina. In: *Molecular and Cellular Basis of Visual Acuity* (S.R. Hilfer & J.B. Sheffield, eds), New York: Springer, pp. 79-105.

[8] Van der Velden, H.A. (1944) Over het aantal lichtquanta dat nodig is voor een lichtprikkel bij het menselijk oog. *Physica* 11:179-189.

[9] Van der Velden, H.A. (1946) The number of quanta necessary for the perception of light of the human eye. *Ophtalmologica* 111:321-331.

[10] Wald, G. (1973) Visual pigments and photoreceptor physiology. In: *Biochemistry and Physiology of Visual Pigments* (H. Langer, ed.), Berlin: Springer, pp. 1-13.

[11] Worthington, C.R. (1974) Structure of photoreceptor membranes. *Ann. Rev. Biophys. Bioengng.* 3:53-81.

CALCIUM BUFFERING IN ROD OUTER SEGMENT OF BOVINE RETINA. EVIDENCE OF A Ca^{2+}ATPASE IN THE DISK MEMBRANES.

Isidoro M.Pepe[1], Isabella Panfoli[2], Carlo Cugnoli[3] and Alessandro Morelli[2]

[1] Istituto di Biofisica dell'Università, Facoltà di Medicina, salita sup della noce 35, 16131 Genova, Italy

[2] Istituto Policattedra di Chimica Biologica dell'Università, Genova, Italy

[3] Istituto di Cibernetica e Biofisica del CNR, Genova, Italy

INTRODUCTION

In the mechanism of vertebrate phototransduction, light closes the plasma membrane channels, which are kept open by the binding with cyclic GMP (cGMP), by activating a cGMP phosphodiesterase. In this way, the "dark current" which flows through the cGMP-gated channels is suppressed, giving rise to an hyperpolarizing signal on the plasma membrane. The recovery of the electrical response back to the basal membrane potential seems to be due to a free Ca^{2+} concentration decrease, which should activate cGMP synthesis for the reopening of the light-sensitive channels (1-7).

Two Ca^{2+} transport systems found in ROS plasma membrane regulate the free calcium concentration in the cytoplasm. One is the cGMP-gated cation channel through which Ca^{2+} ions enter the outer segment of the photoreceptor cell carrying about 10 - 15 % of the total current. Another is the $Na^{+}/Ca^{2+},K^{+}$ exchanger through which Ca^{2+} is extruded from the cell. The presence of a Ca^{2+}ATPase on ROS plasma membrane seems to be ruled out (8-10).

Cytosolic free Ca^{2+} concentrations in the dark was estimated to be about 0.2 μM by fluorescent indicators as fura 2 (11) or about 0.4 μM by aequorin (12).

After bright illumination which suppress the dark current and therefore the calcium influx, extrusion of Ca^{2+} by the Na/Ca,K exchanger occurs at a rate of about 2 μM/sec. The precise level to which calcium concentration can decrease inside the photoreceptor cell is still not known. The lower limit, using the equation of Blaustein and Hodgkin is of the order of 10^{-10}M. However, measurements with fura 2 gave a Ca^{2+} drop to about 0.14 μM after illumination (11).

The mechanisms of Ca^{2+} buffering inside the vertebrate photoreceptor cell are still not clear. The total calcium content of ROS is in the range of 1 - 2 mol/mol of rhodopsin . The overall rhodopsin concentration is about 3 mM (13). This gives a value for the overall calcium concentration in ROS from about 3 to 6 mM. It seems that two calcium buffering systems are present in the outer segment. One of high affinity (k_m of about 0.1 μM) and low capacity (about 200μM), which operates at normal physiological conditions with a ratio of bound to free calcium of about 70 to 1 (12) and which is probably made up of calcium-binding proteins such as arrestin an recoverin (7). The other one of low affinity (k_m between 5μM and 20μM) and high capacity which was found to operate in particular conditions with ROS injected with IBMX and no other ions but calcium in the external medium (12). In this case the ratio of bound to free calcium was 16 to 1. The location of this buffering system has been suggested in the disk membranes.

The calcium sequestered by the disk membranes was found by our group to be of about 2 - 3 mM and probably does not exchange within the time course of the light- response. Therefore, another Ca^{2+} transport system should be considered: the disk membranes that fill ROS seem able to accumulate Ca^{2+} against its gradient, as suggested by the findings of an ATP-dependent Ca^{2+} uptake in bovine ROS disk membranes (14-17) and of a Ca^{2+}-ATPase cytochemically localized inside the disks of toad ROS (18).

In this study we report the characterization of an ATP-dependent Ca^{2+} uptake in ROS disks of bovine retina, sensitive to vanadate, lanthanum and calmodulin. The presence of a Ca^{2+} pump is demonstrated by the isolation on polyacrylamide gel electrophoresis of a Ca^{2+}-activated phosphorylated intermediate located in the disk membranes.

Two types of Ca^{2+}ATPases, PMCA (Plasma Membrane Calcium ATPase) and SERCA (Sarcoplasmic and Endoplasmic Reticulum Calcium ATPase), have been described in mammalian tissues.

The PMCA pumps are calmodulin-binding protein of about 140 KDa that can be activated by digestion with the Ca^{2+}-dependent neutral protease calpain (19-21). The SERCA pumps have molecolar weight in the range of 97-110 KDa, are calmodulin-insensitive and do not undergo proteolysis by calpain. They were found to be inhibited by the tumor-promoting sesquiterpene lactone thapsigargin (22) . Thapsigargin, in turn does not inhibit PMCA pumps. Moreover, PMCA and SERCA pumps show a different trypsinolysis pattern (23, 24).

The results obtained on the 100 KDa intermediate of ROS disks suggest that this Ca^{2+}-pump belongs to the SERCA type. The presence of calcium pump in disk membranes is discussed.

METHODS

Disk Preparations

ROS were isolated from bovine retinae following the method of Schnetkamp and Daemen (25) by sucrose gradient centrifugation. In order to avoid proteolysis, leupeptin (70 μg/ml) was always present while a cocktail of PMSF, phenantroline, iodoacetate and pepstatine was present only in the fist preparations and then eliminated as found without effects. Osmotically intact disks were obtained after bursting ROS for 3 h in 5% Ficoll (Sigma) solution in distilled water containing 5 mM DTT and at ice temperature and purified by collecting them at the 5% Ficoll solution surface after centrifuging for 2 h at 25,000 rpm in a Beckman FW-27 rotor (26). In the experiments with calmodulin, calmodulin-depleted disk membranes were prepared as described by Bachs et al. (27).

Preparation of human red blood cell membranes

RBC membranes were prepared as already described (28). Before assay, membranes were resuspended in a solution containing 20 mM HEPES, 0.05 mM $CaCl_2$, 0.5 mM $MgCl_2$, 130 mM KCl and 2 mM DTT, to a protein concentration of 4-6 mg/ml and immediately used, or extensively washed and finally resuspended in 10 mM Tris, pH 7.5.

Measurements of calcium transport

Measurements of $^{45}CaCl_2$ uptake by disk vescicles were performed at pH 7.2 in a reaction mixture containing: 41 mM

HEPES pH 7.2, 120 mM KCl, 1.5 mM ATP, 1.5 mM $MgCl_2$, 10 μg/ml leupeptin, plus 1.38 μCi of $^{45}CaCl_2$ and cold $CaCl_2$ to a final free Ca^{2+} concentration of 50 μM (final specific activity 0.027 μCi/nmol). Controls contained no ATP or 1.5 mM 5'-adenylylimidodiphosphate (AMP-PNP). About 200 μg of ROS disk proteins were added to each 1.0 ml of reaction mixture to initiate the uptake reaction, then the sample was rapidly filtered at the time indicated through 0.45 μm pore size Millipore filters of mixed cellulose ester (HA type). The filters, presoaked with 4 ml of cold 41 mM Hepes pH 7.2 plus 120 mM KCl, were washed twice with cold 0.25M Sucrose and added to 10 ml of counting scintillant (Packard Pico-Fluor 40) and radioactivity was determined.

Measurements of calcium-stimulated ATP hydrolysis

The release of P_i from ATP was measured spectrophotometrically under the incubation conditions used for the determination of $^{45}CaCl_2$ transport. Samples containing 50 μg of protein (final volume 0.5 ml) were incubated at 37°C for 30 min and the reaction was terminated by adding 50 μl of 50% ice cold TCA. After centrifuging at 4°C for 10 min at 3000 rpm in a Beckman J-6B centrifuge, 400 μl of the supernatant were collected and 100 μl of 5N H_2SO_4 and 100 μl of 2.5% Ammonium Molibdate plus 10 μl of 2.5 mg/ml Fiske and SubbaRow reactive (29) were added, under agitation. Colour was allowed to develop for 10 min at 37°C and absorbance at 660 nm was measured. Free Ca^{2+} concentration less than 10^{-6}M was controlled by a Ca^{2+}-EGTA buffer (30).

Formation of the Ca^{2+}-dependent phosphorylated intermediate

RBC or disk membranes (100 μg), extensively washed with 10mM Tris-HCl pH 7.4, were incubated in a reaction mixture containing 20 mM TRIS-Maleate pH 7.5 and 0.1 mM $CaCl_2$. The reaction was started by the addition of 0.001 nmol of [γ-^{32}P] ATP (specific activity 3000 Ci/mmol) to 50 μl of reaction mixture and it was continued for 30 s at 4°C. The membranes were precipitated with ice-cold 6% trichloroacetic acid containing 1 mM ATP and 10 mM P_i and pellets were washed twice. For the samples to be incubated with hydroxylamine (or TRIS), membranes after the first TCA washing, were resuspended with 0.15 M NH_2OH (or TRIS, for control) at pH 6.6 and, at the end of the incubation for

20 min at room temperature, precipitated again with the same ice-cold trichloroacetic acid-ATP-P_i solution.

The washed precipitates were dissolved in the sample buffer containing 0.25 M Tris-HCl, pH 6.8, 10 mM EDTA, 60% sucrose, 40 mM DTT, 2% SDS and Bromophenol Blue and run onto an acidic 5-10% gradient polyacrylamide slab gel containing 0.1% of SDS at a constant current of 40 mA at 7 °C, according to Sarkadi et al. (31), with minor modifications.

Atomic absorption Spectrometer Measurements

Samples of freshly purified disks were collected by centrifugation at 40,000 rpm in a Beckman FW-27 rotor and dissolved by boiling in 1 % SDS. The total calcium content of disks was determined by flame atomic absorption (Varian) at a wavelength of 422.7 nm, with air/acetylene flame in a solution containing 10 % TCA, 0.1 % SDS and 10 mg/ml lanthanium (final concentration) to avoid interferences.

Protein content was determined by a method based on Coomassie blue staining (32) with BSA as a standard.

RESULTS

Calcium uptake in ROS disk membranes

Intact disk membranes were obtained by bursting rod outer segments in 5% Ficoll. With this method, disks are gently liberated from ROS and therefore, are assumed to have the natural inside-out configuration as *in vivo*. The total calcium concentration contained by the purified disk membranes was measured by atomic absorption spectrometry and resulted to be about 2.1 x 10^{-3}M, which is in the range of values found by several authors (5). This calcium concentration remained practically constant during the time course of the experiments.

Fig 1 shows that disks membranes are able to uptake Ca^{2+} against its gradient from a medium containing $5x10^{-5}$M calcium, when Mg^{2+}-ATP was added, with a maximal initial rate of approximately 1.6 nmol Ca^{2+}/min/mg protein. The uptake was considerably lower and almost saturated whithin the first 5 min in control experiments which did not contain exogenous ATP, or contained 1.5 mM AMP-PNP (5'-

adenylylimidodiphosphate, a non hydrolizable analog of ATP). GTP stimulated only about 25% of the uptake obtained in the presence of ATP.

As continuous irradiation or flashes of light in the visible range, which bleached various amount of disk rhodopsin, had no effect on the Ca^{2+} uptake, all the experiments were carried out in room light.

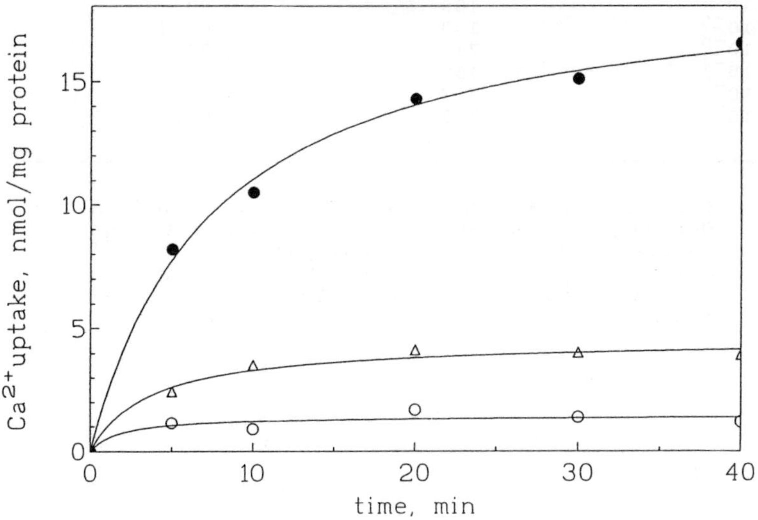

*Figure 1.**Calcium uptake by ROS disks.** Calcium transport was measured as described in Methods. External free Ca^{2+} concentration was $5x10^{-5}M$, pH was 7.3. $^{45}Ca^{2+}$ uptake was determined in the presence of 1.5 mM ATP (●) or 1.5 mM GTP (△) or in the absence of ATP (○)(or adding 1.5 mM AMP-PNP). Each point represents the mean of duplicate measurements. The difference between the two measurements was less than 5%.*

The ATP-dependent Ca^{2+} uptake was completely and rapidly abolished by the addition of the calcium-ionophore A23187 (data not shown). Vanadate or lanthanum, two well known inhibitors of ATPases with phosphorylated intermediates (27-29), inhibited the active calcium uptake. The data of Table 1 were fitted by competitive inhibition curves with K_m (La^{3+}) = 2.4 μM and $K_m(VO_4)^{3-}$ = 50 μM .

The dependence of Ca^{2+} uptake on the external free Ca^{2+} concentration is shown in Fig 2.

TABLE 1

Effect of Vanadate or $LaCl_3$ on the ATP-driven Ca^{2+} uptake by ROS disks.

Vanadate or $LaCl_3$ conc.(M)	+ Vanadate as % of control	+ $LaCl_3$ as % of control
10^{-7}	100	98
10^{-6}	100	69.8
10^{-5}	73	24.1
10^{-4}	43	7.6
10^{-3}	7.7	-
10^{-2}	2.5	0.1

Calcium uptake was measured as described in Methods; vanadate or $LaCl_3$ was added to the samples prior to the addition of $^{45}Ca^{2+}$ solution. External free Ca^{2+} concentration was 5x 10^{-5}M; pH was 7.3. Each point represents the mean of duplicate measurements.

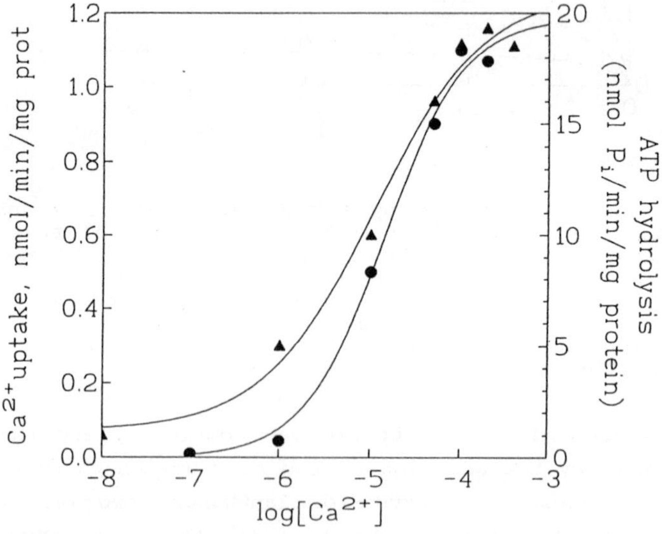

Figure 2. ***Ca^{2+}-dependence of calcium uptake and ATP hydrolysis in ROS disks.*** *Calcium transport was measured as described in Methods after 10 min of incubation. ATP hydrolysis was measured as described in Methods in the absence of Mg^{2+}. Each point represents the mean of duplicate measurements.*

The sigmoid curve fitting the experimental points gave a k_m (Ca^{2+}) of about 13 μM and a Hill coefficient of 1. An inhibition was observed over 200 μM external free Ca^{2+} (data not shown).

The ATP-dependent Ca^{2+} uptake measured in the absence of Mg^{2+} reached saturation and began to decrease after about 10 min of incubation, suggesting that Mg-ATP is the appropriate substrate for the calcium pump. ATP hydrolysis in the presence of Mg^{2+} reached relatively high values (about 60 nmol of P_i/min/mg protein) and was insensitive to changes in calcium ion concentrations, while in the absence of Mg^{2+}, ATP hydrolysis (measured after 10 min incubation) was function of Ca^{2+} ion concentration (Fig 2), reaching half maximal activity at about 10 μM [Ca^{2+}].

The addition of calmodulin to calmodulin-depleted disk membranes enhanced the ATP-dependent Ca^{2+} only up to 40% at 50 μM [Ca^{2+}] as reported in Table 2.

TABLE 2

Effect of calmodulin on ATP-driven Ca^{2+}uptake by calmodulin-depleted disk membranes.

Ca^{2+} conc. (μM)	Control	+Calmodulin
1	0.08	0.14
10	0.48	0.68
50	1.02	1.64
100	1.24	1.81

Calcium uptake was measured as described in Methods in the absence or in the presence of 20 μg/ml calmodulin. Each point represents the mean of duplicate measurements.

A corresponding decrease of k_m(Ca^{2+}) from 34 to 18 μM in the absence or in the presence of calmodulin respectively, was obtained from the sigmoid curves fitting the experimental points.

Isolation of the Ca^{2+}-ATPase phosphoenzyme-intermediate. Effect of thapsigargin, trypsin and calpain.

In order to detect the formation of Ca^{2+}-dependent aspartyl phosphate intermediate (E_1-P), samples of ROS disks and human RBC membranes were labelled with [$-^{32}P$]ATP and run onto an acidic SDS-PAGE. The pH of this gel is low enough (about 6.5) to preserve the phosphoenzymes which would be rapidly hydrolyzed at the high pH of the Laemmli gel system. Fig 3 shows the autoradiographic analysis.

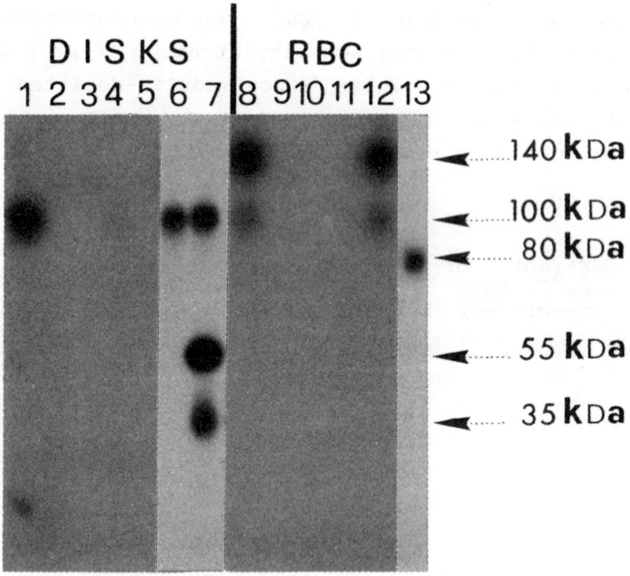

Figure 3. ***Effect of thapsigargin and trypsin on Ca^{2+}induced, phosphorylated intermediate in ROS disks.*** *Autoradiography after acidic SDS-PAGE on slab gel (see Methods for technical details). ROS disks or RBC membranes were incubated with radioactive ATP: in the presence of 50 μM $CaCl_2$ (lane 1,8); resuspended in NH_2OH (100 mM) pH 6.6 after incubation with labelled ATP and 50 μM $CaCl_2$(lane 2,9); in the presence of 0.5 mM EGTA (lane 3,10); in the presence of 5 mM NaCl (lane 4,11); in the presence of 2μM thapsigargin and 50 μM $CaCl_2$(lane 5,12).*
ROS disk and RBC membranes after incubation with radioactive ATP and 50 μM $CaCl_2$: before (lanes 6 and 8) and after 15 min trypsin digestion at 4°C (lanes 7 and 13).

In the presence of 50 μM [Ca^{2+}], a phosphorylated peptide of 100 kDa molecular weight was formed in disk membranes (lane 1) and the well known E_1- P intermediate of human RBC PMCA-pump appeared at 140 kDa (lane 8) (23,33,34). The formation of both phosphopeptides was prevented by the chelation of Ca^{2+} by EGTA (lanes 3, 10), and was decomposed by resuspension with hydroxylamine (lanes 2,9), indicating the acylphosphate nature of the binding (35). NaCl failed to induce any phosphorylated intermediate (lane 4,11).

The effect of thapsigargin, present in a constant ratio of 0.4 μmol/μgr total protein, is shown in lanes 5 and 12. While thapsigargin was ineffective on RBC ATPase (lane 12), as already reported (36), it had an inhibitory effect on the phosphorylated intermediate of disks (lane 5).

Mild proteolysis of ROS disks by trypsin produced two autophosphorylatable fragments of 55 kDa and of 35 kDa (lanes 7), as for the trypsin-digested SERCA-pumps (22). Instead, 15 min trypsin digestion of RBC membranes produced a phosphorylated fragment of 80 kDa (lane 13) as already reported (23,24). On the contrary, 20 min digestion with calpain was not effective on the phosphorylated 100 kDa intermediate of disk membranes (data not shown), while gave two autophosphorylatable bands of about 124 kDa and 80 kDa on on the RBC plasma membrane Ca^{2+}-ATPase, as already reported (33).

Rhodopsin, which was clearly visible in the Coomassie stained band at about 36,000 kDa, was not phosphorylated as its phosphorylation needs a kinase, which was presumably lost during extensive washing of the disk membranes before incubation with radioactive ATP (see Methods).

In order to measure the amount of Ca-ATPase present on disk membranes, different quantities of phosphorylated samples of disks or RBC membranes were run on SDSPAGE. The phosphorylated intermediate bands localized by means of autoradiography were cut off the gel and eluted with a scintillation cocktail. The resulting radioactivity was linear with the protein concentration of the samples. The amount of Ca-ATPase in ROS disks was calculated by comparison with RBC, by assuming the amount of Ca-ATPase of RBC membranes to be 0.1 % of the total membrane protein (37). The result for disks membranes was very similar, that is 0.09 % + 0.001 % of the total disk membrane protein.

DISCUSSION

As the separation of the ROS disks from the disrupted plasma membrane in 5% Ficoll is based upon the different density existing between sealed vescicles and other membranous material, the possibility that our disk preparation could be contaminated by pieces of outer segment plasma membrane that resealed after the osmotic shock in water, cannot be ruled out. However, several experiments seem to exclude the presence of a Ca^{2+}-ATPase on the ROS plasma membrane (9). Contamination from resealed pieces of inner segment membranes seems instead to be excluded, as suggested by the absence of the Na^{+}/K^{+}-ATPase band on gel electrophoresis (see Fig 3). Mitochondrial contamination seems absent in our preparations as no activity was found of the mitochondrial marker succinate dehydrogenase (data not shown). However, it is well-known that only a very small amount of total mitochondrial Ca^{2+} is in an exchangeable dynamic steady-state with the cytosol and that Ca^{2+} transport in mitochondria is not directely due to a ATPase activity but to an electrophoretic transport driven by the internal negative potential (38). Moreover, mitochondrial ATPase does not form covalent phosphorylated intermediates (39).

The results presented here would demonstrate the presence of a Ca^{2+}-ATPase pump acting on the disk membranes of bovine retinal rod outer segments. The enzyme would be of the P-type ATPases (40), as it forms a phosphoprotein intermediate similar to that formed by Ca^{2+}-ATPases during the reaction cycle in which Ca^{2+} ions are pumped through the membrane. The amount of phosphorylated protein resulted to be about 0.09 % of the total disks protein. This value helps to exclude contamination from other retinal tissues, being very similar to the amount of Ca^{2+}-ATPase present in RBC membranes.

It seems that only ATP and Ca^{2+} ions are required for

the formation of the intermediate E-P complex; the presence of Mg^{2+} ions decreased radioactive recovery (data not shown), possibly by accelerating the hydrolysis of the bound phosphate i.e. by favouring the conversion of the enzyme to its non-phosphorylated form.

It is well known that vanadate interferes with the phosphorylation process of P-type ATPases by inhibiting the binding of phosphate with the aspartyl residue of the enzyme, while lanthanum inhibits the hydrolysis of the phosphorylated intermediate (37). In both cases, the reaction cycle would be blocked and calcium uptake inhibited as it is shown to occur in Table 1, suggesting that the ATP-dependent Ca^{2+} uptake measured in disk membranes is due to activity of the Ca^{2+} ATPase.

Active calcium transport by disks shown in Fig 1 is ATP-dependent; its value of about 1.6 nmoles Ca^{2+}/min/mg protein is similar to that found for 5% Ficoll disk preparation by other authors (17). The effect of calmodulin on ATP-driven Ca^{2+}uptake was not striking when compared to that on other preparations. Calcium uptake by disks in the presence of calmodulin reached infact a maximal increase of 40% (Table 2) accompanied with a decrease of $k_m(Ca^{2+})$ of only two fold (from 34 μM to 18 μM), while an increase of calcium uptake of about ten times due to calmodulin with a concomitant decrease of k_m of more than 30 times (from 30 μM to below 1 μM) was found in erythrocyte membrane preparations (37,41-43).

The classification of ROS disk calcium ATPase among those reported in the literature poses some problems. The most part of plasma membrane calcium pumps is sensitive to calmodulin and has a phosphorylated intermediate of molecular weight of about 130 - 140 KDa with the exception of those from liver and neutrophyls, which are insensitive to calmodulin (27,37,44) and have a 70-105 kDa phosphorylated intermediate. On the contrary, the most part of the the so-called SERCA pumps was found to be calmodulin-insensitive and to have molecular weight between 97 and 110 kDa (45,46). Moreover, the 100 KDa band of ROS disks, like the phosphorylated intermediate of Ca^{2+}ATPase of the sarcoplasmic reticulum and unlike that of the plasma membrane (47), was not enhanced by addition of lanthanum (from 40 up to 100 μM upon a wide range of calcium concentrations: data not shown). Most of SERCA pumps were found to be inhibited by thapsigargin (22,36). In our experimental conditions, thapsigargin completely inhibited the formation of the phosphorylated intermediate of ROS disk ATPase and had no effect on RBC pump (see Fig 3).

The Ca^{2+}-dependent neutral protease calpain is known to act selectively on the PMCA pumps by removing the C-terminal fragment and thus producing a calmodulin-like activation of the pump (19-21). The ROS disk Ca^{2+}-ATPase was not susceptible of proteolysis by calpain, as would be expected for a SERCA-pump.

Trypsin digestion pattern on SDS-PAGE has been used as a kind of "fingerprinting" to distinguish between SERCA and PMCA types of pumps (23, 24). In fact, the limited-digestion pattern of PMCA pump yields only an autophosphorylatable fragment of 80 kDa, while that of SERCA pumps produces two autophosphorylatable fragments of about 55 and 35 kDa. As shown in Fig 3, the ROS disk Ca^{2+}-ATPase displays a tryptic digestion pattern typical of SERCA pumps.

In conclusion, the molecular weight of about 100kDa, the low sensitivity to calmodulin, the lack of digestion by calpain, the tryptic fragment pattern and the response to inhibitors such as thapsygargin, suggest that the ROS disk Ca^{2+}-ATPase should be classified among the SERCA type of pumps and not among the PMCA, despite the well known fact that disk membranes are originated from the plasma membrane. Actually, the separation of disks from the plasma membrane is accompanied by their progressively different composition in lipid and protein, which has been suggested to be an evolutionary specialization to attain higher sensitivity (48).

The presence of a Ca^{2+}-ATPase in ROS disks, sufficiently documented by our results and by those from other authors (17,18) poses interesting questions about its physiological role.

The low affinity of the disk pump for Ca^{2+} (K_m of Ca^{2+} uptake is about 13 μM) is consistent with the finding of a low-affinity, high-capacity calcium buffering mechanism observed above 5μM [Ca^{2+}] in the outer segment cytoplasm of vertebrate retina (12) . Ca^{2+} concentration would be kept low in photoreceptor cells not only by the action of the high-affinity low-capacity Na^+:Ca^{2+}exchanger, but also by the disk Ca^{2+}-ATPase that would pump calcium ions into the disks whenever Ca^{2+} concentration would raise over the micromolar range.

Another interesting hypothesis is that the disk Ca^{2+}ATPase would act as an ATP synthesizer which utilizes the concentration gradient of calcium ions (millimolar inside disks and submicromolar in cytoplasm) during normal functioning of the receptor cell. At present, a site of ATP synthesis in the rod cell seems the inner segment which is full of mitocondria, so ATP could reach the outer segment by diffusion through the cilium. As the phototransduction events take place in the outer segment and need fast supply of energy, it is not unlikely that ATP could be synthesized *in situ*, on the disk, at the expense of calcium gradient.

REFERENCES

1 Lamb, T.D. (1986) Transduction in vertebrate photoreceptors: the role of cyclic GMP and calcium. *Trends Neurosci. 9*, 224-228.

2 Stryer, L. (1986) Cyclic GMP cascade of vision. *Ann. Rev. Neurosci. 9*, 87-119.

3 McNaughton, P.A., & Cervetto, L. (1986) The role of calcium concentration in the light response. *Photobiochem. Photobiophys. 13*, 399-414.

4 Chabre, M. & Deterre, P. (1989) Molecular mechanism of visual transduction. *Eur. J. Biochem. 179*, 255-266.

5 McNaughton,P.A. (1990) Light response of vertebrate photoreceptors.*Physiol. Rev. 70*,847-883.

6 Stryer, L. (1991) Visual excitation and recovery. *J. Biol. Chem. 266(17)*, 10711-10714.

7 Kaupp, U.B. and Koch, K.W (1992) Role of cGMP and Ca^{2+} in vertebrate photoreceptor excitation adaptation. *Annu. Rev. Physiol. 54*, 157-175.

8 Yau, K.W. & Nakatani, K. (1985) Light-induced reduction of cytoplasmic free calcium in retinal rod outer segment. *Nature 313*, 579-581.

9 McNaughton, P.A., Cervetto, L. & Nunn, B.J. (1986) Measurements of the intracellular free calcium concentration in salamander rods. *Nature 322*, 261-263.

10 Cervetto, L., Lagnado, L., Perry, R.J., Robinson, D.W. & McNaughton, P.A. (1989) Extrusion of calcium from ROS is driven by both Sodium and Potassium gradient. *Nature 337*, 740-743.

11 Ratto, G.M., Payne, R., Owen, W.G. and Tsien, R.Y. (1988). The concentration of cytosolic free calcium in vertebrate rod outer segment. *J. Neurosci. 8,* 3240-3246.

12 Lagnado, L., Cervetto, L. & McNaughton, P.A.(1992) Calcium homeostasis in the outer segments of retinal rods from the tiger salamander. *J.Physiol. 455*,11-142.

13 Daemen, F.G.M. (1973) Vertebrate rod outer segment membranes. *Biochim. Biophys. Acta 300*, 255-288.

14 Caretta, A. & Cavaggioni, A.J. (1976) On the metabolism of the rod outer segments *J. Physiol. 257*, 687-697.

15 Bownds, D., Gordon-Walker, A., Gaide-Hugueinin, A.C. & Robinson, W. (1971) Characterization and Analysis of Frog Photoreceptor Membranes. *J. Gen. Physiol. 58*, 225-237.

16 Sack, R.A. & Harris, C.M. (1977) Ca^{2+}-dependent ATPase activity of bovine receptor cell outer sement. *Nature 265*, 465-466.

17 Puckett, K.L., Aronson, E.T. & Goldin, S.M. (1985) ATP-dependent uptake activity associated with a disk membrane fraction isolated from bovine retinal rod outer segments. *Biochemistry 24*, 390-400.

18 Davis, W.L., Hagler, H.K., Farmer, G.R., Martin, J.H. & Bridges, G. (1988) Electron microscopic cytochemical localization of Ca ATPase in the rod outer segments of the toad Bufo Marinus. *Anat. Rec. 221 (3)*, 761-768.

19 Carafoli, E. (1991) Physiol. Rev. 71, 129-153.

20 Au,K.S.(1987) Biochim. Biophys. Acta 905, 273-278.

21 Wang, K.W.W., Villalobo, A. & Roufogalis, B.D. (1988) Arch. Biochem. Biophys. 260, 696-704.

22 Thastrup, O., Cullen, P.J., Drobak, B.K., Hanley, M.R. and Dawson, A.P. (1990) Proc.Natl.Acad.Sci. USA 87, 2466-2470.

23 Sarkadi, B., Eynedi, A., Foldes-Papp, Z. and Gardos, G. (1986) J.Biol.Chem. 261, 9552-9557.

24 Eynedi, A., Sarkadi, B., Foldes-Papp, Z., Monostory, S. and Gardos, G. (1986) J.Biol.Chem. 261, 9558-9563.

25 Schnetkamp, P.P. & Daemen, F.J. (1982) Isolation and characterization of osmotically sealed bovine rod outer segments. *Methods Enzymol. 81*, 110-116.

26 Smith, H.G. Jr, Stubbs, G.W. & Litman, B.J. (1985) The isolation and purification of osmotically intact discs from retinal rod outer segment. *Exp. Eye Res. 20*, 211-217.

27 Bachs, O., Famulski, K.S., Mirabelli, F.& Carafoli, E. (1985) ATP-dependent Ca^{2+} transport in vesicles isolated from the bile canalicular region of the hepatocyte plasma membrane. *Eur. J. Biochem. 147*, 1-7.

28 James, P., Vorherr, T., Krebs, J., Morelli, A., Castello, G., McCormick, D.J., Penniston, J.T., De Flora, A. & Carafoli, E. (1989) Modulation of Erythrocyte Ca^{2}+ATPase by selective Calpain Cleavage of the Calmodulin-binding Domain. *J. Biol. Chem. 264*, 8289-8296.

29 Fiske, C.H. and SubbaRow, Y. (1925) *J. Biol. Chem. 66*,

30 Fabiato, A. & Fabiato, F. (1979) Calculator programms for computing the compositions of the solutions containing multiple metals and ligands used for experiments in skinned muscle cells. *J. Physiol. (Paris) 75*, 463-505.

31 Sarkadi, B. Enyedi, A., Føldes-Papp, Z. & Gárdos, G.

(1986) Molecular Characterization of the in situ red cell membrane calcium pump by limited proteolysis. *J. Biol. Chem. 261*, 9552-9557.
32 Esen, A. (1978) A quantitative, semiquantitative and qualitative assay of protein. *Anal. Biochem. 89*, 264-273.
33 James, P., Vorherr, T., Krebs, J., Morelli, A., Castello, G., McCormick, D.J., Penniston, J.T., De Flora, A. & Carafoli, E. (1989) J.Biol. Chem. 264, 8289-8296.
34 Pedersen, P.L. & Carafoli, E.(1987) Trends. Biochem. Sci.12, 146-149.
35 Lichtner, R. & Wolf, H.U. (1980) Biochim. Biophys. Acta 598, 486-493.
36 Lytton, J., Westlin, M,. and Hanley, M.R., (1991) J.Biol.Chem. 266 17067-17071
37 Carafoli,E. (1991) Calcium pump of the plasma membrane. *Physiol. Rev. 71*, 129-153.
38 Carafoli, E. (1987) Intracellular calcium homeostasis. *Ann. Rev. Biochem. 56*, 395-433.
39 Grubmeyer, C., Cross, R.L. & Penefsky, H.S. (1982) Mechanism of ATP hydrolysis by beef heart mitochondrial ATPase. Rate constants for elementary steps in catalysis at a single site.*J. Biol. Chem. 257*, 12092-12100.
40 Pedersen, P.L. & Carafoli, E.(1987) Ion motive ATPases. I. Ubiquity, properties, and significance to cell function. *Trends. Biochem. Sci. 12*, 146-149.
41 Larsen,F.L.& Vincenzi,F.F. (1979) Calcium transport across the plasma membrane: stimulation by calmodulin. *Science 204*,306-308.
42 Muallem,S. and Karlish,S.J.D.(1981) Studies on the mechanism of regulation of the red cell Ca^{2+} pump by calmodulin and ATP. *Biochim.Biophys.Acta 647*,73-86.
43 Jeffery,D.,Roufogalis,B.D. and Katz,S.(1981) The effect of calmodulin on the phosphoprotein intermediate of Mg^{2+} dependent,Ca^{2+} stimulated ATPase in human erythrocyte membranes.*Biochem.J. 194*,481-486.
44 Ochs, D.L. & Reed, P.W. (1984) Ca^{2+}-stimulated, Mg^{2+}-dependent ATPase Activity in Neutrophil Plasma Membrane Vesicles. *J. Biol. Chem. 259*, 102-106.
45 Papp,B., Enyedi, A., Pászty,K., Kovács,T.,Sarkadi,B., Gárdos,G., Magnier,C.,Wuytack,F. & Enouf,J. (1992) Simultaneous presence of two distinct endoplasmic reticulum type calcium-pump isoforms in human cells. *Biochem.J. 288*, 297-302.
46 Lytton,J.,Westlin,M.,Burk,S.E.,Shull,G.E.& MacLennan, D.H. (1992) Functional comparisons between isoforms of the sarcoplasmic or endoplasmic reticulum family of calcium pumps. *J.Biol.Chem. 267*, 14483-14489.
47 Eynedi, A., Sarkadi, B., Foldes-Papp, Z., Monostory, S. and Gardos, G. (1986) Demonstration of two distinct calcium pump in human platelet membrane vescicles.*J.Biol.Chem. 261*, 9558-9563.
48 Boesze-Battaglia, K. and Albert, A.D. (1990) *J.Biol.Chem. 265,* 20727-20730.

cGMP GATED CHANNEL, EXCHANGER AND PHOTOTRANSDUCTION IN ISOLATED ROD OUTER SEGMENTS

G. Rispoli*+, A. Navangione* and V. Vellani*

+University of Washington School of Medicine, Department of Physiology and Biophysics, Seattle, Washington 98195 USA and *University of Ferrara, Institute of General Physiology, Ferrara, ITALY 44100. Mailing address: Ferrara

ABSTRACT

The experiments described in this paper show that the isolated rod outer segment recorded in whole-cell is an ideal preparation to study in situ the guanosine 3':5'-cyclic monophosphate (cGMP) gated channels, the enzymes involved in phototransduction and the Na^+-Ca^{2+},K^+ exchanger. In particular, the role on phototransduction of intracellular Ca^{2+}, Ca^{2+}_i, and the existence of possible modulation of Ca^{2+} movements across the plasma membrane were analyzed. It was found that the reduction of $[Ca^{2+}]_i$ stimulates the GC without affecting the basal PDE activity: this regulation plays a key role in shaping the photoresponse kinetics and the low frequency noise of dark current. The Ca^{2+}_i homeostasis is entirely controlled by the exchangers, which Ca^{2+} extrusion rate was found to be driven exclusively by the electrochemical gradients of the cotransported ions, and by the Ca^{2+} entry through the cGMP gated channels, which were found to be modulated by nucleotide triphosphates.

INTRODUCTION

The electrical signal that initiates the visual process in vertebrates is generated by the modulation of a light sensitive circulating current (called dark current) that flows into the outer segment (OS) and out of the inner segment of the retinal photoreceptors. The current entering the OS depolarizes the cell because it is carried mostly by Na^+ and Ca^{2+} through plasma membrane channels opened by cGMP. The ionic gradients are maintained by an electrogenic Na^+-Ca^{2+},K^+ exchanger in the OS[1,2] and by a Na^+-K^+ pump in the inner segment. The intracellular cGMP concentration, $[cGMP]_i$, is set in the dark by the balance between a constant hydrolysis operated by a dark active Phosphodiesterase (PDE) and a constant synthesis operated by a Guanylyl Cyclase[3] (GC). Light causes the activation of a G-protein coupled enzymatic cascade which ends with the increase in the basal activity of PDE, causing a fall in $[cGMP]_i$, the closure of the cGMP gated channels, the reduction of the cationic inward current and the hyperpolarization of the cell. The hyperpolarization reduces the rate of transmitter release at the synaptic terminal and that is the message that is relayed to the second order retinal neurons. The closure of the cGMP gated channels decreases Ca^{2+} influx, while the exchangers continue the Ca^{2+} extrusion, thus $[Ca^{2+}]_i$ falls during light stimulation. This $[Ca^{2+}]_i$ fall triggers several mechanisms taking part in dark state recovery and light adaptation of photoreceptor,[4,5,6,7,8,9,10] one of which is GC stimulation. The enzymatic cascade, the cGMP gated channels and the Na^+-Ca^{2+},K^+ exchangers are entirely compartmentalized in the OS: this particular architecture allowed the development of a new preparation to study photoreceptor physiology, consisting in the whole-cell recording from OS detached from the rest of the retina[11] (Fig. 1). This method isolates the OS from the biochemical influence of the inner segment, and it allows the control through the patch pipette of the membrane potential and of the intracellular milieu. The electrical recordings are greatly simplified since (depending upon the pipette solution) it is possible the selective recording of the cGMP-activated current and/or the exchange current, with the transduction cascade enabled to function or not. The intracellular perfusion of the OS with a minimal internal solution containing millimolar amounts of adenosine 5'-triphosphate (ATP) and guanosine 5'-triphosphate (GTP) supports normal phototransduction (Fig. 3), showing that the enzymatic cascade, the cGMP gated channels and the exchangers are functionally intact. The whole-cell patch pipette then functions as an artificial inner segment, because it provides and maintains the driving force for the permeating ions and supplies the nucleotides needed for the development of a sustained inward dark current and its light sensitivity. The intracellular nucleotides level can be entirely controlled by the patch pipette (since the mitochondria are not present in the OS), allowing also the study of the GC and its Ca^{2+} regulation in situ. The importance of Ca^{2+}_i on the light response of photoreceptor raises also the possibility that its homeostasis does not rely entirely on the number of channels open and on the ionic gradients that drive exchanger activity, but rather that the Ca^{2+} movements across the plasma membrane are indeed regulated by additional mechanism/s. In fact, almost in all cellular systems are found channels which conductance is modulated through several pathways.[12] One of the most common regulation pathway of Ca^{2+} permeation occurs through a phosphorylation,[13] and more and more channels are found to be modulated by a cGMP dependent kinase.[14,15,16] On the other side, the Na^+-Ca^{2+} exchanger of several tissues is ATP regulated, as the cardiac one.[17] The OS preparation allows the control (through the patch pipette) over the intracellular nucleotides level, as well as the selective recording of the cGMP-activated current or the exchange current. Thus, the effect of nucleoside triphosphates on the Ca^{2+} permeability of the light sensitive channels and on the

Ca^{2+} extrusion by the exchanger can be conveniently studied with this preparation. Brief accounts of some of this work have appeared previously.[5,7,18,19]

METHODS

Animals and Preparation. Whole-cell recordings were obtained from mechanically isolated OS from rods of the nocturnal lizards *Gekko gecko* as described elsewhere.[11] All manipulations were done under infrared illumination, using image converters. The retina was "peeled" from an eyecup piece and was gently stretched in ~200 µl of Ringer to obtain the OS. The fluid drop containing the OS was transferred to the recording chamber and held by surface tension between two cover slips. The chamber, that was open on three sides, was mounted on a modified inverted microscope and the OS was viewed with an infrared-sensitive TV camera (illumination wavelength >850 nm).

Patch Pipettes, apparatus and recording procedure. Patch pipettes were prepared from 100 µl Drummond micropipette glass in the conventional manner[20] and fire polished to a pipette resistance of 7-12 MΩ. The OS was stimulated with uniform unpolarized green light (520 nm) having a measured unattenuated intensity of $1.7 \cdot 10^5$ photons/($mm^2 \cdot sec$). Neutral density filters were used to reduce the intensity of the unattenuated light. Where indicated, the light stimuli were generated by an ultrabright LED diode (555 nm) driven by a computer controlled current-to-voltage converter. The original routines used to generate light stimuli were written in Basic. The LED was coupled to the fluorescence lamp port of the inverted microscope; the light emerging from LED passed through a neutral density filter (logarithmic attenuation factor: 2.0), a diffusion filter (45WF, Olympus), a 10x objective, a dicroic mirror (35267, Olympus, 550 nm) housed in the exciter filter insertion slot and finally light was focused on the OS through the objective. The LED was positioned so that the spot of light had the desired intensity when the OS was viewed in sharp focus by the TV camera. The OS current was recorded with an Axopatch 1D amplifier (Axon Instruments, Burlingame, CA, USA) or a List EPC-7 patch clamp (Medical System Corp., Great Neck, NY). A fire polished patch pipette was pressed against the surface of an OS to obtain a gigaseal (10-40 GΩ). Breakthrough from cell-attached to whole-cell recording was promoted by suction impulses delivered by mouth. The access resistance ranged from 25 to 60 MΩ; the pipette voltage was clamped at -20 mV.

Solutions and perfusion. The composition of the normal Ringer was: 160 mM NaCl, 3.3 mM KCl, 1 mM $CaCl_2$, 1.7 mM $MgCl_2$, 1.7 mM SO_4, 2.8 mM Hepes and 10 mM Dextrose. The 1 mM Ca^{2+} was replaced with 10 mM Ca^{2+} in the high Ca^{2+} Ringer and with 2 mM EGTA (ethylene glycol-bis-(β-aminoethyl ether)- *N,N,N',N'*-tetraacetic acid) in the 0 Ca^{2+} Ringer.

Isotonic solutions contained 166 mM of the monovalent cation indicated in the text and 2 mM EGTA, while Isotonic Ca^{2+} contained 119 mM Ca^{2+} only. Pipettes were filled with one of the "intracellular" solutions indicated in the text or with one of the solutions reported in Table 1. The nucleotides used were: guanosine 3':5'-cyclic monophosphate, sodium salt; guanosine 5'-triphosphate, sodium salt; adenosine 5'-triphosphate, dipotassium salt. All solutions were buffered to pH≈7.4 with 5 mM N-2-hydroxyethylpiperazine-*N'*-2-ethanesulphonic acid (Hepes) and NaOH, KOH or tetraethylammonium hydroxide (TEA-OH) depending upon whether the solution contained high Na^+, high K^+ or 0 Na^+ and 0 K^+, respectively. Osmolality of all solution was adjusted with sucrose to ~320 mOs/Kg when needed. All chemicals were obtained from Sigma Chemical Co. (St. Louis, MO), while BAPTA (1,2-bis(*O*-aminophenoxy)ethane-*N,N,N',N'*-tetraacetic acid) tetrapotassium salt was from Molecular Probes (Eugene, OR) and from Sigma. The external solution was changed rapidly (≤50 msec) during whole-cell recording by moving laterally a multibarrelled perfusion pipette in front of the OS.

REAGENT	*I*	*II*	*III*	*IV*	*V*
ATP	0	0	5	5	5
GTP	0	0	1	1	1
cGMP	0	0.05	0	0	0
BAPTA	0	10	10	0	0
EDTA	0	1	1	0	0
K^+	139	139	139	139	139
Na^+	10	10	10	10	10
Mg^{2+}	0.5	6.61	6.61	1.5	6.05
Ca^{2+}	0	0	0	20	0
Aspartate	122	69	69	118	113
Cl^-	25	25	25	53	24

Table 1. Composition of dialysis solutions (mM).

RESULTS

Fig. 1 and 2 illustrates a family of recordings from OS using different intracellular and extracellular media. The traces selected represent the typical behavior of the OS in each of the conditions indicated (see figure legend and Table 1); the breakthrough from cell attached to whole-cell recording occurs at the arrow (zero time). No stable dark current was recorded upon dialyzing the OS with a standard internal solution containing 0 Ca^{2+} and 0 nucleotides (solution I of Table 1; Fig. 1, trace *a*). Under these conditions, the input resistance measured in whole cell configuration was about as large as the seal resistance measured in cell attached, which was up to 40 GΩ. The incorporation of cGMP in the dialysis solution (solution II of Table 1) caused the development of an inward current which size at the steady state increased with [cGMP]$_i$ (Fig. 1, trace *b*). This current became light sensitive if GTP was added to the dialysis solution (solution III of Table 1; Fig. 1, trace *c*). The main OS current source - bathed in a solution containing Na^+, Na^+_o, and dialyzed with millimolar amounts of Ca^{2+} and K^+ - was the exchanger, which electrogenic activity gave an inward current (Fig. 2, traces *a* and *b*). This current was affected by the gradients of Na^+, Ca^{2+} and K^+ as expected for a Na^+-Ca^{2+} exchanger which cotransports K^+ also. In the forward mode of operation (Ca^{2+} and K^+ extruded and Na^+

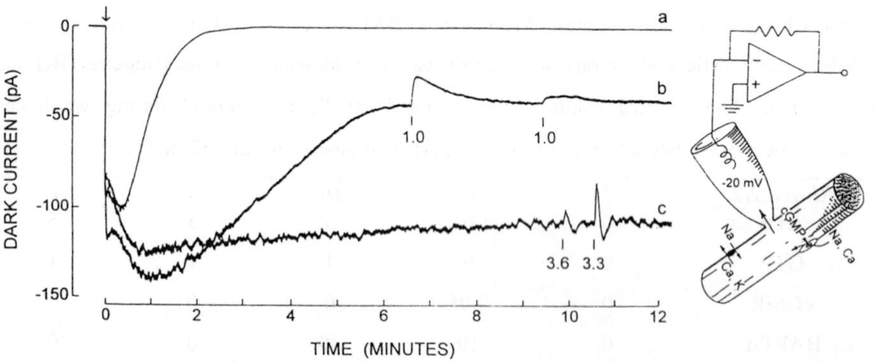

Fig. 1. Left panel: chart record of the cGMP-activated current under various internal nucleotides conditions. The intensity of the light flash (delivered at the vertical bar) is reported as the logarithmic attenuation factor of the neutral density filters used. Right panel: cartoon of the recording of a detached rod outer segment recorded in whole cell.

imported) the current was inward, and attained maximum amplitude using Isotonic Na^+_o, 30 mM Ca^{2+}_i and 124 mM K^+_i solutions (Fig. 2, trace *a*). Adding 1 mM Ca^{2+}_o reduced the steady-state exchange current of about one half (Fig. 2, trace *b*), giving and apparent dissociation constant for Ca^{2+}_o of about 1 mM. In the absence of K^+_i the exchange current was about zero (trace *c*), showing that K^+_i was required for forward mode operation. Furthermore, the exchange current was detected only when Ca^{2+} and K^+ were present on the same side of the membrane and Na^+ was present on the opposite side. This conclusion was based on a series of experiments where several combinations of intracellular and extracellular solutions satisfying this requirement were tested. Recording *d* in Fig. 2 was obtained using the same solutions as in *a*, but placed on the opposite sides of the membrane. The current was outward (reverse mode, Ca^{2+} and K^+ imported and Na^+ extruded), showing that Ca^{2+}_i is not required for reverse exchange functioning. The intracellular applications of ATP and/or GTP did not change the exchange current level (solution IV of Table 1), indicating that the exchange process is entirely

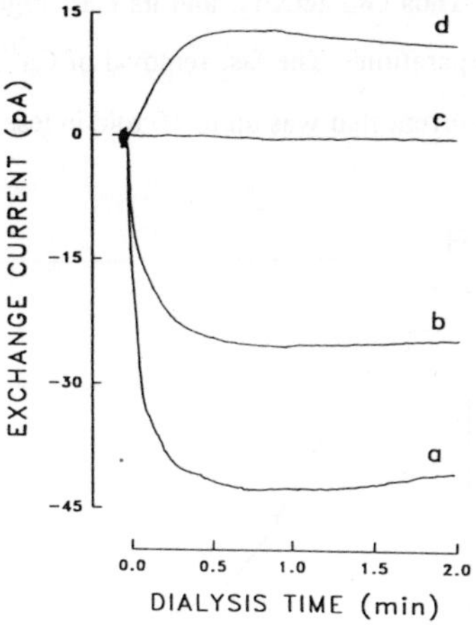

Fig. 2. Chart record of exchange currents under various internal and external ionic conditions. Forward exchange currents were obtained in the presence of K^+_i, Ca^{2+}_i and Na^+_o, while the reverse mode ones in the presence of K^+_o, Ca^{2+}_o and Na^+_i. Traces (mean current ± SEM at the steady-state, n = number of OS averaged): *a*, - 50.6 ± 2.7 pA, n = 13; *b*, - 24.1 ± 7.5 pA, n = 23; *c*, 0 exchange current, n = 48 in total (input resistance ≈ 5 - 40 GΩ), the trace reported is from an OS recorded with Isotonic Ca^{2+}_o and Isotonic Na^+_i; *d*, 5.9 ± 0.3 pA, n = 14.

regulated by ionic gradients, and not by intracellular factors. This view is also supported by experiments showing that the exchanger functionality in OS and excised patches outside-out are similar (data not shown). The electrical recordings from OS dialyzed with a standard

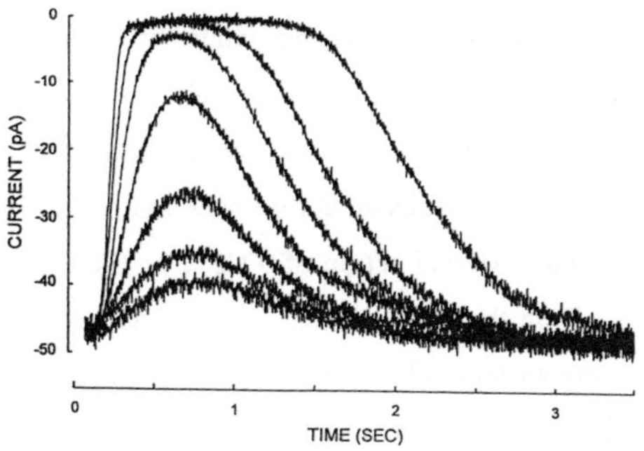

Fig. 3. Response family recorded from OS dialyzed with 5 mM ATP and 1 mM GTP. Flash intensities (delivered by the ultrabright LED described in the method section) were as follows: 0.2, 0.4, 0.9, 1.8, 3.5, 7, 200 photons·μm^{-2}.

internal solution containing ATP and GTP only (solution V of Table 1) were similar to the ones obtained from intact photoreceptors by having similar dark current level, light sensitivity, stimulus-response relationship, response kinetics (Fig. 3), adaptation behavior and light-sensitive noise characteristics. Since the channel responds to a cGMP change with a time constant smaller than one msec, the dark current can be regarded as a real time recording of the enzyme activity that regulates $[cGMP]_i$. The remarkable similarity between the recordings from OS and the intact photoreceptor (see, for example, Forti et al.[21]) shows that the integrity of the phototransduction machinery is preserved in the "OS preparation". The presence of a stable dark current when GTP is incorporated in the dialysis solution without cGMP shows also that the OS contains a functionally intact GC. Thus GC activity, and its Ca^{2+} regulation, can be conveniently studied in situ with the "OS preparation". The fast removal of Ca^{2+}_o (t≤50 msec) caused a rapid and large increase in the dark current that was up to 10 fold in less than

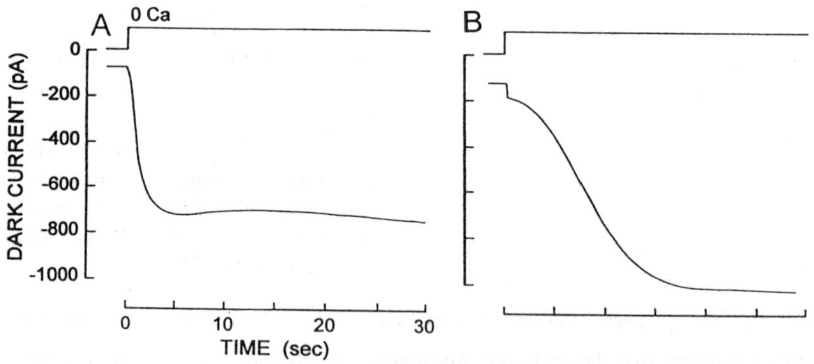

Fig. 4. 0 Ca^{2+} response in OS dialyzed with triphosphates (5 mM ATP and 1 mM GTP; *A*) and with triphosphates and Ca^{2+} chelators (*B*).

5 sec[22,5] (Fig. 4*A*). The Ca^{2+} entry is abolished when Ca^{2+}_o is removed, while the exchanger (which turnover rate is accelerated in 0 Ca^{2+}_o: Fig. 2) keeps extruding Ca^{2+} until it reaches the equilibrium. The $[Ca^{2+}]_i$ concentration that equilibrates the exchanger is several times smaller than its physiological level,[23] thus the reduction of Ca^{2+}_o causes an abrupt decrease of $[Ca^{2+}]_i$. In order to decide if this current increase is due to an external action of 0 Ca^{2+}_o on the cGMP gated channels or to an internal action of Ca^{2+} fall on the enzymes that control $[cGMP]_i$, the Ca^{2+}_i changes were slowed down by incorporating an high affinity Ca^{2+} buffer in the dialysis solution (solution III of Table 1). The Ca^{2+} buffer unmasked two components in the response to 0 Ca^{2+}_o: an initial small increase, which had the same timing of the solution change, followed by a larger and slower increase (Fig. 4*B*). All the current in 0 Ca^{2+}_o was carried through the cGMP gated channels, rather than through a leak, since it was completely suppressed by turning on the cGMP specific PDE with light. The different kinetics of the two components indicate that the fast one is due to an external action on the channels while the second one, which was slowed down by the Ca^{2+} buffer, is due to an internal action on the enzymes that control the $[cGMP]_i$. The two possibilities are that 0 Ca^{2+}_o either stimulates the basal activity of the GC or inhibit the basal activity of the PDE. Since the synthesis of cGMP requires the

presence of GTP, it was investigated first the effect of triphosphates depletion on the 0 Ca^{2+}_o response. The slow component of the 0 Ca^{2+}_o response was absent in cGMP-only dialyzed OS (Fig. 5*B*: the cGMP was included to provide a stable dark current which was not maintained if the dialysis solution lacked GTP), while the fast component was still present. However, the basal activity of PDE might require nucleotide triphosphates: if this is the case, the absence of the slow component of the 0 Ca^{2+}_o response in cGMP-only dialyzed OS could be due to the lack of basal PDE activity. To test this possibility, a cGMP-only dialyzed OS was extracellularly perfused with a membrane permeable PDE inhibitor (as 40 μM IBMX in Fig. 5*A*): the PDE inhibitor caused a large increase in current,[24] indicating that the basal activity of

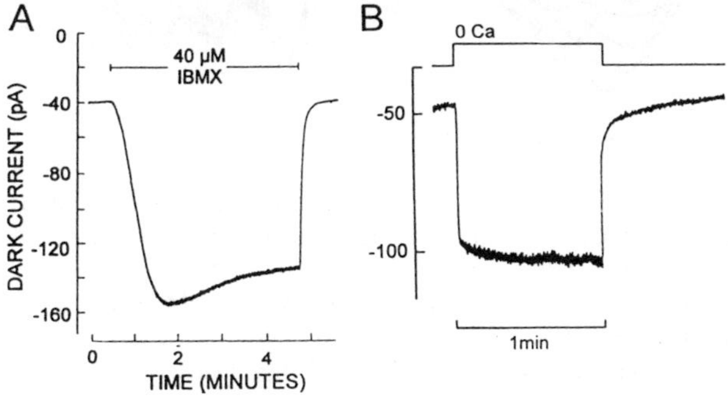

Fig. 5. The effect of IBMX (*A*) and the 0 Ca^{2+} response (*B*) in cGMP-only dialyzed OS.

PDE does not require triphosphates. Consequently, a fall in Ca^{2+}_i (induced by the 0 Ca^{2+}_o perfusion) does not inhibit the basal activity of the PDE, but rather it stimulates the basal activity of the GC when the GC substrate is available. The GC regulation by Ca^{2+}_i gives a possible mechanism for photoresponse recovery. In fact, the closure of the channels by light causes a decrease in Ca^{2+}_i influx, while the exchanger continues the extrusion, causing a fall in the $[Ca^{2+}]_i$ and GC stimulation. The increased activity of the GC raises the $[cGMP]_i$, causing the reopening of the channels previously closed by light. The channel reopening increases the Ca^{2+} influx, which reestablishes the $[Ca^{2+}]_i$ and in turn the GC activity that were present before the light stimulation. If this interpretation is correct, the kinetics of the photoresponse is expected to be critically linked to the kinetics of the Ca^{2+}_i changes. In fact, if the Ca^{2+}_i changes were slowed down by incorporating in the dialysis solution an high affinity Ca^{2+} buffer, the photoresponses were dramatically slowed down (Fig. 6*A*) compared to the control[25] (Fig. 3). We conclude therefore that the Ca^{2+} regulation of the GC is a negative feedback mechanism that is designed to clamp the dark current to a constant level. According to this scheme, a fluctuation in the $[cGMP]_i$ causes a fluctuation in the dark current which in turn induces a fluctuation in the $[Ca^{2+}]_i$ and GC activity. Because of the negative feedback, the GC fluctuation will be in a direction to counteract the $[cGMP]_i$ fluctuation, and this compensation is fast for the high gain of the feedback. If the GC is made to detect a variation in the Ca^{2+} influx with a delay, it is expected a much slower compensation of the cGMP fluctuations and an increase in

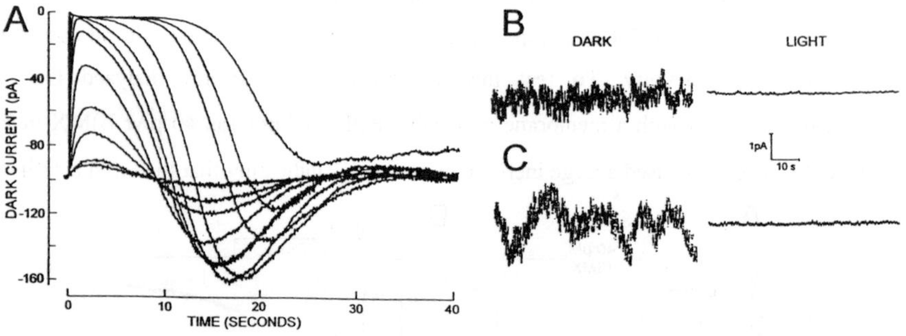

Fig. 6. *A*, response family recorded from OS dialyzed with 5 mM ATP, 1 mM GTP and Ca^{2+} chelators. Flash intensities (controlled by the neutral density filters as described in the methods section) were: 0.9, 1.7, 3.4, 8.5, 17, 34, 85, 170, 340, 854, 1704, 3400 photons·μm^{-2}. Low frequency noise in OS dialyzed with 5 mM ATP + 1 mM GTP (*B*, left panel) and with triphosphates + Ca^{2+} chelators (*C*, left panel); the fluctuations were light suppressed ($1.7 \cdot 10^5$ photons·mm^{-2}·sec^{-1}; *B* and *C*, right panel).

their size. In fact, the recordings from OS dialyzed with ATP, GTP and BAPTA had slower and wider fluctuations in current (Fig. 6*B*) than the ones from OS dialyzed without BAPTA (Fig. 6*C*). This low frequency noise was completely suppressed by light (Fig. 6*B* and *C*, right panel), showing that it was indeed originated by fluctuations in $[cGMP]_i$. Furthermore, no low frequency noise was recorded from cGMP-only dialyzed OS (Fig. 1, trace *b*). It is then concluded that the $[cGMP]_i$ fluctuations do not arise from intrinsic fluctuations of basal PDE activity, but from the endogenous buffering capacity for Ca^{2+}_i and possibly by G-protein-induced fluctuations in PDE activity. The importance of Ca^{2+}_i in photoreceptor physiology raises questions about the possible existence of mechanism/s that might regulate the Ca^{2+} movements across the plasma membrane. The putative modulation of the extrusion rate is excluded in the present work, although a modulation might still occur in the Ca^{2+} influx. The size of the first component of the 0 Ca^{2+}_o response was larger in cGMP-only dialyzed OS (Fig. 5*B*) in respect to the one recorded from triphosphate dialyzed OS (Fig. 4*B*). Since Ca^{2+}_i is not expected to change during the time it takes to remove Ca^{2+}_o (that is, 50 msec or less), the size of the first component is essentially given by the variation of the current flowing through the same number of cGMP gated channels that were open before the solution change. This suggests that $Ca^{2+}{}_o$ blocks the cGMP-gated channel more in the presence of triphosphates than in their absence. This prediction is confirmed by the fact that there is an instantaneous decrease in current upon switching a cGMP-only dialyzed OS to 10 mM $Ca^{2+}{}_o$ (Fig. 7*A*), while this solution change gave even an instantaneous increase in current in the presence of triphosphates (Fig. 7*B*). The striking difference between the two recording conditions in the first component of the 0 and 10 mM $Ca^{2+}{}_o$ response could be explained by a direct modification of the channel by triphosphates or by an indirect action mediated, for instance, by a phosphorylation. The latter hypothesis was checked by testing different kinase inhibitors supposedly able to make the first component in the presence of triphosphates similar to the one recorded from cGMP-only dialyzed OS. However, the first component was unaffected by the incorporation in the triphosphate dialysis solution of 550 μM H7 or 740 μM H8 or 680 nM Staurosporin. On the other side, this result does not rule out the presence of a phosphorylation, because the above compounds did not cause changes in light response suggestive of rhodopsin kinase inhibition,

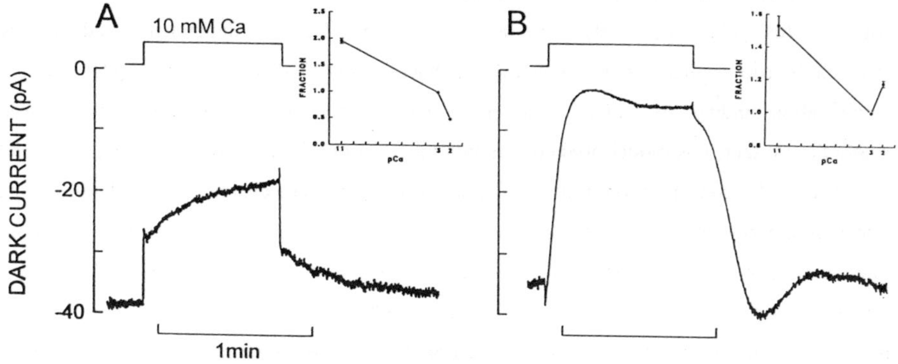

Fig. 7. 10 mM Ca^{2+} response in OS dialyzed with 5 mM ATP, 1 mM GTP and Ca^{2+} chelators (*A*) and with only cGMP (*B*). Insets, plot of the instantaneous fractional current change (first component) vs. the extracellular Ca^{2+} concentration in each of the recording conditions (averaged data from 2 to 6 OS, 1 to 3 measurements on each OS).

raising the possibility that rod kinases are, in general, insensitive to the inhibitors used.

DISCUSSION

Several reports have demonstrated that the cGMP gated channel is more selective for Ca^{2+} than for Na^+.[26,27,28,29] Ca^{2+} is supposed to block the channel by occupying one or more binding sites within the channel for which cations (mainly Na^+) compete in order to permeate. The stronger binding of Ca^{2+} to this/these binding site/s in respect to Na^+ prolongs its dwell time, making its flux smaller than the Na^+ one.[28,30,31] The removal of $Ca^{2+}{}_o$ abolishes the Ca^{2+} influx while it increases the current carried by Na^+,[26,27,32] resulting in a net increase in current (Fig. 4*B*). The instantaneous current amplitude (first component) recorded just after switching the OS to high Ca^{2+}_o was smaller than the current amplitude in normal Ringer in cGMP-only dialyzed OS (Fig. 7*B*), while just the opposite occurred in triphosphate dialyzed OS (Fig. 7*A*). This shows that triphosphates affects the Ca^{2+} movements across the channel: the most simple hypothesis is that the Ca^{2+} binding within the channel is less strong in the presence of triphosphates, thus allowing Na^+ and Ca^{2+} itself to permeate the channel more easily. Furthermore, the first component of the 0 Ca^{2+} response in cGMP-only experiments was larger than the one in the presence of triphosphates, confirming that Ca^{2+} blocks more the channel in the former conditions and its removal allows a larger fraction of the current to enter the channel. The plot of the instantaneous fractional current change vs. the $[Ca^{2+}]_o$ can be regarded as the plot of the membrane conductance vs. the ratio between $[Na^+]_o$ and $[Ca^{2+}]_o$. This plot goes through a minimum in the presence of triphosphates (Fig. 7*A*, inset), suggestive of an anomalous mole fraction dependence of the membrane conductance, while it is monotonic in the absence of triphosphates (Fig. 7*B*, inset). The anomalous mole fraction dependence induced by intracellular triphosphates indicates that the channel can not be described by a model having a single cation binding site flanked by an energy barrier on each side, where ions permeate in single file. In fact, this model predicts a monotonic relationship between the total current and the extracellular concentration of one of the permeating ions. The channel can not be described either by a series of binding sites and barriers in which ions permeate in single file without interacting with each other, because this scheme can be regarded as a sequence of two barriers-one well models, each one of them giving the aforementioned monotonic relationship. A simple model to account for the anomalous mole fraction dependence is the one with two binding sites that are close enough for bound cations to interact by repulsion.[33,34] According to this scheme, an increase in Ca^{2+}_o raises the probability that both binding sites are occupied by cations, and the consequent ionic repulsion shortens the dwell time for Ca^{2+}, which now carry a larger current and reduces to a smaller extent the flux of the other ions. Other groups found a monotonic relationship between the cGMP-activated current recorded from an excised patch and the Ca^{2+} concentration at its extracellular side, and the channel was sketched by a single binding site model.[29,35] This is consistent with the results obtained from OS dialyzed with cGMP-only. Furthermore, the channel properties just after the excision of the patch were different than later in the experiment.[36] These results suggest that the channel in transducing rods behaves differently than in excised patches.[29] The triphosphates-induced modulation of

Ca^{2+} permeation might simply occur via a nucleotide binding site at the intracellular side of the channel. This binding site might be a protein kinase target (possibly cyclic nucleotides dependent) as suggested by other groups,[36,37] or even a binding site for a second messenger synthesized or released through a G-protein and/or a phosphorylation pathway. The data so far presented might be explained also by a population of two or more different cGMP gated channels (instead of a single channel type being in different states in consequence of a modulation). For instance, only one population might have an high Ca^{2+} permeability and be switched on (directly or not) by triphosphates.

REFERENCES

1. K.-W. Yau and K. Nakatani (1984). Electrogenic Na-Ca exchange in retinal rod outer segments. *Nature* 311, 661-663.
2. A. L. Hodgkin, P. A. McNaughton and B. J. Nunn (1987). Measurement of sodium-calcium exchange in salamander rods. *J. Physiol.* 391, 347-370.
3. N. D. Goldberg, A. Ames III, J. E. Grander and T. F. Walseth (1983). Magnitude of increase in retinal cGMP metabolic flux determined by ^{18}O incorporation into nucleotide a-phosphoryls corresponds with intensity of photic stimulation. *J. Biol. Chem.* 258, 9213-9219.
4. V. Torre, H. R. Matthews and T. D. Lamb (1986). Role of calcium in regulating the cyclic GMP cascade of phototransduction in retinal rods. *Proc. Natl. Acad Sci.* USA 83, 7109-7113.
5. G. Rispoli, W. A. Sather and P. B. Detwiler (1988). Effect of triphosphate nucleosides on the response of detached outer segments to low external calcium. *Biophys. J.* 53, 390a.
6. K. W. Koch and L. Stryer (1988). Highly cooperative feedback control of retinal rod Guanylate Cyclase by calcium ion. *Nature* 334, 64-66.
7. G. Rispoli and P. B. Detwiler (1991). Voltage dependence of light-sensitive and Na:Ca exchange current in detached lizard retinal rod outer segments. *Biophys. J.* 59, 534a.
8. Y.-T. Hsu and R. S. Molday (1993). Modulation of the cGMP gated channel of rod photoreceptor cells by calmodulin. *Nature* 361, 76-79.
9. S. Kawamura (1993). Rhodopsin phosphorylation as a mechanism of cyclic GMP phosphodiesterase regulation by S-modulin. *Nature* 362, 855-857.
10. L. Lagnado and D. A. Baylor (1994). Calcium controls light-triggered formation of catalytically active Rhodopsin. *Nature* 367, 273-277.
11. G. Rispoli, W. A. Sather and P. B. Detwiler (1993). Visual transduction in dialyzed detached rod outer segments from lizard retina. *J. Physiol.* 465, 513-537.
12. B. Hille (1993). *Ionic channels of excitable membranes*. Sunderland, Mass. (USA), Sinauer.
13. L. K. Kaczmarek (1988). The regulation of neuronal calcium and potassium channels by protein phosphorylation. *Advances in Second Messenger and Phosphoprotein Research*. P. Greengard and G. A. Robinson. New York, Raven Press. 24, pp. 113-138.
14. D. Paupardin-Tritsh, C. Hammond, H. M. Gerschenfeld, A. C. Nairn and P. Greengard (1986). cGMP-dependent protein kinase enhances Ca^{2+} current and potentiates the serotonin-induced Ca^{2+} current increase in snail neurones. *Nature* 323, 812-814.
15. D. B. Light, J. D. Corbin and B. A. Stanton (1990). Dual ion-channel regulation by cyclic GMP and cyclic GMP-dependent protein kinase. *Nature* 344, 336-339.
16. P. F. Mery, S. M. Lohmann, U. Walter and R. Fischmeister (1991). Ca current is regulated by cGMP-dependent protein kinase in mammalian cardiac myocytes. *Proc. Natl. Acad. Sci.* USA. 88, 1197-1201.
17. J. P. Reeves (1992). Molecular aspects of sodium-calcium exchange. *Arch. Biochem. and Biophys.* 292, 329-334.
18. G. Rispoli, W. A. Sather and P. B. Detwiler (1991). Ca regulation of guanylate cyclase is responsible for low frequency noise in retinal rods. *Biophys. J.* 59, 434a.

19. G. Rispoli, A. T. Fineberg and P. B. Detwiler (1990). Null current measurements provide estimate of intracellular Ca in rod outer segments. *Biophys. J.* 57, 368a.
20. O. P. Hamill, A. Marty, E. Neher, B. Sakmann and F. J. Sigworth (1981). Improved patch-clamp tecniques for high-resolution current recording from cells and cell-free membrane patches. *Pflügers Arch.* 391, 85-100.
21. S. Forti, A. Menini, G. Rispoli and V. Torre (1989). Kinetics of phototransduction in retinal rods of the newt Triturus cristatus. *J. Physiol.* 419, 265-295.
22. T. D. Lamb, P. A. McNaughton and K. W. Yau (1981). External and internal actions on the response of salamander retinal rods to altered external calcium concentration. *J. Physiol.* 403, 473-494.
23. L. Cervetto, L. Lagnado, R. J. Perry, D. W. Robinson and P. A. McNaughton (1989). Extrusion of calcium from rod outer segments is driven by both sodium and potassium gradients. *Nature* 337, 740-743.
24. G. Rispoli, P. G. Gillespie and P. B. Detwiler (1990). Comparative effects of phospho-diesterase inhibitors on detached rod outer segment function. *Sensory transduction*. A. Borsellino, L. Cervetto and V. Torre. New York, Plenum Press. 157-167.
25. T. D. Lamb, H. R. Matthews and V. Torre (1986). Incorporation of calcium buffers into salamander retinal rods: a rejection of the calcium hypothesis of phototransduction. *J. Physiol.* 372, 315-349.
26. M. Capovilla, A. Caretta, A. Cavaggioni, L. Cervetto and R. T. Sorbi (1983). Ionic movements through light-sensitive channels of toad rods. *J. Physiol.* 343, 295-310.
27. A. L. Hodgkin, P. A. Mc Naughton and B. J. Nunn (1985). The ionic selectivity and calcium dependence of the light-sensitive pathway in toad rods. *J. Physiol.* 358, 447-468.
28. A. Menini, G. Rispoli and V. Torre (1988). The ionic selectivity of the light-sensitive current in isolated rods of the tiger salamander. *J. Physiol.* 402, 279-300.
29. A. L. Zimmerman and D. A. Baylor (1992). Cation interactions within the cyclic GMP-activated channel of retinal rods from tiger salamander. *J. Physiol.* 449, 759-783.
30. L. W. Haynes and K. W. Yau (1985). Cyclic GMP-sensitive conductance in outer segment membrane of catfish cones. *Nature* 317, 61-64.
31. L. W. Haynes, A. R. Kay, and K. W. Yau (1986). Single cyclic GMP-activated channel activity in excised patches of rod outer segment membrane. *Nature* 321, 66-70.
32. K. W. Yau and D. A. Baylor (1989). Cyclic GMP-activated conductance of retinal photoreceptor cells. *Ann. Rev. Neurosci.* 12, 289-327.
33. W. Almers and E. W. McCleskey (1984). The non selective conductance in calcium channels of frog muscle: Calcium selectivity in a single-file pore. *J. Physiol.* 353, 585-608.
34. P. Hess and R. W. Tsien (1984). Mechanism of ion permeation through calcium channels. *Nature* 309, 453-456.
35. A. Menini (1990). Currents carried by monovalent cations through the cGMP-activated channels in excised patches from salamander rods. *J. Physiol.* 424, 167-185.
36. S. Gordon, D. L. Brautigan and A. L. Zimmermann (1992) Protein phosphatases modulate the apparent agonist affinity of the light-regulated ion channel in retinal rods. *Neuron* 9, 739-748.
37. M.-Y. Liu, J. Li, K.-W. Yau (1994). Phosphorylation of N terminal domain of human rod cyclic nucleotide-gated channel by protein kinase C and the cGMP dependent protein kinase, *Inv. Ophthal. Visual Sci.* 35, no. 4, 1474p.

SIGNALING IN VERTEBRATE PHOTORECEPTORS: A TALE OF TWO INTRACELLULAR MESSENGERS

U. B. Kaupp

Institut für Biologische Informationsverarbeitung, Forschungszentrum Jülich, Postfach 1913, 52425 Jülich, Germany

INTRODUCTION

Rod and cone photoreceptor cells transduce the absorption of light into a brief, hyperpolarizing voltage pulse by closing cation channels in the plasma membrane. In recent years there have been many striking advances in understanding the physico-chemical basis of photo-electrical excitation in photoreceptor cells. An intracellular messenger, guanosine 3',5'-cyclic monophosphate (cGMP), carries the message of the absorption event to ion channels in the surface membrane of the photoreceptor outer segment. Light decreases the cGMP concentration by activating an enzyme cascade that hydrolytically destroys the intracellular messenger. A negative feedback loop involving Ca^{2+} and a guanylyl cyclase (GC) controls the recovery of the photoreceptor cell from the light response. Light also initiates a sequence of events that changes the sensitivity and response kinetics of photoreceptors. These cellular processes - collectively called adaptation - enable the photoreceptor cell to adjust its sensitivity to the ambient illumination. Although we know much less about adaptation, recent experiments suggest that Ca^{2+} ions play an important role by controlling the formation and degradation of cGMP.

cGMP is continuously synthesized and degraded by an enzymatic cycle that involves several proteins. Similarly, Ca^{2+} homeostasis of the cell is controlled by a cycle which involves Ca^{2+} influx through the cGMP-gated channel and subsequent extrusion of Ca^{2+} from the cell by a Na/Ca, K-exchanger. The cGMP- and the Ca^{2+}-cycles communicate with each other through a network of enzymatic reactions. A change in one cycle also causes a change in the other, and vice versa.

OVERVIEW OF VISUAL TRANSDUCTION: THE cGMP-ENZYME CASCADE

Vertebrate photoreceptor cells consist of two major parts, an outer segment and an inner segment, that are connected by a thin rudimentary cilium. The inner segment houses subcellular organelles such as the nucleus, mitochondria, and endoplasmic reticulum, all important for protein synthesis and energy production. The outer segment houses the molecular components for light absorption and generation of an electrical signal. It consists of a stack of flattened membrane sacks, called disks, that are covered by the surface membrane. The majority of the visual pigment *rhodopsin* is located at high concentration in the disk membrane; the ion channels that are indirectly regulated by light are exclusively located in the plasma membrane. A highly amplifying cascade carries the message of light absorption from rhodopsin in the disk membrane across the cytosolic

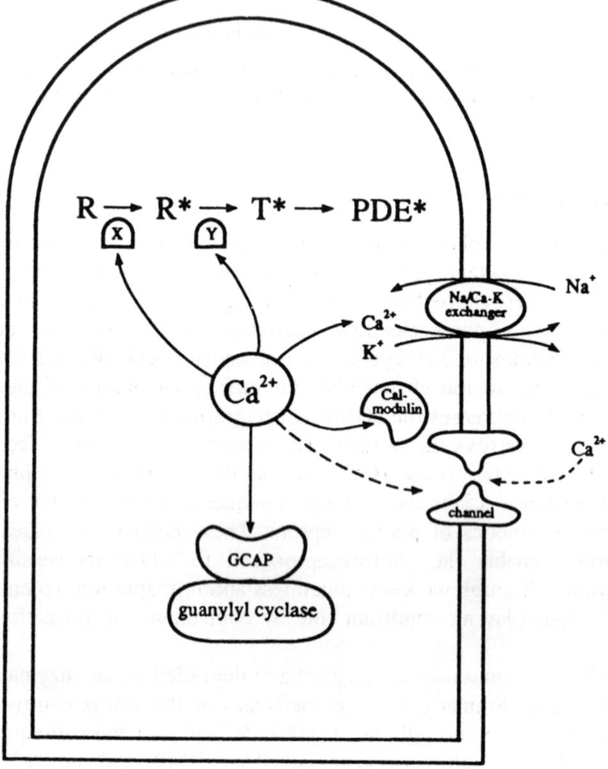

Figure 1 - Molecular components of the cGMP- and the Ca^{2+}-cycle in a vertebrate rod outer segment. Internal Ca^{2+} is controlled by Ca^{2+} influx through the cGMP-gated channel and Ca^{2+} efflux via the Na/Ca, K-exchanger. cGMP is hydrolyzed via the light-activated enzyme cascade and synthesized by the guanylyl cyclase (GC). Interaction between a modulator protein (GCAP) and guanylyl cyclase leads to stimulated cGMP synthesis at low Ca^{2+} concentrations.

space to the ion channels in the plasma membrane. In the dark, the concentration of cGMP is relatively high and a few percents of the cGMP-gated channels are kept open. Upon illumination, the cGMP concentration decreases and ion channels close. The decrease in cGMP concentration is accomplished by a light-stimulated cascade of enzymes that involves rhodopsin (R), transducin (T), and a phosphodiesterase (PDE) (see Figure 1 and Table).

Light isomerizes the chromophore of rhodopsin, 11-*cis* retinal, and triggers a series of conformational changes in the protein part of the molecule. One intermediate of the photochemical cycle of rhodopsin, called metarhodopsin II (R^*), is formed within a few milliseconds and persists for several minutes. Newly formed R^* activates another protein, called *transducin* (T). Transducin, a heterotrimeric protein that is composed of one α, β, and γ-subunit each, belongs to the large family of G-proteins. The GDP-bound form of transducin is inactive, whereas the GTP-bound form is active. R^* catalyzes the exchange of GDP for GTP at the α-subunit of transducin. Activated transducin dissociates into $T_{\beta\gamma}$ and T_α-GTP (T^*), which in turn removes two inhibitor molecules from the *phosphodiesterase* (PDE), and thereby starts cGMP hydrolysis. One activated PDE molecule catalyzes the hydrolysis of a thousand or so cGMP molecules per second. The overall amplification of the cascade can be as large as 1,000,000 but the simultaneous onset of adapting processes allow one R^* to hydrolyze only several ten thousands of cGMP molecules.

cGMP hydrolysis is stopped by termination of the enzyme cascade. Each component of the enzyme cascade is inactivated independently. Termination of the first activation step $R^* \rightarrow T^*$ requires two other proteins, a *rhodopsin kinase* and *arrestin*. The specific kinase can incorporate up to nine phosphate groups into the C-terminal region of R^*. Arrestin binds to phosphorylated R^* with much higher affinity than T and thereby uncouples $R^* \rightarrow T^*$ activation. T_α-GTP exhibits an endogenous GTPase activity that removes a terminal phosphate group from GTP whereby T_α-GDP is formed. T_α-GDP then associates with $T_{\beta\gamma}$ to yield functional transducin that again can be activated by R^*. Finally, the catalytic subunits of the PDE^* recombine with the inhibitory subunits *I* to give inactive PDE.

Recovery of the photoreceptor cell from light stimulation is accomplished by synthesis of new cGMP which is initiated by a light-stimulated drop of the intracellular Ca^{2+} concentration.

ROLE OF Ca^{2+} IN RECOVERY AND LIGHT ADAPTATION

Background light decreases the sensitivity of photoreceptor cells by controlling the gain of the transduction pathway. Thus, more photons must be captured in the presence of background illumination to generate the same response amplitude as in the dark-adapted state. The gain control prevents saturation of the light response and enables the cell to register changes over a wide range of light intensities. This physiological phenomenon is called light adaptation. Light adaptation seems to be a universal property of all vertebrate photoreceptors (Yau, 1991). The underlying mechanism involves a light-induced change in the internal Ca^{2+} level. Ca^{2+} serves as the intracellular messenger of

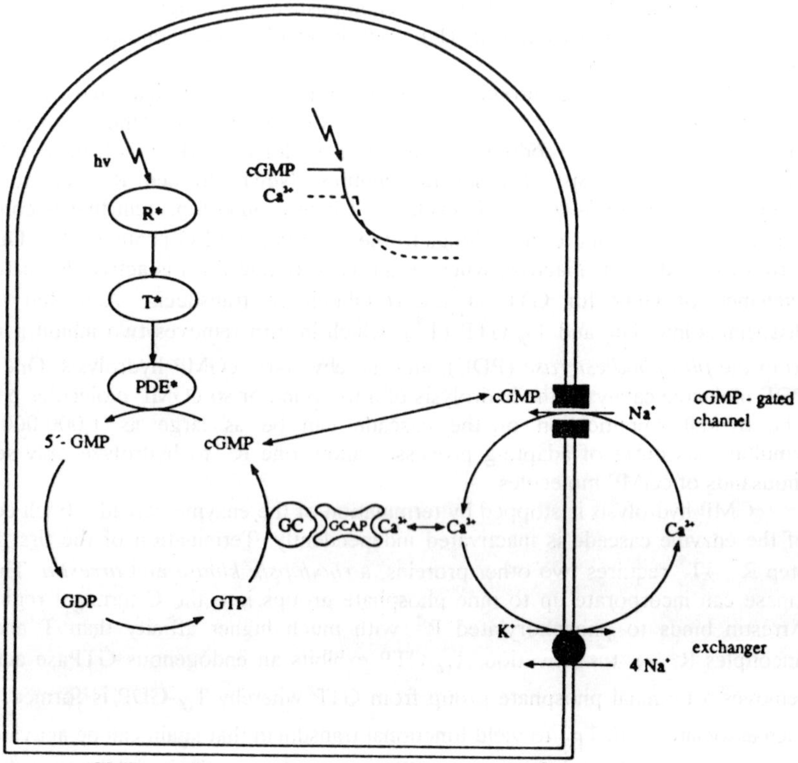

Figure 2 - Sites of Ca^{2+} activation in rod photoreceptors. Abbreviations are described in the text. X and Y refer to novel, unidentified Ca^{2+}-binding proteins. Y may be identical with recoverin.

light adaptation but also promotes the recovery of the cell from preceding light stimulation.

The Ca^{2+} concentration in the cell is determined by a delicate balance between Ca^{2+} entry through the cGMP-gated channels and extrusion through the Na/Ca, K-exchanger (Figure 1). In light, the Ca^{2+} influx is blocked by closure of the cGMP-gated channels while the $Na^+/Ca^{2+}-K^+$ exchange molecules continue to export Ca^{2+} from the cell and thereby decrease the internal Ca^{2+} concentration.

Ca^{2+} seems to act on several different targets of the cGMP-signaling pathway. A fall in Ca^{2+} (1) increases the affinity of the cGMP-gated channel for cGMP, (2) stimulates the synthesis of cGMP by a guanylyl cyclase, (3) accelerates deactivation of R^* by phosphorylation, and (4) lowers the catalytic rhodopsin activity generated by light (Figure 2).

Interaction of Ca^{2+} with the cGMP-gated channel

The cGMP-gated channel is also permeable for divalent cations and allows Ca^{2+} and Mg^{2+} to enter the cell in the dark. Under physiological conditions approximately 70 % of the inward dark current is carried by Na^+, 15 % by Ca^{2+}, 5 % by Mg^{2+} via the cGMP-gated channel, and 8 % by the Na^+ flux through the Na/Ca, K-exchanger (Nakatani & Yau, 1988). The permeability for divalent cations is important for the normal physiology of the cell. Ca^{2+} is part of several feedback mechanisms that regulate the synthesis of cGMP by a GC (Koch & Stryer, 1988) and the degradation by the light-stimulated PDE, thereby providing a route for recovery from the light response and for light adaptation. Ca^{2+} ions also block the cGMP-gated channel by inhibiting the flow of more permeant Na^+ and K^+ ions. The major single-channel conductance in rod cells in the absence of Ca^{2+} or Mg^{2+} is 20-25 pS (Haynes et al., 1986, Zimmerman & Baylor, 1986) and decreases below 0.1 pS in the presence of 1 mM Ca^{2+} or Mg^{2+} (Bodoia & Detwiler, 1985). A single-channel conductance of ≤ 0.1 pS reduces current noise generated by the open-closed fluctuations of the channel and thus improves the reliability of single photon detection (Yau & Baylor, 1989). While traversing the channel pore, Ca^{2+} competes with other cations for common binding site(s) in the pore. Divalent ions bind more strongly and hence occupy the channel for a longer time, thereby impeding flow of monovalent ions.

Ca^{2+} controls ion permeation and channel gating at three distinct sites. Ca^{2+} ions block the channel at micromolar concentrations ($K_i \sim 3$ μM) from the extracellular site and at millimolar concentrations from the intracellular site (Root & MacKinnon, 1993; Eismann et al. 1994). A negative glutamate residue in the pore motif of the channel polypeptide forms the high affinity Ca^{2+}-binding site at the entrance to the narrow constriction of the aqueous pore. When the glutamate is replaced by a neutral amino acid residue, the blockage by external Ca^{2+} is abolished.

The channel activity is also controlled by a Ca^{2+}/calmodulin-dependent mechanism (Hsu & Molday, 1993). Calmodulin binds to the channel and modulates the open probability. At elevated Ca^{2+} levels, the affinity of the cGMP-gated channel for cGMP is smaller than at low Ca^{2+} concentrations. Thus, the channel tracks internal Ca^{2+} and tends to close when Ca^{2+} level becomes too high.

Control of guanylyl cyclase activity by Ca^{2+}

In the dark, when internal Ca^{2+} is relatively high, guanylyl cyclase is inactive. The light-stimulated decrease in internal Ca accelerates the synthesis of cGMP by a guanylyl cyclase (Koch & Stryer, 1988). Ca^{2+} does not control guanylyl cyclase activity directly, but by means of a modulator protein that can stimulate basal guanylyl cyclase activity. When the Ca^{2+} concentration drops after light stimulation, probably Ca^{2+} dissociates from the modulator protein, thereby activating the guanylyl cyclase. The Ca^{2+} dependence of GC activation is very steep with a Hill coefficient n ≥·3, suggesting that the modulator protein can accommodate several Ca^{2+} ions.

A novel Ca^{2+}-binding protein, called *recoverin* had been previously implicated in the activation of guanylyl cyclase (Dizhoor et al. 1991; Lambrecht & Koch, 1991). Recent evidence, however, undermines a role of recoverin in Ca^{2+}-dependent cGMP synthesis (Gray-Keller et al. 1993). Instead another, genetically related Ca^{2+}-binding protein of similar size, designated *g*uanylyl-*c*yclase *a*ctivating *p*rotein (GCAP) has been identified and shown to modulate guanylyl cyclase activity in a Ca^{2+}-dependent manner (Gorczyca et al., 1994; Dizhoor et al., 1994; Palczewski et al., 1994).

Control of the cGMP-signaling pathway by Ca^{2+}

Ca^{2+} affects the cGMP-signaling pathway by two distinct mechanisms. A decrease in Ca^{2+} shortens the lifetime of catalytically active PDE (Kawamura & Murakami, 1991). The effect requires the presence of recoverin and ATP. Perhaps recoverin accelerates deactivation of R^* by stimulating the phosphorylation by rhodopsin kinase (Kawamura, 1993).

Finally, Ca^{2+} controls light-triggered formation of catalytically active rhodopsin (Lagnado & Baylor, 1994). Lowering intracellular Ca^{2+} reduces the catalytic rhodopsin activity equivalent to a fourfold reduction in the number of rhodopsin molecules available for activation. The effect is cooperative and saturable and probably involves an unknown Ca^{2+}-binding protein.

Acknowledgement - The author's work was supported by the Deutsche Forschungsgemeinschaft and Fonds der Chemie.

References

[1] Bodoia, R.D. and Detwiler, P.B. (1985) Patch-clamp recordings of the light-sensitive dark noise in retinal rods from the lizard and frog. J. Physiol., *367*:183-216.

[2] Dizhoor, A.M., Ray, S., Kumar, S., Niemi, G., Spencer, M., Brolley, D., Walsh, K.A., Philipov, P.P., Hurley, J.B., and Stryer, L. (1991) Recoverin: A calcium sensitive activator of retinal rod guanylate cyclase. Science, *251*:915-918.

[3] Dizhoor, A.M., Lowe, D.G., Olshevskaya, E.V., Laura, R.P., and Hurley, J.B. (1994) The human photoreceptor membrane guanylyl cyclase, RetGC, is present in outer segments and is regulated by calcium and a soluble activator. Neuron, *12*, 1345-1352.

[4] Eismann, E., Müller, F., Heinemann, S.H., and Kaupp, U.B. (1994) A single negative charge within the pore region of a cGMP-gated channel controls rectification, Ca^{2+} blockage, and ionic selectivity. Proc. Natl. Acad. Sci. USA, *91*, 1109-1113.
[5] Gorczyca, W.A., Gray-Keller, M.P., Detwiler, P.B., and Palczewski, K. (1994) Purification and physiological evaluation of a guanylate cyclase activating protein from retinal rods. Proc. Natl. Acad. Sci. USA, *91*, 4014-4018.
[6] Gray-Keller, M.P., Polans, A.S., Palczewski, K., and Detwiler, P.B. (1993) The effect of recoverin-like calcium-binding proteins on the photoresponse of retinal rods. Neuron, *10*:523-531.
[7] Haynes, L.W., Kay, A.R., and Yau, K.-W. (1986) Single cyclic GMP-activated channel activity in excised patches of rod outer segment membrane. Nature, *321*:66-70.
[8] Hsu, Y.T. and Molday, R.S. (1993) Modulation of the cGMP-gated channel of rod photoreceptor cells by calmodulin. Nature, *361*:76-79.
[9] Kawamura, S. (1993) Rhodopsin phosphorylation as a mechanism of cyclic GMP phosphodiesterase regulation by S-modulin. Nature, *362*:855-857.
[10] Kawamura, S. and Murakami, M. (1991) Calcium-dependent regulation of cyclic GMP phosphodiesterase by a protein from frog retinal rods. Nature, *349*:420-423.
[11] Koch, K.-W. and Stryer, L. (1988) Highly cooperative feedback control of retinal rod guanylate cyclase by calcium ions. Nature, *334*:64-66.
[12] Lagnado, L. and Baylor, D.A. (1994) Calcium controls light-triggered formation of catalytically active rhodopsin. Nature, *367*:273-277.
[13] Lambrecht, H.-G. and Koch, K.-W. (1991) A 26 kd calcium binding protein from bovine rod outer segments as modulator of photoreceptor guanylate cyclase. EMBO J., *10*:793-798.
[14] Nakatani, K. and Yau, K.-W. (1988) Calcium and light adaptation in retinal rods and cones. Nature, *334*:69-71.
[15] Palczewski, K., Subbaraya, I., Gorczyca, W.A., Helekar, B.S., Ruiz, C.C., Ohguro, H., Huang, J., Zhao, X., Crabb, J.W., Johnson, R.S., Walsh, K.A., Gray-Keller, M.P., Detwiler, P.B., and Baehr, W. (1994) Molecular cloning and characterization of retinal photoreceptor guanylyl cylcase-activating protein. Neuron, *13*, 395-404.
[16] Root, M.J. and MacKinnon, R. (1993) Identification of an external divalent cation-binding site in the pore of a cGMP-activated channel. Neuron, *11*:459-466.
[17] Yau, K.-W. (1991) Calcium and light adaptation in retinal photoreceptors. Curr. Opinion Neurobiol., *1*:252-257.
[18] Yau, K.-W. and Baylor, D.A. (1989) Cyclic GMP-activated conductance of retinal photoreceptor cells. Annu. Rev. Neurosci., *12*:289-327.
[19] Zimmerman, A.L. and Baylor, D.A. (1986) Cyclic GMP-sensitive conductance of retinal rods consists of aqueous pores. Nature, *321*:70-72.

MOLECULAR BIOLOGY OF CYCLIC NUCLEOTIDE-GATED ION CHANNELS

U. B. Kaupp

Institut für Biologische Informationsverarbeitung, Forschungszentrum Jülich, Postfach 1913, 52425 Jülich, Germany

INTRODUCTION

Cyclic nucleotides, *i.e.* cAMP and cGMP, control the activity of a variety of ion channels by diverse mechanisms. Cyclic nucleotide-sensitive channels fall into distantly related families. Members of one family, designated cyclic nucleotide-*gated* (CNG) channels, obligatory require cAMP or cGMP to open, although they share common structural features with members of the superfamily of voltage-gated channels. CNG channels have been best characterized in sensory cells, but they also exist in other neurons and in non-neuronal tissue. In photoreceptors of the vertebrate retina and in olfactory sensory neurons (OSN), CNG channels serve as final targets of signalling pathways that transduce the respective stimulus into electrical signals.

Their specific function in other cell types is less clear. Some CNG channels are highly permeable to Ca^{2+} and may serve as Ca^{2+} entry pathways under the control of membrane receptors which are able to change the concentration of cyclic nucleotides. This review describes recent progress in understanding the physiological and structural properties of CNG channels.

DIVERSITY AND DISTRIBUTION OF CNG CHANNELS

For a while, CNG channels were considered as a curiosity of nature, perhaps confined to vertebrate rod and cone photoreceptor cells (Fesenko et al., 1985; Haynes & Yau, 1985) and OSNs (Nakamura & Gold, 1987), where they have been first discovered. One of the major advances in the past years has been the discovery that similar CNG channels are present in other neurons of the retina and in the central nervous system (Ahmad et al., 1994; Schaad et al., 1995), but also in non-neuronal tissues (Ahmad et al., 1992; Biel et al., 1993; Biel et al., 1994; Weyand et al., 1994) (see also Table 1).

Except for a CNG channel in spermatozoa (Weyand et al., 1994), relatively little is known about the cellular localization of cloned CNG channel genes in non-neuronal tissues.

CNG channels differ in their ion selectivity, ligand selectivity and sensitivity, and their interaction with Ca^{2+} ions. They are either cation-selective (Menini, 1990; Torre & Menini, 1994) or K^+-selective (Delgado et al., 1991; Hatt & Ache, 1994). A few prefer cAMP over cGMP (Kolesnikov et al., 1991; Delgado et al., 1991) but the majority is specific for cGMP. The photoreceptor CNG channels are at least one order of magnitude less sensitive to cyclic nucleotides than CNG channels expressed in OSN (Fesenko et al., 1985; Haynes & Yau, 1985; Nakamura & Gold, 1987; Altenhofen et al., 1991).

What is the molecular basis for this functional diversity; and how is the tissue distribution of each of these channel proteins? Most importantly, what is the physiological significance of this spezialisation?

Table 1 - Tissue distribution of CNG channel α-subunits

***r*CNGCα**	***c*CNGCα**	***o*CNGCα**
rod photoreceptor	cone photoreceptor	olfactory sensory neuron
bipolar cells	spermatozoa	aorta
ganglion cells	kidney	
kidney	heart	
pineal gland	pineal gland	

A number of cyclic nucleotide-modulated (CNM) channels has been recently characterized that disclose a bewildering potpourri of physiological properties (Fig. 1b). This functional heterogeneity suggests that CNM channels fall into several genetic subfamilies. A familial relationship between CNG- and CNM-channels has been established in a few cases. The primary structure of several K^+ channel genes, including the *ether-á-go-go (eag)* channel in *Drosophila* and its vertebrate homologues [Refs. 23-27] reveal a low but significant sequence similarity to CNG channels and most notably carry a cyclic nucleotide-binding motif [28]. The pore motif signature of *eag* is typical of a K^+-selective channel, characterized by two additional amino acid residues (Tyr-Gly; [29]) that are missing in the non-selective CNG channels. The strong voltage-dependent activation of the *eag* channel is consistent with an archetypical voltage-sensor (S4) motif that, unlike the S4 motif of CNG channels, is not compromised by flanking glutamate residues. The *Drosophila eag* channel is modulated by cAMP and seems to be permeable to Ca^{2+} [30]. Interestingly, the rat homologue of *eag* is neither Ca^{2+} permeable nor modulated by cyclic nucleotides [24].

Understanding of the diversity and molecular composition of CNG channels has benefited greatly from the cloning of their components. CNG channels form heterooligomeric complexes consisting of several distinct yet homologous subunits. Rod (Kaupp et al., 1989) and cone (Weyand et al., 1994; Bönigk et al., 1993) photoreceptors and OSN (Dhallan et al., 1990; Ludwig et al., 1990; Goulding et al., 1992) each express a unique form of one principal subunit - designated here α-subunit, or *r*CNGCα, *c*CNGC α and *o*CNGCα. When heterologously expressed, these α-subunits form functional channels by themselves; and they are responsible for many of the key channel characteristics. Additional subunits which are homologous to the α-subunits have been discovered in rod photoreceptors (Chen et al., 1993; Körschen et al., 1995) and OSN (Liman & Buck, 1994; Bradley et al., 1994).

These subunits do not form functional channels by themselves, but impart to the respective α-subunits specific properties which are more characteristic of the native channels (see below).

The three α-subunit genes seem to be expressed in other tissues as well. For example, cDNA clones encoding *r*CNGCα have also been isolated from kidney (Ahmad et al., 1992), retinal ganglion cells (Ahmad et al., 1994), bipolar cells (I Dabin, DS Roy, CI Barnstable *ARVO Abstract* 1995, 1302-205), and pineal gland (Schaad, 1995); those encoding *c*CNGCα have been isolated from testis (Weyand et al., 1994), kidney (Biel et al., 1994), and pineal gland (W Bönigk, F Müller, UB Kaupp, unpublished observation) (see Table 1).

In some of these tissues α-subunits have been identified by amplification of short cDNA sequences. Therefore, these subunits could also represent splice variants of those genes expressed in vertebrate photoreceptors or OSNs. Besides expression of splice variants, functional diversity of CNG channels may be further enhanced by using different β-subunits and by post-translational processing of channel polypeptides (Bönigk et al., 1993; Molday et al., 1991).

A *Drosophila* CNG channel gene that is highly homologous to the vertebrate α-subunits has recently been cloned (DmCNGC; (Baumann et al., 1994)). DmCNGC is expressed in eyes and antennae, but has not yet been localized at the cellular level. Whether the native *Drosophila* CNG channel exists as heterooligomer is not known.

Cyclic nucleotide-activated currents have also been measured in the ventral photoreceptor of *Limulus* (Bacigalupo et al., 1991), in hair cells of the inner ear (Kolesnikov et al., 1991), *Drosophila* muscle (Delgado et al., 1991), retinal bipolar cells (Nawy & Jahr, 1991), pinealocytes (Dryer & Henderson, 1991), taste cells (MS Herness, *Soc Neurosci Abstr* 1993, 84.1), extra-ocular photoreceptors of molluscs (Gotow et al., 1994), and hyperpolarizing photoreceptors of *Pecten* (M de Pilar Gomez, E Nasi, *Biophys Soc Abstr* 1995, A18). Whether these CNG channels are composed of subunits encoded by one of the already known genes or by related genes still needs to be determined.

SUBUNIT COMPOSITION OF CNG CHANNELS

The first CNG channel was purified from bovine rod photoreceptors as a polypeptide with a relative molecular mass (M_r) of ~ 63K that, when reconstituted into liposomes or planar lipid bilayers, gave rise to cGMP-stimulated channel activity (Cook et al., 1987; Hanke et al., 1988).

Heterologous expression of cloned cDNA encoding this polypeptide (α-subunit) resulted in channel activity largely similar to that of the native channel, nourishing the belief that CNG channels perhaps are homooligomers (Kaupp et al., 1989). However, native CNG channels and heterologously expressed α-subunits differ in characteristic ways with respect to ion selectivity, ligand sensitivity, gating kinetics and pharmacology (Nakamura & Gold, 1987; Dhallan et al., 1990; Altenhofen et al., 1991; Goulding et al., 1992; Chen et al., 1993; Nizzari et al., 1993).

These differences suggest that α-subunits require additional factor(s) or subunit(s) to display native properties. Chen et al. (1993) recently reported the cloning of a second 'subunit' of the human rod CNG channel, that, when co-expressed with the α-subunit, restores sensitivity to the channel blocker l-*cis*-diltiazem and flickery gating behavior. This second 'subunit' appears to exist in two splice variants with a predicted M_r of ~ 70K and ~ 102K (Chen et al., 1993). However, polypeptides in the size range of 70K - 102K have not been detected in purified rod channel preparations (Cook et al., 1987; Molday et al., 1990); instead the 63K channel polypeptide copurifies with a ~ 240K component (Cook et al., 1987; Molday et al., 1990). The 240K component is a calmodulin-binding protein that confers to the reconstituted 63K protein a Ca^{2+}/calmodulin-mediated decrease in ligand sensitivity (Hsu & Molday, 1993). Cloning, functional expression, Western blot analysis and immunohistochemical localization now identify the 240K polypeptide as the genuine β-subunit of the CNG channel from bovine rod photoreceptor (*r*CNGCβ; (Körschen et al., 1995)). The 240K polypeptide has an unusual bipartite structure that has not yet been discovered in other channel proteins.

The cytoplasmic N-terminal part is essentially identical to a glutamic acid-rich protein (GARP; (Sugimoto et al., 1991)), whereas the C-terminal part is highly homologous to the previously cloned second 'subunit' from human rods (Chen et al., 1993), and comprises the predicted membrane-spanning segments (S1-S6), the cGMP-binding site, the pore region, and a calmodulin-binding site (see Fig. 1b and below). Although the ligand sensitivity between homooligomeric (α) and heterooligomeric (α + β) channels is rather similar, the ligand binding site of the β-subunit is involved in cGMP-binding, because it is specifically labeled by a photoaffinity analog (Brown et al., 1993).

The isolation of clones encoding the GARP-part alone and Northern blot analysis suggest that at least the GARP-part may be also expressed separately (Sugimoto et al., 1991; Körschen et al., 1995). cDNAs encoding the human second 'subunit' may either represent partial clones or shorter splice variants that are expressed, however, in other retinal cells.

The bipartite structure of the bovine rod β-subunit raises several questions. What is the function of the GARP-part in the 240K protein? It could provide a means for the channel to interact with other proteins, e.g. Na^+/Ca^{2+}-K^+ exchanger (Bauer & Drechsler, 1992) or it may be required for targeting of CNG channels to the rod outer segment. It needs to be shown whether β-subunits in other cells are built up as a modular system consisting of two distinct parts that can be alternatively expressed as single or separate polypeptides and thereby increase the functional diversity of the CNG channel family.

A second subunit has also been identified for the olfactory CNG channel (Liman & Buch, 1994; Bradley et al., 1994). When co-expressed with the α-subunit, heterooligomeric channels display ligand sensitivities that are more characteristic of the native channel, in particular with respect to cAMP. While *r*CNGCα and *r*CNGCβ are profoundly dissimilar, the second subunit of the olfactory CNG channel is much more similar to all α-subunits and even to DmCNGC than to *r*CNGCβ (Fig. 1a). It is therefore tentatively designated *o*CNGCα_2. This raises the interesting possibility that OSN may express a third subunit that is more akin to *r*CNGCβ.

STRUCTURAL AND FUCTIONAL SEQUENCE MOTIFS

CNG channel α-subunits carry two short sequence domains - a pore-forming (*P*) region and a residual voltage-sensor (S4) motif - that are reminiscent of similar motifs in *Shaker*-type K^+ channels. It has been tacitly assumed that CNG channels and this K^+ channel subfamily share a common overall topology, characterized by a hydrophobic domain carrying six membrane-spanning segments. The topology of *r*CNGCα and DmCNGC has now been examined by a gene fusion approach (Henn et al., 1995) using different bacterial reporter enzymes that are active either in the cytoplasm or periplasm. CNG channels are found to have indeed six membrane-spanning segments, suggesting that CNG and *Shaker*-type K^+ channels, albeit distant relatives within the gene superfamily, share a common overall topology. The topology model is also consistent with the intracellular localization of N- and C-terminus (Molday et al., 1991), and N-glycosilation of Asn 327 in *r*CNGCα (Fig. 1b; (Wohlfart et al., 1992).

The *P* region, located between segments S5 and S6, has been identified by constructing chimera in which the *P* motif and thereby ion selectivity was exchanged between *r*CNGCα and *o*CNGCα (Goulding et al., 1993), and by site-directed mutagenesis (Heginbotham et al., 1992; Root & MacKinnon, 1993; Eismann et al., 1994; Sesti et al., 1995). The *P* region comprises a glutamate (residue 363 in *r*CNGCα of bovine) that is crucially important for Ca^{2+} blockage (see below), ionic selectivity and gating (Root & MacKinnon, 1993; Eismann et al., 1994; Sesti et al., 1995). This negatively charged residue is conserved in all cloned α-subunits including *o*CNGCα_2, but is replaced by a glycine residue in the putative pore region of *r*CNGCβ. Because CNG channels form oligomeric complexes composed of four subunits, the high-affinity blocking site probably consists of several glutamate residues. In homomeric channels, composed only of α-subunits, probably four glutamates form two equivalent, independent sites that can be protonated (Root & MacKinnon, 1994). Protonation accounts for the occurrence of sub-conductance levels (Goulding et al., 1992; Root & MacKinnon, 1994) and pH-dependence of current amplitude (Menini & Nunn, 1990; Tanaka, 1993). A similar set of conserved but non-equivalent glutamate residues exists in the *P* region of Ca^{2+} channels (Kim et al., 1993; Yang et al., 1993), and the mechanism and structural basis for Ca^{2+} selectivity and permeation seem to be similar in CNG- and Ca^{2+} channels (Frings et al., 1995).

The canonical voltage-sensor (S4) motif in K^+ channels contains up to seven positively charged amino acid residues (mainly arginine). Transmembrane segment S4 of all CNG channels contains a voltage-sensor-like motif, characterized by fewer arginine residues and several adjacent glutamate residues. The function of the S4-motif in CNG channels is unknown.

A region comprising 80-100 residues in the C-terminal part has been identified as cyclic nucleotide-binding domain by sequence comparison with other cyclic nucleotide-regulated proteins and by mutagenesis (Kaupp et al., 1989; Altenhofen et al., 1991; Goulding et al., 1994). In analogy to the catabolite-gene-activator protein (CAP), a bacterial cAMP-binding protein whose structure has been solved (McKay & Steitz, 1981), the binding site could consist of an eight-stranded β-roll and three α-helices.

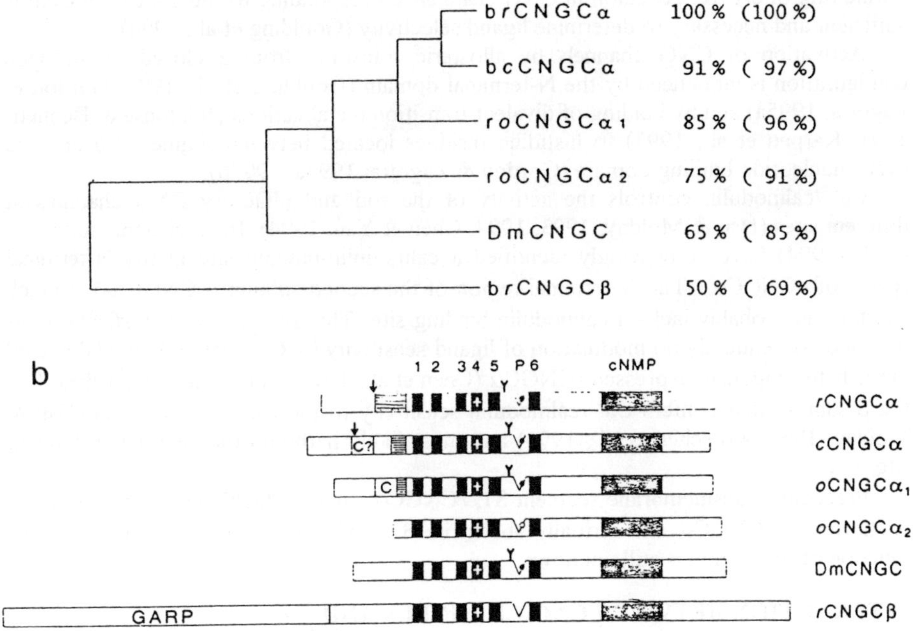

Figure 1 - (a) Evolutionary tree of cyclic nucleotide-binding domain of CNG channel from bovine rod and cone α-subunits (b*r*CNGCα, b*c*CNGCα), from both rat olfactory α-subunits (r*o*CNGCα1, r*o*CNGCα2), from *Drosophila* (DmCNGC), and from bovine rod β-subunit (b*r*CNGCβ). Numbers indicate the percentage of identity (similarity, considering identical and conservatively exchanged amino acids) and refer to b*r*CNGCα. **(b)** Structural features of CNG channel subunits. Transmembrane segments 1-6 are represented by black boxes; cyclic nucleotide-binding domains (cNMP) by grey boxes; highly charged regions by hatched boxes; *P*, pore motif; *C*, calmodulin-binding region, GARP, glutamic acid-rich part; *Y*, identified (*r*CNGC α; (Wohlfart *et al.*, 1992)) or consensus glycosilation sites; white cross, residual voltage-sensor motif in segment 4; arrows, site of post-translational cleavage in *r*CNGCα (Molday *et al.*, 1991) and putative site in *c*CNGCα; filled circles, glutamate residue and open circles, aspartate residue in *P* region.

The C-helix which follows the β-roll in CAP forms two hydrogen bonds with the purine ring of cAMP. A region in CNG channels corresponding to the C-helix in CAP is sufficient and necessary to determine ligand selectivity (Goulding et al., 1994).

Activation of CNG channels by allosteric transition from a closed to an open configuration is influenced by the N-terminal domain (Goulding et al., 1994; Gordon & Zagotta, 1995a) and by binding of divalent transition metal cations (Ildefonse & Bennett, 1991; Karpen et al., 1993) to histidine residues located between segment S6 and the cyclic nucleotide-binding domain (Gordon & Zagotta, 1995a, 1995b).

Ca^{2+}/calmodulin controls the activity of the rod and olfactory CNG channels at different sites (Hsu & Molday, 1993; 1994; Chen & Yau, 1994). In an elegant study, Liu et al. (1994) have convincingly identified a calmodulin-binding site in the N-terminal region of *o*CNGCα_1. The N-terminal region of the second subunit (*o*CNGCα_2) is much shorter and probably lacks a calmodulin-binding site. This site is absent in *r*CNGCα as well and consequently no modulation of ligand sensitivity by Ca^{2+}/calmodulin is detected in the heterologously expressed *r*CNGCα (Chen et al., 1994; Körschen et al., 1995). It is the β-subunit that confers Ca^{2+}/calmodulin sensitivity to the rod CNG channel (Hsu & Molday, 1993; Körschen et al., 1995) and most likely harbours the calmodulin-binding site.

Preceding transmembrane segment S1, *r*CNGCα carries a highly charged region that is shorter in *c*CNGCα and virtually absent in *o*CNGCα (Fig. 1b, striped boxes). The function of this region is still unknown.

INTERACTION BETWEEN CNG CHANNELS AND Ca^{2+}

Ca^{2+} ions permeate and at the same time block CNG channels (for summary see McNaughton, 1990; Zufall et al., 1994). Ca^{2+} entry through CNG channels in photoreceptors and OSN is physiologically important because Ca^{2+} takes part in several feedback mechanisms that regulate the synthesis and degradation of cyclic nucleotides (Pugh & Lamb, 1990; Yau, 1991; Koch, 1994; Shepherd, 1994).

The intracellular Ca^{2+} concentration ($[Ca^{2+}]_i$) in photoreceptors and OSN is changed by the activity of CNG channels. In vertebrate photoreceptors, CNG channels are open in the dark and close in response to light stimulation. $[Ca^{2+}]_i$ in the dark is determined by the balance of a steady Ca^{2+} entry through open CNG channels and Ca^{2+} extrusion *via* a Na^+/Ca^{2+}-K^+ exchanger. When Ca^{2+} entry is blocked by the closure of CNG channels in light, continued extrusion lowers $[Ca^{2+}]_i$ which serves as key signal of light adaptation (Pugh & Lamb, 1990; Fain & Matthews, 1990). In contrast to photoreceptors, CNG channels in OSN are closed at rest and open upon stimulation by odorants, leading to an increase of $[Ca^{2+}]_i$ (Restrepo et al., 1990; 1993). A rise in $[Ca^{2+}]_i$ both facilitates and terminates the response of OSN to odorants. It activates a depolarizing Cl^- current (Kleene & Gesteland, 1991; Kleene, 1993; Kurahashi & Yau, 1993) and thereby greatly amplifies the response (Lowe & Gold, 1993). Subsequently, Ca^{2+} stimulates the degradation of cAMP via a Ca^{2+}/calmodulin-activated phosphodiesterase (Borisy et al., 1992) and causes CNG channel to close by a mechanism also involving calmodulin (Chen & Yau, 1994).

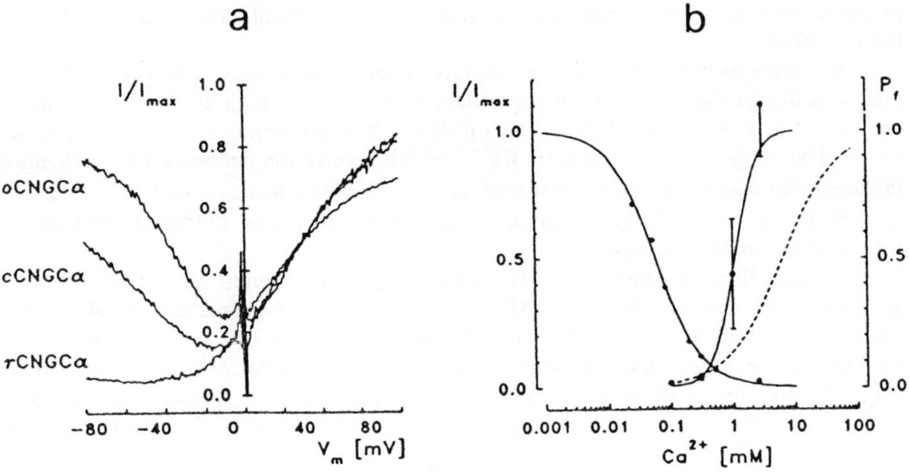

Figure 2 - Ca^{2+} blockage and Ca^{2+} permeation of CNG channels. **(a)** Fraction of unblocked current I/I_{max} measured in outside-out membrane patches of heterologously expressed *r*CNGCα, *c*CNGCα and *o*CNGCα in the presence of 52 μM $[Ca^{2+}]_o$. I/I_{max} values of 1 and 0 mean that the channel is completely unblocked or blocked, respectively. The voltage dependence of blockage is determined by two processes, access of Ca^{2+} to the blocking site and dissociation from this site to the cytoplasmic face of the channel; i.e. permeation. At positive voltage, blockage by external Ca^{2+} is very similar for all three channel types. However, at negative voltage, Ca^{2+} blockage is greatly relieved in *c*CNGCα and *o*CNGCα, but not in *r*CNGCα. As a consequence, Ca^{2+} entry at hyperpolarized voltages would be much larger in cells that express *c*CNGC or *o*CNGC.
(b) Comparison of Ca^{2+} dependence of channel blockage (I/I_{max}; open circles) and fractional Ca^{2+} current (P_f, closed circles) for heterologously expressed *o*CNGCα. A P_f value of 1 indicates that the entire current is carried by Ca^{2+} ions. The increase in P_f with $[Ca^{2+}]_o$ is very steep, suggesting that several Ca^{2+} ions interact with the intrapore binding site to achieve a sizeable flux rate. The interrupted line predicts the P_f value according to the Goldman-Hodgkin-Katz formalism (for detail see Frings *et al.*, 1995).

While passing through CNG channels, $[Ca^{2+}]_o$ blocks the current carried by Na^+ or K^+ ions in a voltage-dependent fashion (see Fig. 2a; (McNaughton, 1990; Zufall et al., 1994)). Blockage reduces current noise generated by the channel itself and, in photoreceptor cells, thus improves the reliability of single-photon detection (Yau & Baylor, 1989).

Ca^{2+} interacts with CNG channels at three different sites either directly or indirectly. Firstly, it blocks the channel from the extracellular side with high affinity by binding to the glutamate residue 363 in the *P* region (Root & MacKinnon, 1993; Eismann et al., 1994). The blockage by extracellular Ca^{2+} and the relative ion permeability are changed in heterooligomeric ($\alpha + \beta$) compared to homooligomeric (α) rod CNG channels, suggesting that *r*CNGCβ takes part in lining the aqueous pore and contributes to the Ca^{2+} binding site (Körschen et al., 1995).

Secondly, divalent cations at 50-100-fold higher concentrations block CNG channels also from the cytoplasmic side (Menini et al., 1988; Colamartino et al., 1991; Zimmermann & Baylor, 1992; Picones & Korenbrot, 1995). Blockage by intracellular Ca^{2+} or Mg^{2+} involves a low-affinity binding site ($K_i \sim 1$ mM) distinct from the glutamate site in the pore region (Root & MacKinnon, 1993; Eismann et al., 1994). While maximal blockage by $[Ca^{2+}]_i$ seems to be similar in rod and cone CNG channels (Picones & Korenbrot, 1995), blockage by $[Ca^{2+}]_o$ is roughly 8-fold more efficient in rod than in cone photoreceptor channels.

Finally, Ca^{2+}/calmodulin decrease the apparent ligand sensitivity of CNG channels (Hsu & Molday, 1993; Chen & Yau, 1994). This modulation provides a feedback mechanism that helps terminating the cellular response and restoring the resting state. It seems to be more important for OSN than for rod photoreceptors because modulation by Ca^{2+}/calmodulin lowers ligand sensitivity only twofold in *r*CNGC (Hsu & Molday, 1993; Chen et al., 1994), but up to 20-fold in *o*CNGC (Chen & Yau, 1994).

CNG channels greatly differ in Ca^{2+} permeability and in voltage-dependent blockage by $[Ca^{2+}]_o$ blockage (Perry & McNaughton, 1991; Frings et al., 1995; Picones & Korenbrot, 1995). Heterologously expressed *c*CNGCα and *o*CNGCα are severalfold more permeable to Ca^{2+} than *r*CNGCα (Frings et al., 1995). In fact, Ca^{2+} permeation of *o*CNGCα is so pronounced that at millimolar $[Ca^{2+}]_o$, the current is almost entirely carried by Ca^{2+} ions (Fig. 2b; Frings et al., 1995). These differences imply profound consequences for cellular Ca^{2+} homeostasis, and probably represent an evolutionary adjustment to the specific ways by which CNG channels and Ca^{2+} participate in the cellular responses of rod and cone photoreceptors and of OSN. Obviously, a channel that is open at rest, allowing Ca^{2+} „trickle" into the cell (as the rod CNG channel), must be designed differently than a channel that is closed at rest and, upon opening, allows Ca^{2+} to rush into the cell (as the olfactory CNG channel).

What has been previously considered as the archetypical function of CNG channels, namely the generation of receptor currents in sensory neurons may be restricted to vertebrate photoreceptors. In OSN, CNG channels only provide an initial Ca^{2+} current that triggers the depolarizing Cl^- current. CNG channels in non-neuronal tissue may also subserve functions entirely different from that in vertebrate photoreceptors. For example, the CNG channel in sperm, which is identical to that expressed in cone photoreceptors, might play a crucial role in chemotaxis or in the acrosome reaction (Weyand et al., 1994). Both cellular processes are strongly dependent on extracellular Ca^{2+} and cyclic nucleotides (Ward & Kopf, 1993) and CNG channels by virtue of their high Ca^{2+} permeability could provide a pathway for Ca^{2+} entry controlled by cyclic nucleotides.

In the cone synapse, CNG channels in addition to voltage-activated Ca^{2+} channels control exocytosis (Rieke & Schwartz, 1994). Release of neurotransmitter from synaptic vesicles is highly Ca^{2+}-dependent and CNG channels extend the range of membrane voltage at which this process can operate.

I suggest that the primary if not exclusive function of CNG channels in many cells is to increase $[Ca^{2+}]_i$ rather than to change the membrane voltage by conducting Na^+ or K^+ ions. A rise in $[Ca^{2+}]_i$ stimulates a host of cellular responses such as adaptation, exocytosis, cell motility, chemotaxis, fluid transport or axon guidance. In general, CNG channels represent an alternative pathway for Ca^{2+} entry in electrically excitable and non-excitable cells besides voltage-gated Ca^{2+} channels, IP_3 receptors, and the so-called calcium release-activated calcium current, or I_{CRAC} (Putney, 1993).

CONCLUSIONS

Cyclic nucleotide-*sensitive* channels probably form a large, heterogenous class of ion channels that falls into several distantly related subfamilies. A number of cyclic nucleotide-*modulated* (CNM) channels has recently been characterized that disclose a bewildering potpourri of physiological properties. A familial relationship between CNG- and CNM-channels has been established in a few cases (Warme et al., 1991; Guy et al., 1991; Anderson et al., 1992; Sentenac et al., 1992; Brüggemann et al., 1993; Warme & Ganetzky, 1994; Ludwig et al., 1994). CNM channel polypeptides carry structural elements similar to those present in CNG channels. It will be a challenge for biophysically oriented work to elucidate on the molecular level how these distinct features generate such a rich functional diversity.

The most important task will be to clearly define the specific functions of CNG channels in excitable and non-excitable tissues and to relate it to their spatial localization and molecular composition. I trust that this field will continue to make significant contributions to our understanding of cellular signalling.

Acknowledgements - I thank many members of the Institute for critical reading of the manuscript and helpful discussions. Work reported from the author's laboratory was supported by the Deutsche Forschungsgemeinschaft.

REFERENCES

[1] Ahmad, I., Korbmacher, C., Segal, A.S., Cheung, P., Boulpaep, E.L., and Barnstable, C.J. (1992) Mouse cortical collecting duct cells show nonselective cation channel activity and express a gene related to the cGMP-gated rod photoreceptor channel. Proc. Natl. Acad. Sci. USA, *89*:10262-10266.

[2] Ahmad, I., Leinders-Zufall, T., Kocsis, J.D., Shepherd, G.M., Zufall, F., and Barnstable, C.J. (1994) Retinal ganglion cells express a cGMP-gated cation conductance activatable by nitric oxide donors. Neuron, *12*:155-165.

[3] Altenhofen, W., Ludwig, J., Eismann, E., Kraus, W., Bönigk, W., and Kaupp, U.B. (1991) Control of ligand specificity in cyclic nucleotide-gated channels from rod photoreceptors and olfactory epithelium. Proc. Natl. Acad. Sci. USA, *88*:9868-9872.

[4] Anderson, J.A., Huprikar, S.S., Kochian, L.V., Lucas, W.J., and Gaber, R.F. (1992) Functional expression of a probable Arabidopsis thaliana potassium channel in Saccharomyces cerevisiae. Proc. Natl. Acad. Sci. USA, *89*:3736-3740.

[5] Bacigalupo, J., Johnson, E.C., Vergara, C., and Lisman, J.E. (1991) Light-dependent channels from excised patches of Limulus ventral photoreceptors are opened by cGMP. Proc. Natl. Acad. Sci. USA, *88*:7938-7942.

[6] Bauer, P.J. and Drechsler, M. (1992) Association of cyclic GMP-gated channels and Na-Ca-K exchangers in bovine retinal rod outer segment plasma membranes. J. Physiol., *451*:109-131.

[7] Baumann, A., Frings, S., Godde, M., Seifert, R., and Kaupp, U.B. (1994) Primary structure and functional expression of a Drosophila cyclic nucleotide-gated channel present in eyes and antennae. EMBO J., *13*:5040-5050.

[8] Biel, M., Altenhofen, W., Hullin, R., Ludwig, J., Freichel, M., Flockerzi, V., Dascal, N., Kaupp, U.B., and Hofmann, F. (1993) Primary structure and functional expression of a cyclic nucleotide-gated channel from rabbit aorta. FEBS Lett., *329*:134-138.

[9] Biel, M., Zong, X., Distler, M., Bosse, E., Klugbauer, N., Murakami, M., Flockerzi, V., and Hofmann, F. (1994) Another member of the cyclic nucleotide-gated channel family, expressed in testis, kidney, and heart. Proc. Natl. Acad. Sci. USA, *91*:3505-3509.

[10] Bönigk, W., Altenhofen, W., Müller, F., Dose, A., Illing, M., Molday, R.S., and Kaupp, U.B. (1993) Rod and cone photoreceptor cells express distinct genes for cGMP-gated channels. Neuron, *10*:865-877.

[11] Borisy, F.F., Ronnett, G.V., Cunningham, A.M., Juilfs, D., Beavo, J., and Snyder, S.H. (1992) Calcium/calmodulin-activated phosphodiesterase expressed in olfactory receptor neurons. J. Neurosci., *12*:915-923.

[12] Bradley, J., Li, J., Davidson, N., Lester, H.A., and Zinn, K. (1994) Heteromeric olfactory cyclic nucleotide-gated channels: A new subunit that confers increased sensitivity to cAMP. Proc. Natl. Acad. Sci. USA, *91*:8890-8894.

[13] Brown, R.L., Gerber, W.V., and Karpen, J.W. (1993) Specific labeling and permanent activation of the retinal rod cGMP-activated channel by the photoaffinity analog 8-p-azidophenacylthio-cGMP. Proc. Natl. Acad. Sci. USA, *90*:5369-5373.

[14] Brüggemann, A., Pardo, L.A., Stühmer, W., and Pongs, O. (1993) Ether-à-go-go encodes a voltage-gated channel permeable to K and Ca and modulated by cAMP. Nature, *365*:445-448.

[15] Chen, T.-Y., Illing, M., Molday, L.L., Hsu, Y.-T., Yau, K.-W., and Molday, R.S. (1994) Subunit 2 (or ß) of retinal rod cGMP-gated cation channel is a component of the 240-kDa channel-associated protein and mediates Ca-calmodulin modulation. Proc. Natl. Acad. Sci. USA, *91*:11757-11761.

[16] Chen, T.-Y., Peng, Y.-W., Dhallan, R.S., Ahamed, B., Reed, R.R., and Yau, K.-W. (1993) A new subunit of the cyclic nucleotide-gated cation channel in retinal rods. Nature, *362*:764-767.

[17] Chen, T.-Y. and Yau, K.-W. (1994) Direct modulation by Ca-calmodulin of cyclic nucleotide-activated channel of rat olfactory receptor neurons. Nature, *368*:545-548.

[18] Colamartino, G., Menini, A., and Torre, V. (1991) Blockage and permeation of divalent cations through the cyclic GMP-activated channel from tiger salamander retinal rods. J. Physiol., *440*:189-206.

[19] Cook, N.J., Hanke, W., and Kaupp, U.B. (1987) Identification, purification, and functional reconstitution of the cyclic GMP-dependent channel from rod photoreceptors. Proc. Natl. Acad. Sci. USA, *84*:585-589.

[20] Delgado, R., Hidalgo, P., Diaz, F., Latorre, R., and Labarca, P. (1991) A cyclic AMP-activated K channel in Drosophila larval muscle is persistently activated in dunce. Proc. Natl. Acad. Sci. USA, *88*:557-560.

[21] Dhallan, R.S., Yau, K.-W., Schrader, K.A., and Reed, R.R. (1990) Primary structure and functional expression of a cyclic nucleotide-activated channel from olfactory neurons. Nature, *347*:184-187.

[22] Dryer, S.E. and Henderson, D. (1991) A cyclic GMP-activated channel in dissociated cells of the chick pineal gland. Nature, *353*:756-758.

[23] Eismann, E., Müller, F., Heinemann, S.H., and Kaupp, U.B. (1994) A single negative charge within the pore region of a cGMP-gated channel controls rectification, Ca blockage and ionic selectivity. Proc. Natl. Acad. Sci. USA, *91*:1109-1113.

[24] Fain, G.L. and Matthews, H.R. (1990) Calcium and the mechanism of light adaptation in vertebrate photoreceptors. Trends Neurosci., *13*:378-384.

[25] Fesenko, E.E., Kolesnikov, S.S., and Lyubarsky, A.L. (1985) Induction by cyclic GMP of cationic conductance in plasma membrane of retinal rod outer segment. Nature, *313*:310-313.

[26] Frings, S., Seifert, R., Godde, M., and Kaupp, U.B. (1995) Profoundly different calcium permeation and blockage determine the specific function of distinct cyclic nucleotide-gated channels. Neuron - in press.

[27] Gordon, S.E. and Zagotta, W.N. (1995a) A histidine residue associated with the gate of the cyclic nucleotide-activated channels in rod photoreceptors. Neuron, *14*:177-183.

[28] Gordon, S.E. and Zagotta, W.N. (1995b) Localization of regions affecting an allosteric transition in cyclic nucleotide-activated channels. Neuron, *14*:857-864.

[29] Gotow, T., Nishi, T., and Kijima, H. (1994) Single K channels closed by light and opened by cyclic GMP in molluscan extra-ocular photoreceptor cells. Brain Res., *662*:268-272.

[30] Goulding, E.H., Ngai, J., Kramer, R.H., Colicos, S., Axel, R., Siegelbaum, S.A., and Chess, A. (1992) Molecular cloning and single-channel properties of the cyclic nucleotide-gated channel from catfish olfactory neurons. Neuron, *8*:45-58.

[31] Goulding, E.H., Tibbs, G.R., Liu, D., and Siegelbaum, S.A. (1993) Role of H5 domain in determining pore diameter and ion permeation through cyclic nucleotide-gated channels. Nature, *364*:61-64.

[32] Goulding, E.H., Tibbs, G.R., and Siegelbaum, S.A. (1994) Molecular

mechanism of cyclic-nucleotide-gated channel activation. Nature, *372*:369-374.

[33] Guy, H.R., Durell, S.R., Warmke, J., Drysdale, R., and Ganetzky, B. (1991) Similarities in amino acid sequences of Drosophila eag and cyclic nucleotide-gated channels. Science, *254*:730.

[34] Hanke, W., Cook, N.J., and Kaupp, U.B. (1988) cGMP-dependent channel protein from photoreceptor membranes: Single-channel activity of the purified and reconstituted protein. Proc. Natl. Acad. Sci. USA, *85*:94-98.

[35] Hatt, H. and Ache, B.W. (1994) Cyclic nucleotide- and inositol phosphate-gated ion channels in lobster olfactory receptor neurons. Proc. Natl. Acad. Sci. USA, *91*:6264-6268.

[36] Haynes, L.W. and Yau, K.-W. (1985) Cyclic GMP-sensitive conductance in outer segment membrane of catfish cones. Nature, *317*:61-64.

[37] Heginbotham, L., Abramson, T., and MacKinnon, R. (1992) A functional connection between the pores of distantly related ion channels as revealed by mutant K channels. Science, *258*:1152-1155.

[38] Henn, D.K., Baumann, A., and Kaupp, U.B. (1995) Probing of the transmembrane topology of cyclic nucleotide-gated channels by a gene fusion approach. Proc. Natl. Acad. Sci. USA - in press.

[39] Hsu, Y.-T. and Molday, R.S. (1993) Modulation of the cGMP-gated channel of rod photoreceptor cells by calmodulin. Nature, *361*:76-79.

[40] Hsu, Y.-T. and Molday, R.S. (1994) Interaction of calmodulin with the cyclic GMP-gated channel of rod photoreceptor cells. J. Biol. Chem., *269*:29765-29770.

[41] Ildefonse, M., Bennett, N. (1991) Single-channel study of the cGMP-dependent conductance of retinal rods from incorporation of native vesicles into planar lipid bilayers. J. Membr. Biol., *123*:133-147.

[42] Karpen, J.W., Brown, R.L., Stryer, L., and Baylor, D.A. (1993) Interactions between divalent cations and the gating machinery of cyclic GMP-activated channels in salamander retinal rods. J. Gen. Physiol., *101*:1-25.

[43] Kaupp, U.B., Niidome, T., Tanabe, T., Terada, S., Bönigk, W., Stühmer, W., Cook, N.J., Kangawa, K., Matsuo, H., Hirose, T., Miyata, T., and Numa, S. (1989) Primary structure and functional expression from complementary DNA of the rod photoreceptor cyclic GMP-gated channel. Nature, *342*:762-766.

[44] Kim, M.-S., Morii, T., Sun, L.-X., Imoto, S., and Mori, Y. Structural determinants of ion selectivity in brain calcium channel. FEBS Lett., *318*:145-148.

[45] Kleene, S.J. and Gesteland, R.C. (1991) Calcium-activated chloride conductance in frog olfactory cilia. J. Neurosci., *11*:3624-3629.

[46] Kleene, S.J. (1993) Origin of the chloride current in olfactory transduction. Neuron, *11*:123-132.

[47] Koch, K.-W. (1994) Calcium as modulator of phototransduction in vertebrate photoreceptor cells. Rev. Physiol. Biochem. Pharmacol., *125*:149-192.

[48] Kolesnikov, S.S., Rebrik, T.I., Zhainazarov, A.B., Tavartkiladze, G.A., and Kalamkarov, G.R. (1991) A cyclic AMP-gated conductance in cochlear hair cells. FEBS Lett., *290*:167-170.

[49] Körschen, H.G., Illing, M., Seifert, R., Williams, A., Gotzes, S., Colville, C.,

Müller, F., Dosé, A., Godde, M., Molday, L., Kaupp, U.B., and Molday, R.S. (1995) A 240K protein represents the complete ß-subunit of cyclic nucleotide-gated channel from rod photoreceptors. Neuron - in press -
[50] Kurahashi, T. and Yau, K.-W. (1993) Co-existence of cationic and chloride components in odorant-induced current of vertebrate olfactory receptor cells. Nature, *363*:71-74.
[51] Liman, E.R. and Buck, L.B. (1994) A second subunit of the olfactory cyclic nucleotide-gated channel confers high sensitivity to cAMP. Neuron, *13*:611-621.
[52] Liu, M., Chen, T.-Y., Ahamed, B., Li, J., and Yau, K.-W. (1994) Calcium-calmodulin modulation of the olfactory cyclic nucleotide-gated cation channel. Science, *266*:1348-1354.
[53] Lowe, G. and Gold, G.H. (1993) Nonlinear amplification by calcium-dependent chloride channels in olfactory receptor cells. Nature, *366*: 283-286.
[54] Ludwig, J., Margalit, T., Eismann, E., Lancet, D., Kaupp, U.B. (1990) Primary structure of cAMP-gated channel from bovine olfactory epithelium. FEBS Lett., *270*:24-29.
[55] Ludwig, J., Terlau, H., Wunder, F., Brüggemann, A., Pardo, L.A., Marquardt, A., Stühmer, W., and Pongs, O. (1994) Functional expression of a rat homologue of the voltage gated ether á go-go potassium channel reveals differences in selectivity and activation kinetics between the Drosophila channel and its mammalian counterpart. EMBO J., *13*:4451-4458.
[56] McKay, D.B. and Steitz, T.A. (1981) Structure of catabolite gene activator protein at 2.9 Å resolution suggestes binding to left-handed B-DNA. Nature, *290*:744-749.
[57] McNaughton, P.A. (1990) Light response of vertebrate photoreceptors. Physiol. Rev., *70*:847-883.
[58] Menini, A. (1990) Currents carried by monovalent cations through cyclic GMP-activated channels in excised patches from salamander rods. J. Physiol., *424*:167-185.
[59] Menini, A. and Nunn, B.J. (1990) The effect of pH on the cyclic GMP-activated conductance in retinal rods. In Sensory Transduction, edited by Borsellino A et al. New York: Plenum Press;175-181.
[60] Menini, A., Rispoli, G., and Torre, V. (1988) The ionic selectivity of the light-sensitive current in isolated rods of the tiger salamander. J. Physiol., *402*:279-300.
[61] Molday, L.L., Cook, N.J., Kaupp, U.B., and Molday, R.S. (1990) The cGMP-gated cation channel of bovine rod photoreceptor cells is associated with a 240-kDa protein exhibiting immunochemical cross-reactivity with spectrin. J. Biol. Chem., *265*:18690-18695.
[62] Molday, R.S., Molday, L.L., Dose, A., Clark-Lewis, I., Illing, M., Cook, N.J., Eismann, E., and Kaupp, U.B. (1991) The cGMP-gated channel of the rod photoreceptor cell characterization and orientation of the amino terminus. J. Biol. Chem., *266*:21917-21922.
[63] Nakamura, T. and Gold, G.H. (1987) A cyclic nucleotide-gated conductance in olfactory receptor cilia. Nature, *325*:442-444.
[64] Nawy, S. and Jahr, C.E. (1991) cGMP-gated conductance in retinal bipolar cells

is suppressed by the photoreceptor transmitter. Neuron, *7*:677-683.
[65] Nizzari, M., Sesti, F., Giraudo, M.T., Virginio, C., Cattaneo, A., and Torre, V. (1993) Single-channel properties of cloned cGMP-activated channels from retinal rods. Proc. Roy. Soc. London B, *254*:69-73.
[66] Perry, R.J. and McNaughton, P.A. (1991) Response properties of cones from the retina of the tiger salamander. J. Physiol., *433*:561-587.
[67] Picones, A. and Korenbrot J.I. (1995) Permeability and interaction of Ca with cGMP-gated ion channels differ in retinal rod and cone photoreceptors. Biophys. J., *69*:120-127.
[68] Pugh, E.N.Jr. and Lamb, T.D. (1990) Cyclic GMP and calcium: The internal messengers of excitation and adaptation in vertebrate photoreceptors. Vision Res., *30*:1923-1948.
[69] Putney, J.W.Jr. (1993) Excitement about calcium signaling in inexcitable cells. Science, *262*:676-678.
[70] Restrepo, D., Miyamoto, T., Bryant, B.P., and Teeter, J.H. (1990) Odor stimuli trigger influx of calcium into olfactory neurons of the channel catfish. Science, *249*:1166-1168.
[71] Restrepo, D., Okada, Y., and Teeter, J.H. (1993) Odorant-regulated Ca gradients in rat olfactory neurons. J. Gen. Physiol., *102*:907-924.
[72] Rieke, F. and Schwartz, E.A. (1994) A cGMP-gated current can control exocytosis at cone synapses. Neuron, *13*:863-873.
[73] Root, M.J. and MacKinnon, R. (1993) Identification of an external divalent cation-binding site in the pore of a cGMP-activated channel. Neuron, *11*:459-466.
[74] Root, M.J. and MacKinnon, R. (1994) Two identical noninteracting sites in an ion channel revealed by proton transfer. Science, *265*:1852-1856.
[75] Schaad, N.C., Venecek, J., Rodriguez, I.R., Klein, D.C., Holtzclaw, L., and Russel, J.T. (1995) Vasoactive intestinal peptide elevates pinealocyte intracellular calcium concentrations by enhancing influx: Evidence for involvement of a cyclic GMP-dependent mechanism. Molec. Pharmacol., *47*:923-933.
[76] Sentenac, H., Bonneaud, N., Minet, M., Lacroute, F., Salmon, J.-M., Gaymard, F., and Grignon, C. (1992) Cloning and expression in yeast of a plant potassium ion transport system. Science, *256*:663-665.
[77] Sesti, F., Eismann, E., Kaupp, U.B., Nizarri, M., and Torre, V. (1995) The multi-ion nature of cGMP-gated channel from vertebrate rods. J. Physiol. - in press
[78] Shepherd, G.M. (1994) Discrimination of molecular signals by the olfactory receptor neuron. Neuron, *13*:771-790.
[79] Sugimoto, Y., Yatsunami, K., Tsujimoto, M., Khorana, H.G., and Ichikawa, A. (1991) The amino acid sequence of a glutamic acid-rich protein from bovine retina as deduced from the cDNA sequence. Proc. Natl. Acad. Sci. USA, *88*:3116-3119.
[80] Tanaka, J.C. (1993) The effects of protons on 3',5'-cGMP-activated currents in photoreceptor patches. Biophys. J., *65*:2517-2523.
[81] Torre, V. and Menini, A. (1994) Selectivity and single-channel properties of the cGMP-activated channel in amphibian retinal rods. In Handbook of Membrane Channels: Molecular and Cellular Physiology, ed. Peracchia C. San Diego: Academic

Press; 345-358.
[82] Ward, C.R. and Kopf, G.S. (1993) Molecular events mediating sperm activation. Developmental Biol., *158*:9-34.
[83] Warmke, J., Drysdale, R. and Ganetzky, B. (1991) A distinct potassium channel polypeptide encoded by the Drosophila eag locus. Science, *252*:1560-1562.
[84] Warmke, J.W. and Ganetzky, B. (1994) A family of potassium channel genes related to eag in Drosophila and mammals. Proc. Natl. Acad. Sci. USA, *91*:3438-3442.
[85] Weyand, I., Godde, M., Frings, S., Weiner, J., Müller, F., Altenhofen, W., Hatt, H., and Kaupp, U.B. (1994) Cloning and functional expression of a cyclic-nucleotide-gated channel from mammalian sperm. Nature, *368*:859-863.
[86] Wohlfart, P., Haase, W., Molday, R.S., and Cook, N.J. (1992) Antibodies against synthetic peptides used to determine the topology and site of glycosylation of the cGMP-gated channel from bovine rod photoreceptors. J. Biol. Chem., *267*:644-648.
[87] Yang, J., Ellinor, P.T., Sather, W.A., Zhang, J.-F., and Tsien, R.W. (1993) Molecular determinants of Ca selectivity and ion permeation in L-type Ca channels. Nature, *366*: 158-161.
[88] Yau, K.-W. and Baylor, D.A. (1989) Cyclic GMP-activated conductance of retinal photoreceptor cells. Annu. Rev. Neurosci., *12*:289-327.
[89] Yau, K.-W. (1991) Calcium and light adaptation in retinal photoreceptors. Curr. Opinion Neurobiol., *1*:252-257.
[90] Zimmermann, A.L. and Baylor, D.A. (1992)Cation interactions within the cyclic GMP-activated channel of retinal rods from the tiger salamander. J. Physiol., *449*:759-783.
[91] Zufall, F., Firestein, S., and Shepherd. G.M. (1994) Cyclic nucleotide-gated ion channels and sensory transduction in olfactory receptor neurons. Annu. Rev. Biophys. Biomol. Struct., *23*:577-607.

DISTRIBUTION OF CYCLIC NUCLEOTIDES BINDING SITES DURING DEVELOPMENT IN FROG AND RODENT RETINA

Antonio Caretta

Istituto di Fisiologia Umana, Universita' di Parma
Via Gramsci, 14, 43100 Parma, Italy

INTRODUCTION

Cyclic nucleotides partecipate in the regulation of the nervous cell metabolism in different transduction systems.

Cyclic nucleotides directly regulate ion channel in olfactory cells [1], while in taste cells cAMP is responsible via phosphorylation of proteins for the closure of the K channel [2]. In vision cGMP gates directly ion channels of the rod outer segments [3,4]. In the vertebrate retina cAMP regulates gap junction permeability amongst horizontal cells [5].

In previous papers [6,7], making use of fluorescent analogues of cyclic nucleotides, namely 8-thioacetamido-fluorescein-cAMP and -cGMP, we described the localization of cAMP and cGMP binding sites in the frog retina. cGMP exclusively stained the membrane of the rod outer segments while cAMP stained both the inner nuclear layer and the outer plexiform layer.

In this study the observations were extended in a more systematic way, reporting the distribution of the cyclic nucleotide binding sites during the development of both frog and mouse retina. The histological features of mouse retina are well understood during development [8-11]. Moreover, the results can be compared to the ones obtained with the rd /rd strain, which carries an autosomal recessive photoreceptor degeneration, laying on chromosome 5, characterized by an early onset and a rapid progression [12,13].

The data presented in this paper support the hypotesis of a distribution of cyclic nucleotides binding sites in the retina dependent on developmental modulation.

METHODS

Tadpoles, adult frogs, albino mice at embryonic day (E)16, postnatal (P)0, P3, P5, P7, P8, P10, P12, P14 and adult, and rd-rd mice at P7, P15, P21 were deeply anesthetized and killed by cervical dislocation. Eyes (the heads for tadpoles and E16 mice) were rapidly removed and frozen by plunging into liquid nitrogen cooled pentane. Coronal sections were cut at 10-20 μm on a cryostat. The sections were air dried and stained with SAF-cAMP or SAF-cGMP nucleotides in 130 mM KCl, 20 mM Tris-HCl , 0.5 mM EDTA, 1mM IBMX, pH 7.6. The sections were covered with a dilute solution of the fluorescent nucleotide (1 μM was found to be the most convenient for viewing and photography), a coverslip was applied on it and fluorescence was observed with a Zeiss epifluorescent microscope. Adiacent sections were stained with cresyl violet in order to identify the labelled structure. Photographs were taken with a 25x objective, using a 400 ASA Ilford film, exposure 15 sec. 8-thioacetamido-cAMP (SAF-cAMP) and 8-thioacetamido-cGMP (SAF-cGMP) were synthesized as previously described [6]. After 2 hours incubation on a section no hydrolysis could be observed on TLC. The controls were carried out in the presence of 1000 fold excess of 8-Br-cAMP or 8-Br-cAMP (as chasers) in the fluorescent solution.

Microphotometric measurement of fluorescence

Fluorescence was measured with a sensitive photomultiplier (Ortec mod 9201) on the fluorescence microscope and expressed in arbitrary units. The light source was a 50 watt xenon lamp, the excitation interference filter was at 475 nm, the emission interference filter was at 530 nm. A narrow exciting beam was used (100x objective, effective N.A. 0.15) giving a 20 mm diameter field. The emitted fluorescence was collected with the same objective (100x, N.A. 1.35). The photon signals of the photomultiplier were amplified and integrated (time constant 0.2 sec) to give an analog reading. The fluorescence of the fields, the width of the fluorescent band, together with contiguous reference fields without dots (F_0), were scored and the fluorescence of the band (F_d) was obtained :

$$F_d = \frac{F_n - F_0}{n}$$

F_d was averaged over many fields and divided by the mean volume of the dots to give f_d, the fluorescence per unit volume. The mean volume was derived from the average radius of the dots estimated on photographs (3200 ASA Kodak film, exposure time 8 sec, 100 x objective) and assuming a spheric shape. Further assuming equal fluorescence quantum yield, the concentration of SAF-cAMP in the dots was derived using an experimentally determined proportionality factor K = 48 between the concentration of SAF-cAMP [L] in standard solutions and the fluorescence f_s per unit volume : $[L] = Kf_s$.

f_s was derived from the fluorescence F_s observed in the field with the relationship :

$$f_s = \frac{F_s}{s\pi r^2 h}$$

where r = 10 mm is the field radius, h = 40 mm is the thickness of the fluorescent layer and s = 0.85 is a factor experimentally determined from the integrated distribution of the fluorescence of a thin layer along the optical axis.

RESULTS

Frog

In tadpoles, SAF-cAMP labelling can be observed from the two-legs stage onwards. It is present in the outer plexiform layer and inner segment layer as in adult frogs. Dot like labelling in inner retinal layers has not been observed in tadpoles or frogs. SAF-cGMP labelling on photoreceptor outer segments can be observed in four-legs tadpoles, much weaker than in the adult frog.

Mouse

The first detectable signal of the SAF-cAMP binding sites in the mouse retina can be observed since the late stage of the embryo development (E20). A marked fluorescent band lies on a fairly thick and undifferentiated neuroblastic layer (ventricular layer). The developing ventricular layer lacks entirely an internal organization, although at this age an active cell migration has been observed by some authors. The labelling, however, does not seem to be in proximity of the cell body, but rather on the surface of the layer, where the bulbous protrusions, namely the future inner segments, are developing.

At birth the pattern of labelling resembles the one observed at E20; yet the fluorescent band has become thicker.

As a result of the cell migration and development, at P5 the histological pattern reveals a noticeable degree of differentiation consisting in the presence of a second layer (inner nuclear layer) separated from the outer nuclear layer by the outer plexiform layer. The SAF-cAMP staining at this age still occurs, but greatly strenghtened, as a single dense band on the surface of the inner segment layer.

At P7-8 the retinal labelling is characterized by the appearance of the SAF-cGMP binding sites and the splitting of the cAMP band, together with the presence of bright dots in the ganglion cell layer. SAF-cGMP labelling is concentrated in the area corresponding to the developing photoreceptors, which, in the early stage, is very weakly labelled. The retina now is almost fully developed, only the photoreceptor layer appear to be thinner, as its development has just begun.

More interesting is the presence of a new band labelled with SAF-cAMP in the proximity of the outer plexiform layer, yielding a very dense and bright labelling. Moreover, small and isolated bright dots (2-3 μm diameter) show up in the ganglion cell layer; they are mainly distributed in isolated groups and with a density of about 10 +/- 3 for every 500 μm. The brightness of the dots increases over a period of 4-5 days and then slowly fades away, entirely disappearing by P25.

At P11 SAF-cAMP labels new structures as bright dots at the level of the inner nuclear layer. Their brightness increases up to P20, remains constant for some months, and then slowly disappear by the 9th month.

During aging the SAF-cAMP labelling pattern in the outer plexiform layer undergoes some modification. The brightness and the thickness of the fluorescent band increases up to the third month, then it begins a slow decrease, and, although it doesn't entirely disappear, in the one year old mouse it is present as a faint and thin band.

No modification can be seen in SAF-cGMP staining of photoreceptor outer segments during aging.

Retinal degenerative mice

The distribution of the SAF-cAMP and SAF-cGMP binding sites was also studied in a strain of mice carrying the rd/rd mutation. Retina of the rd/rd mice has the normal histological features during the early postnatal days; at P7, in effect, the rd/rd retina is still undistinguishable from the normal one. The first evident feature of the degeneration is the lack of the development of the outer segments, as they are phagocited by the pigment epithelium. Consequentely SAF-cGMP is not effective in staining the retina, neither at this stage nor later. On the other hand, cAMP binding sites are present in the inner segment and outer plexiform layers, which do not appear to be significantly modified. The brightness of the cAMP band increases over few days up to P14 and very rapidly disappears by P18. As a contrast, the labelling of the ganglion cell and inner nuclear layers has the same age dependent modifications as observed in the normal mice.

Microphotometric measurements

In order to estimate the local concentration of SAF-cyclic nucleotides and its age related changes, we made photometric measurements. In normal mice the intensity of the SAF-cAMP band increases from P1 to P6, then, on appearance of the outer plexiform layer band, its intensity decreases slightly from P6 to P9. From P9 to P90 its intensity increase again and then it remains constant until 1 year. The outer plexiform layer labelling intensity increases from P7 to P20; from P90 its intensity decreases and in 1 year old mice its value is about 30% of the P90 value.

In rd/rd mice we have not been able to measure the intensity of the rod inner segment labelling because of the disturbing effect of the pigment epithelium.

In the outer plexiform layer the intensity of labelling increases from P7 to P12. At this age the intensity is close to that of normal mice (70% of the normal value). Then the intensity of the labelling decreases and at P20 labelling becomes undetectable.

Biochemical characterization

SAF-cGMP and SAF-cAMP stainings were measured for their specificity in the presence of 8-Br-cAMP or 8-Br-cGMP; SAF-cGMP binding shows the same sensitivity to competition from other nucleotides as previously described for frog rod outer segments [7]. SAF-cAMP labelling was displaced by other cyclic nucleotides (8Br-cAMP > cAMP > cIMP > 8Br-cGMP > cGMP). Cyclic nucleotides substituted in position 6 were ineffective. All other nucleotides tested (ATP, GTP, GDP, ADP, GMP, AMP, Adenosine) were ineffective.

Binding experiments to different retinal fractions showed that SAF-cGMP binding occurs only to the rod outer segment fractions, in agreement with microscopy observations and previous experiments. SAF-cAMP binding is mainly restricted to the membrane fraction sedimented at 35% sucrose w/w. The binding curves presented different slopes and affinity. In case of SAF-cAMP binding, it was possible to follow the time course : $t_{1/2}$ = 6 min. Binding of SAF-cAMP was not affected by low and high ionic strength washings of membranes, while it was abolished by trypsin or urea treaments. It was greatly reduced and abolished by mild detergent treatments.

DISCUSSION

This paper first report the localization and distribution of highly concentrated cyclic nucleotides binding sites in the frog and mouse retina during the development.

In frogs, SAF-cAMP and cGMP labelling in the outer layers of the retina is present like in mice. On the other hand, no punctuate labelling could be observed in the inner retinal layers, possibly suggesting functional differences between amphibian and mammalian retinas.

In mice, the highly concentrated cyclic nucleotide binding sites are detected at birth on the surface of the neuroblastic layer (venticular layer). The photoreceptors and the horizontal cells have already undergone the last mythosis, which occurs at E15 and are developing the vitreal processes and the inner segments as bulbous protrusions. It is possible that the binding sites are lodged in the precursors of the inner segments. This hypothesis is supported by the fact that the labelling increases as the bulbous protrusions first, and the inner segments afterwards get larger during the early postnatal development. It is, however to be pointed out that bulbous protrusions are already present, although less developed since E15, when SAF-cAMP is not effective in labelling the retina. It can be concluded that these sites do appear during the late development of the inner sgments and do not stain cells in active mythosis.

At P7-8 SAF-cAMP stains the outer plexiform layer. The labelling is contemporary to the late maturation of synapses between photoreceptors and bipolar cells. The development of the outer plexiform layer begins with the formation of diad configuration of synaptic element, and at P8 their complexity has markedly increased. By P10 the triad configuration forms. Therefore, it cannot be ruled out any relationship between cAMP binding sites and synapse elements during development, although the density of the sites increases steadily from P8 to P14, the latter date representing the full maturation of the outer plexiform layer.

The experiments performed on the rd/rd mice provide further information supporting this hypothesis. In this defective strain, cAMP binding sites develop in the external layer of the retina with the same time course and density as in the normal mice until P8. From this date onwards they steadily decline, entirely disappearing by P14.

In the rd/rd mice only the photoreceptor cells are affected by the degeneration, which starts at P8 and is almost complete by P14. The SAF-cGMP binding sites localized in the photoreceptors cells do not develop, while SAF-cAMP binding sites start to appear, but fade away as photoreceptors degenerate. SAF-cAMP labelling in the inner nuclear and ganglion cell layer seems to be not dependent on photoreceptors activity, since the same time course as in normal mice can be observed.

REFERENCES

1. R.S. Dhallan, K. Yau, K.A. Schrader, R.R. Reed (1990). Primary structure and functional expression of a cyclic nucleotide-activated channel from olfactory neurons. *Nature* 347, 184-187.

2. P. Avenet, F. Hofmann, Lindemann (1988). Transduction in taste receptor cells requires cAMP dependent protein kinase. *Nature*, 331, 351-354.

3. A. Caretta, A. Cavaggioni and R.T. Sorbi (1979). Cyclic GMP and the permeability of the disks of the frog photoreceptors. *J. Physiol*. London, 259, 171-178.

4. E.E. Fesenko, S.S. Kolesnikov, A.L. Lyubarsky (1985) .Induction by cGMP of cationic conductance in plasma membrane of retinal outer segment. *Nature* 313, 310-313.

5. M. Piccolino, J.Neyton, H.M. Gerschenfeld (1984) Decrease in gap junction permeability induced by dopamine and 3'5'-adenosine monophosphate in horizontal cells of turtle retina. *J. Neurosci.* 4, 2477-2488.

6. A. Caretta, A. Cavaggioni and R.T. Sorbi (1985) Binding stoichiometry of a fluorescent cGMP analogue to membranes of rod outer segments. *Eur.J. Biochem.* 153, 49-53.

7. A. Caretta and H. Saibil (1989). Visualization of cyclic nucleotide binding sites in the vertebrate retina by fluorescent microscopy. *J. Cell. Biol.* 108, 1517-1522.

8. L.D. Beazley, V.H. Perry, B. Baker and J.E. Darby (1987) An investigation into the role of ganglion cells in the regulation of division and death of other retinal cells. *Dev. Brain Res.* 33, 169-184

9. J.C. Blank, A.M. Adinolfi and R.N. Lolley (1974). Synaptogenesis in the photoreceptor terminal of the mouse retina. *J. Comp. Neurol.* 156, 81-94.

10. J.W. Hinds and P.L. Hinds (1979). Differentiation of photoreceptors and horizontal cells in the embryonic mouse retina: an electron microscopy, serial section analysis. *J. Comp. Neurol.* 187, 495-512.

11. J.W. Hinds and P.L. Hinds (1983). Development of retina amacrine cells in the mouse embryo: evidence for two modes of formation. *J. Comp. Neurol.* 213, 1-23.

12. J.C. Blank, A.M. Adinolfi and R.N. Lolley (1974). Photoreceptors degeneration and synaptogenesis in retinal-degenerative (rd) mice. *J. Comp. Neurol.* 156, 95-106.

13. J.C. Blank and D. Bok (1977) An autoradiographic analysis of postnatal cell proliferation in the normal and degenerative mouse retina. *J. Comp. Neurol.* 174, 317-328.

Light sensitivity of rod photoreceptors: dependency on gain and on capture efficiency.

Gian Michele Ratto

Inst. of Neurophysiology, Via San Zeno 51 Pisa
and *Technobiochip*, Marciana Italy.
Phone: (39) 50 559723. E-mail: fengari@vm.cnuce.cnr.it

Introduction

Light sensitivity is the photoreceptor's capability of responding to light, and it is maximally expressed in the receptors devoted to low-light vision: the retinal rods. Both lower vertebrate and mammalian rods have exquisite sensitivities and, when dark adapted, they are capable of producing a measurable signal in response to the absorption of a single photon[1,2]. It is important to understand the factors that define photoreceptors performances, since their sensitivity is likely to shape and limit the overall performances of the visual system.

The electrical responses evoked by light flashes can be recorded by a suction electrode: the rod outer segment is sucked inside a tight glass pipette so that a large fraction of the dark current flowing between outer and inner segment, is collected and recorded[3,2]. The dark adapted photoreceptor is exposed to brief light stimuli of variable intensities and a fraction of the incoming light is absorbed by the photo pigment. The activated rhodopsin triggers a chain of enzymatic reactions that ends in the closure of membrane channels. The resulting suppression of some of the membrane current is recorded by the suction electrode (see references 4, 5 and 6 for reviews on the vertebrate photoreceptor's transductive process).

Rods light sensitivity depends both on the efficiency with which light is captured by the cell and on the gain of the mechanism that converts the photochemical event in the electrical response. Modulating either one or the other factor does modify the photoreceptor sensitivity. This paper will show how the dark adapted light sensitivity is determined both by the amplitude of the response to a single photoisomerisation and by the capture efficiency of the photoreceptor.

The intensity-response curve

Photoreceptors respond to flashes of increasing intensities with a suppression of the light-sensitive current of increasing amplitude. Figure 5 shows the responses of two rodent photoreceptors to flashes of various intensities. When the amplitude of the normalised response is plotted against the logarithm of the flash intensity, the data points fall on a characteristic sigmoidal curve. Lamb, McNaughton and Yau[7] found a simple and elegant argument predicting this curve. According to their model the absorption of a single photon ends in the complete closure of the channels placed on a section of length Δ centred around the photoisomerisation site (see figure 1). The average response $R(I)$ to a flash of intensity I (photons per flash per unit of length) will be:

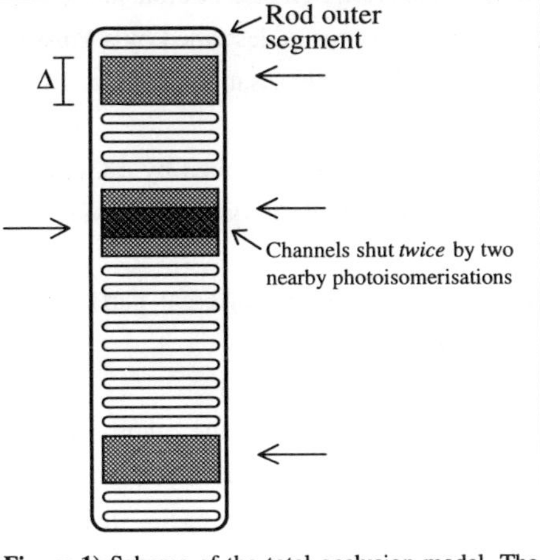

Figure 1) Scheme of the total occlusion model. The elongated shape is the Outer Segment (OS) containing the stacked disks. Each absorbed photon (horizontal arrows) shuts the light sensitive channels along a length Δ of the OS. The total response is given by the summed contributions of all photoisomerisations. Notice how two nearby photoisomerisations can produce a comparably small event due to the non-linear summation (grey area).

$$1) \qquad R(I) = L \cdot \langle r(I) \rangle$$

where L is the outer segment length and $\langle r(I) \rangle$ is the average response of each outer segment section. Following the total occlusion model the channels on each section of the rod outer segment will either be shut, if at least one photoisomerisation occurred in the neigh borough of width Δ, or they will be left undisturbed. Therefore

$$2) \qquad \langle r(I) \rangle = \frac{R_{Max}}{L} \cdot \left(1 - e^{-\Delta\beta I}\right)$$

where R_{Max} is the saturating response, Δ is the length of the occluded region and β is the fraction of incoming light which is absorbed by the photoreceptor. The term $1 - e^{-\Delta\beta I}$ is the probability that one or more events occur in the neighbour of amplitude Δ. Finally the intensity-response curve is:

$$3) \qquad R(I) = R_{Max} \cdot \left(1 - e^{-\Delta\beta I}\right)$$

From this equation we can compute the parameter I_2 which is the light intensity required to shut half of the light-sensitive current and it is a good indicator of the rod dark adapted light sensitivity:

$$4) \qquad I_2 = \frac{\ln(2)}{\Delta\beta}$$

A consequence of this scheme is that the amplitude of a flash response depends also on the location where the photons are absorbed. For example if two or more photons are captured by the same disk, the resulting response will be identical to the single photon response. Of course the total occlusion model idealises the real situations and a further factor that has to be taken in account when predicting the response amplitude is the variable size of each single photon event. This variability does not change the shape of the intensity-response curve, since this relationship estimates the size of an average flash response, but it becomes important when studying the variance of the response[8,1].

These arguments demonstrate that the variability of a response to a non-saturating flash is given by the sum of three separate contributions:

a) The variable number of photoisomerisations: this is often referred as the photon shot noise, and it is inherent to the quantic nature of light. This noise source is described by a Poisson distribution, therefore the variance of the number of photoisomerisation is equal to its mean value.

b) The intrinsic variability of a single photon response. This is well described by a Gaussian distribution and its variance has to be determined by performing the quantal analysis (see following section).

c) The variable amount of superimposition of responses occurring in different parts of the photoreceptor.

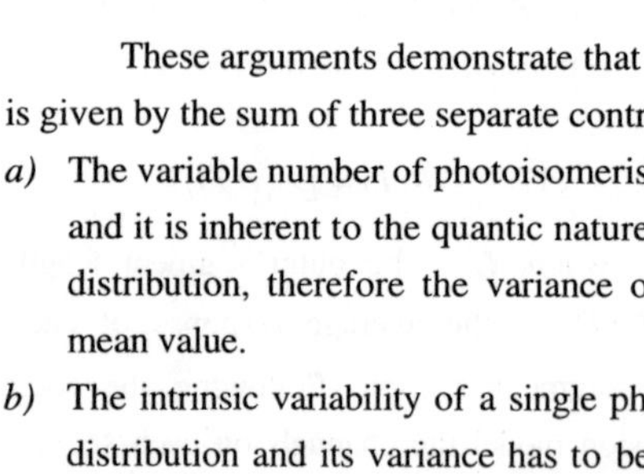

Figure 2) Large symbols: intensity-response plot for two albino rat rods. The number of photoisomerisations occurring at the dimmest flash intensity has been computed from each cell quantum analyses. The crosses have been computed by the Monte Carlo simulation employing the single photons response amplitude and variance obtained from the quantal analysis of the dim flash responses of the two cells. The smooth line is a best fit to the saturating exponential of equation 3.

This scheme of the photoreceptor response can be implemented in a Monte Carlo simulation. In each numerical experiment the response amplitude is computed as the end result of three random processes:

a) Given the flash intensity and the capture efficiency of the photoreceptor, the number of photoisomerisations actually occurring is computed as a variable regulated by a Poisson distribution with a given mean value.

b) The amplitude of the response to each photoisomerisation is computed as a Gaussian variable with mean and variance obtained by the quantal analysis.

c) The events are summed by placing them

on randomly chosen locations along the outer segment.

Figure 2 shows the intensity-response curve from two cells plus the values obtained by the numerical simulation. The continuous line is a saturating exponential fitted to the simulation data points. The standard deviation of the response can be computed by employing the same method, as it will be shown in figure 6.

Recently Pugh and Lamb completed the description of a model of the photoresponse activation[9] based on the present knowledge of the phototransduction cascade. Interestingly the intensity-response relationship they predicted has the same form of equation 3.

The single photon response

In the previous section we established a firm link between the cell sensitivity and its single photon response. The current response to a single photon event has been observed first in amphibian rod photoreceptors[1] and more recently in monkey and rodents rod photo-receptors[2,10,11]. Panel *A* of figure 3 shows a number of responses to dim flashes eliciting in average about 0.8 photoisomerisations.

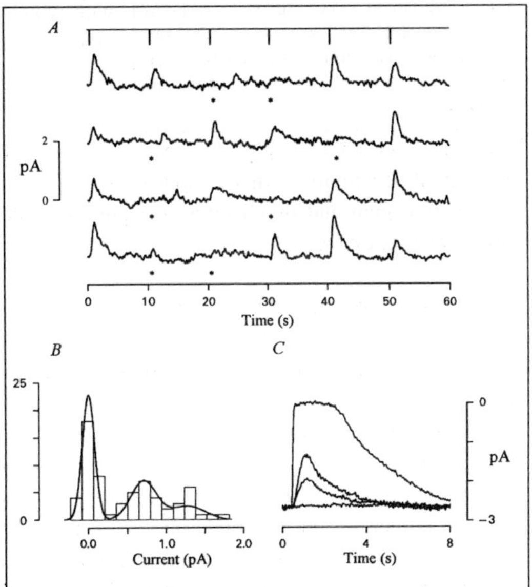

Figure 3) ***A***: responses of a dark adapted rat rod held at room temperature to 24 dim flashes (3.5 Φ/μm^{-2}). The asterisks indicates flashes that failed in evoking a response. ***B***: amplitude histogram of the responses to 64 flashes. The continuous line is the sum of three Gaussians. ***C***: Saturating response and responses to failures, one and two photoisomerisations. Modified from reference 10.

A computerised procedure measures the amplitude of the response to each event (for the numerical methods involved see references 1, 2 and 11). The amplitude histogram is plotted in panel *B*. The continuous line is the sum of three Gaussians, the three peaks corresponding to the responses evoked by failures and by one and two photoisomerisations.

From the histogram is possible to establish size criteria for the events belonging to each of these three groups, and the variance of the single photon response. These data are used in

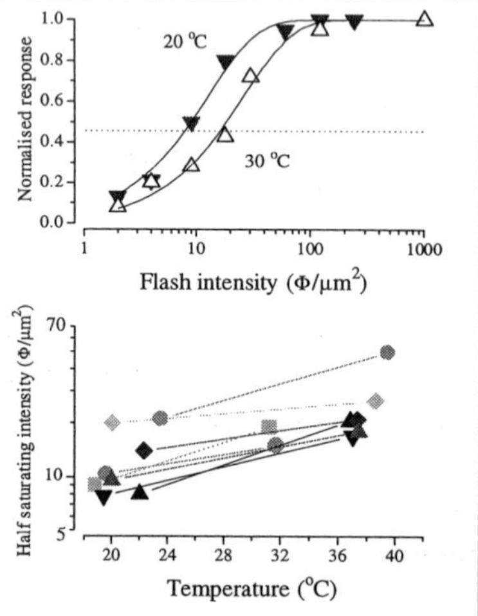

Figure 4) Upper panel: Intensity response relationship for a cell kept at two different temperatures. The smooth lines are best fit to equation 3. Lower panel: the effect of temperature on I_2.

the Monte Carlo simulations. Panel *C* shows, together with the saturating response, the averaged responses for failure, one and two photoisomerisations: the last response is smaller than twice the single photon response because of the non-linear summation showed in figure 1.

Single photon response: dependency on temperature

Equation 4 shows that the sensitivity, defined as the light intensity required to elicit an half saturating response, is function both of the efficiency with which photons are absorbed (β) and by the gain of the cascade converting the photoisomerisation in channels closure (Δ). Agents modulating either one or the other factor modify the photoreceptor's sensitivity. As an example changes in temperature affect light sensitivity both by modifying the photo-transduction gain and by changing the photoreceptor capture efficiency[11].

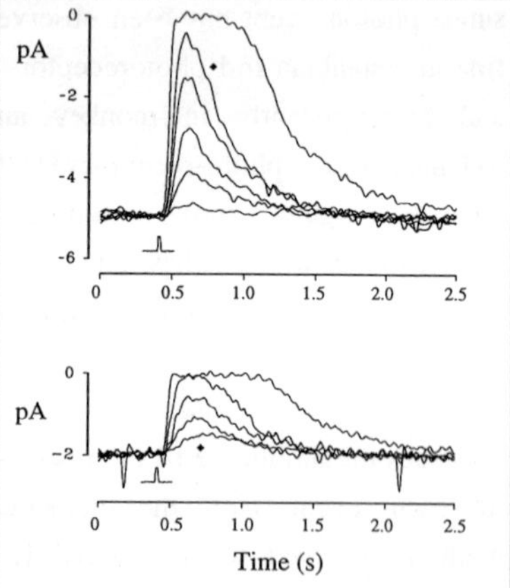

Figure 5) Responses recorded from P35 (upper panel) and P13 rat rods. The flash intensities were 55, 210, 300, 450, 1150 and 7600 $\Phi/\mu m^{-2}$ (upper panel) and 1140, 2300, 6300, 14000 and 44000 $\Phi/\mu m^{-2}$ (lower panel). Modified from reference 11.

The upper panel of figure 4 shows the response-intensity curves for a rod held at 20 and 30 °C. At higher temperature the data points shift to the right indicating a decrease in sensitivity. In this cell I_2 decreased with temperature from 9.4 $\Phi/\mu m^{-2}$ to 18.2 $\Phi/\mu m^{-2}$, with an overall decrease in sensitivity of almost 0.3 log units. The decrease in sensitivity is due either to a decrease in the single photon response or to a decrease in rhodopsin quantum efficiency. As a matter of fact the observed change in sensitivity is due to the occurrence of both events.

The single photon response was determined by performing the quantal analysis on a number of cells held at room and body temperature. We found that, in average, the normalised single photon response decreased from 19% of the dark current at room temperature to 10% at body temperature. In the same cells I_2 increased from 11.4 $\Phi/\mu m^{-2}$ to 30.0 $\Phi/\mu m^{-2}$. Therefore the decrease in gain (53%) is not sufficient to fully account for the observed decrease in sensitivity (38%).

This discrepancy is due to a reduction in the rhodopsin in quantum efficiency. This can be seen easily by looking at the number of failures at both temperatures. This is a measure of the capability of the rod at capturing light and in these cells the failure probability increased in average from 0.44 to 0.55. The sensitivities changes measured at different temperatures are then due both to variations in the gain of the phototransductive pathway and to changes in the rhodopsin quantum efficiency.

Development of sensitivity in mammalian rods

The light sensitivity of neonatal mammals is very low. The dark adapted sensitivity for the detection of large stimuli is 50 times lower in a one month old baby than in an adult. Since in the early post-natal life the entire mammalian visual system goes through a complex process of maturation there are certainly many factors contributing to this increase in sensitivity. However it is clear that any change occurring at the very beginning of visual elaboration, *i.e.* at the photoreceptors level, would certainly affect all the subsequent stages of visual perception.

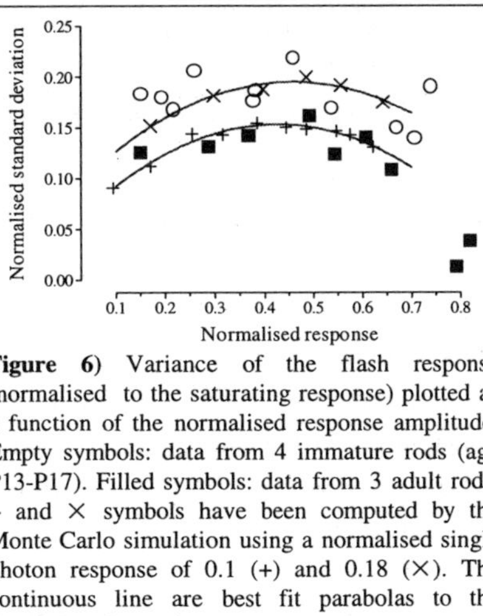

Figure 6) Variance of the flash response (normalised to the saturating response) plotted as a function of the normalised response amplitude. Empty symbols: data from 4 immature rods (age P13-P17). Filled symbols: data from 3 adult rods. + and × symbols have been computed by the Monte Carlo simulation using a normalised single photon response of 0.1 (+) and 0.18 (×). The continuous line are best fit parabolas to the simulation data point.

By recording photoreceptors isolated from rats retinae dissected at different postnatal ages, it has been found that sensitivity is very low in immature photoreceptors and gradually increases during the post-natal development[(11)], paralleling the changes in sensitivity occurring during the maturation of the visual system. Figure 5 shows a series of flash responses for a 35 days old (P35, upper panel) and a P13 albino rat. The kinetic of the responses of the two cells is quiet similar. The current is smaller in the younger animal, reflecting the shorter length of the immature outer segment. In spite of the similitude of the responses there is a large difference in sensitivity. As an example the intensity of the saturating flash increases from 7600 $\Phi/\mu m^{-2}$ at P35 to 44000 $\Phi/\mu m^{-2}$ at P13. The traces indicated with a diamond symbol have been elicited by flashes of similar intensity: it is apparent how the sensitivity of the younger photoreceptor is markedly smaller. The half saturating intensities I_2 were 303 $\Phi/\mu m^{-2}$ for the P35 rod and 3000 $\Phi/\mu m^{-2}$ for the P 13 cell.

But what is the cause for this large increase in sensitivity during the post-natal maturation? As we have seen it must be due either to a change in gain *i.e.* change in size of the single photon response, or to a change in capture efficiency of the photoreceptor. The immature rods are more than 1 log unit less sensitive than adults, if this were due to a lower gain of the immature cascade the single photon response would be at least 10 times smaller than in the adult. Symmetrically the response variance would be much smaller. Figure 6 shows the response variance as a function of the response amplitude. The variance of the immature receptors is larger than the adults', indicating that the single photon response is no smaller in the less sensitive cells, but is actually larger. The figure also shows data resulting from Monte Carlo simulations computed for two different sizes ot the single photon response.

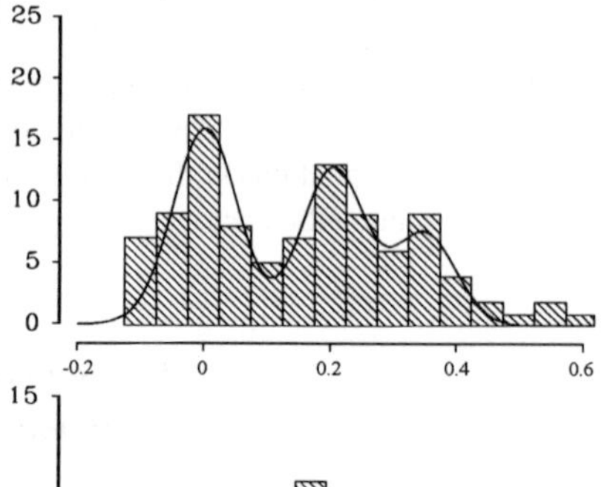

Figure 7) Amplitude histograms for the dim flash responses from an adult (upper panel) and a P13 rod. I_2 for the two cells were 53 and 3000 $\Phi/\mu m^{-2}$ respectively. The normalised single photon responses was 4% (adult) and 21% (P13). Modified from 11.

In agrement with the simulation, the quantal analysis shows that immature photoreceptors have a larger single photon response. In seven adult cells the single photon response was 0.50 pA or 13.9% of the dark current while in 4 immature cells (younger than P17) it was 0.58 pA or 23.7% of the dark current. Therefore the gain of the phototransductive pathway is about two times larger in immature photoreceptors, when expressed as fraction of closed channels (Fig. 7).

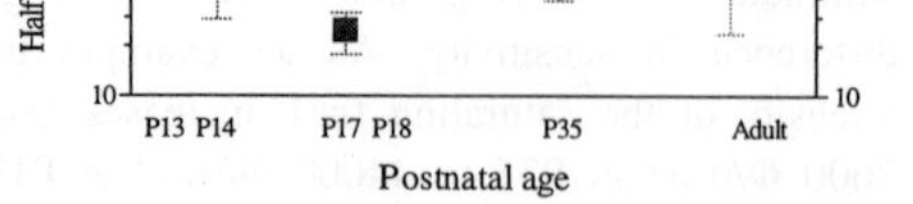

Figure 8) Development of light sensitivity. The diamond symbols indicates sensitivity at various ages in normal photoreceptors. Square symbols indicates sensitivity after treatment with 9-cis-retinal. The bars indicate the standard deviation.

Since the neo-natal photoreceptors have normal gain their low sensitivity must be due to a very low capture efficiency because of a low density of functional pigment in the immature cells. Rhodopsin is constituted by two components: a protein (opsin) which is coupled to a molecule of the light-sensitive cromophore (11-*cis* retinal). Therefore the very low concentration of functional pigment can be due either to a very low concentration of opsin incorporated in the disks, or to a cromophore deficit. If this were the case it would be possible to restore normal sensitivity in immature rods, just by esogenously adding the cromophore. Figure 8 shows that the treatment with cromophore induces a large sensitivity increase in the immature photoreceptors: these experiments prove that neonatal photoreceptors have a full complement of protein and that the cromophore is missing. Why would that be the case? Retinoids are known for being important messenger during the morphological development. Possibly, during the early post-natal development, the retina protects itself from the retinoids' morphogenic action by inhibiting their entire methabolic pathway. This inhibition would be removed once that the retina reaches maturity, and, only at this stage, retinal would be allowed of fully playing its role in the photoreceptors. Of course these speculations are carrying us away from the original purpose of this paper, but they demonstrate how the study of the dark-adapted sensitivity is not only usefull for the comprehension of the mechanisms of phototransduction, but it is also an important tool to further explore the cellular basis of visual perception.

References

1) D.A. Baylor, T.D. Lamb and K.W. Yau (1979). Responses of retinal rods to single photons. *J. Physiol.* **288**, 613-634.

2) D.A. Baylor, B.J. Nunn and J.L. Schnapf (1984). The photocurrent, noise and spectral sensitivity of rods of the monkey *Macaca fascicularis. J. Physiol.* **357**, 575-607.

3) D.A. Baylor, T.D. Lamb and K.W. Yau (1979). The membrane current of single rod outer segments. *J. Physiol.* **288**, 589-611.

4) L. Stryer (1986). Cyclic GMP cascade of vision. *Annual Rev. of Neurosci.* **9**, 87-119.

5) P.A. McNaughton (1990). Light response of vertebrate photoreceptors. *Physiol. Rev.* **70**, 847-883.

6) Y. Koutalos and K-W Yau (1993). A rich complexity emerges in phototransduction. *Current Opinion in Neurobiology* **3**, 513-519.

7) T.D. Lamb, P.A. McNaughton and K.W. Yau (1981). Spatial spread of activation and background desensitization in toad rod outer segments. *J. Physiol.* **319**, 463-496.

8) P.G Lillywhite and S.B. Laughlin (1979). Transducer noise in a photoreceptor. *Nature* **277**, 569-572.

9) T.D. Lamb and E.N. Jr Pugh (1992). A quantitative account of the activation steps involved in phototransduction in amphibian photoreceptors. *J. Physiol.* **449**, 719-758.

10) D.W. Robinson, G.M. Ratto, L. Lagnado and P.A. McNaughton (1993). Temperature dependence of the light response in rat rods. *J. Physiol.* **462**, 465-481.

11) G.M. Ratto, D.W. Robinson, B. Yan and P.A. McNaughton (1991). Development of the light response in neonatal mammalian rods. *Nature* **351**, 654-657.

LIGHT SENSITIVITY IN PINEAL PHOTORECEPTORS OF TELEOSTS. A COMPARISON WITH THE PROPERTIES OF RETINAL PHOTORESPONSES

Claudia Kusmic

Dipartimento di Fisiologia e Biochimica. Universita' di Pisa.

Via S. Zeno 31. 56127 Pisa. Italy

Within the last decades, many investigators have assessed the existence and the functional importance of extra-retinal photoreception, in both invertebrates and vertebrates. Phototactic responses as well as pigment adaptation of the skin to light are still present in fish and amphibians even after the removal of the eyes [1,2]. In addition, extra-retinal photoreceptors seem to mediate the effect of light on the "biological clock" by neural and neuroendocrine mechanisms in all classes from mollusks to higher vertebrates [3,4]. It follows that extra-retinal photoreceptors deserve attention just because they are able to record the most evident phenomenon of the environment, namely the daily alternation between light and dark. Pineal organ is an important component of the biological clock and, as a part of the photoneuroendocrine system, is involved in the regulation of many autonomic functions in response to changes in photoperiodic environment [5]. In addition, light influences the synthesis of melatonin and other indolaminergic compounds with nocturnal peak and low day-time levels [6-9]. The light-dependent wavering in the level of melatonin production is highly conserved throughout all vertebrate classes, despite the fact that pineal cellular types and the control of their activities by light underwent remarkable modifications during the phylogenesis [10,11]. In higher vertebrates photic information is collected by retinal photoreceptors and transmitted to the pineal organ throughout a complex neural pathway involving the hypothalamus, the suprachiasmatic nucleus, the superior cervical ganglion and some fibers of the sympathetic system [12]. In lower vertebrates light directly acts on the pineal organ for the presence of true photoreceptors in the organ itself [13,14]. Pineal photoreceptors transmit processed light information to several brain areas by modulating the impulse frequency of second order neurons as well as modulating the endocrine melatonin signal. It is conceivable to attribute special operating properties to pineal photoreceptors considering that the organ is a permanent target of direct sun rays, a very different condition from retinal photoreceptors of the lateral eye, centered on the horizontal plane. In addition, a stronger indication that pineal photoreceptors might not completely conform to the adaptation parameters of retinal photoreceptors [15-16] derives from the evidence that the voltage-dependent synthesis of indolamines has a steady relation with the intensity of daylight [17]. This is not the case if pineal responses should undergo a similar time-dependent relaxation to that observed on light adapting retinal photoreceptors during prolonged illumination.

Several previous studies defined some of the electrical and the spectral properties of pineal photoreceptors of teleosts [18-24]. The present work is aimed to analyze the response of pineal photoreceptors to brief and prolonged illumination and the results are discussed in comparison with those obtained in retinal cones of the same species.

Methods. Adult teleosts (*Salmo irideus, Carassius auratus* and *Tinca tinca*) were used for the experiments. Animals were dark adapted for a time between three hours and overnight before enucleation and isolation of the pineal organ. Under 0.25% MS 222 anesthesia, animals were decapitated, enucleated and their eyes immersed in filtered saline solution, containing (in mM): NaCl 153; KCl 2.6; $MgCl_2$ 1; $CaCl_2$ 1.8; HEPES 10; glucose 10; pH 7.35.

In the experiments on pineal organ the skin and the fronto-parietal bone covering the pineal organ were removed under a dissecting microscope and the organ was isolated, freed of connective tissue and placed in the perfusion chamber put on the stage of an inverted microscope.

Enucleated eyes were hemisected and retinas, freed from pigment epithelium, were laid on a nylon net-filter (mesh opening 80 μm) with the photoreceptor layer upside down. The vitreal side of retinas was then gently squeezed against the net filter in order to dissociate photoreceptors and allow them to fall into a small drop of saline solution. This procedure left most photoreceptors undamaged. The aliquot containing isolated photoreceptors was transferred to the perfusion chamber on the stage of an inverted microscope.

Specimens were observed under infrared illumination (>850 nm) with the aid of a TV camera and monitor. Both retinal photoreceptors and pineal organs were illuminated with either white or monochromatic light of different intensities obtained by interposing interference and neutral density narrow band filters (Balzers B40, 12.5 nm bandwidth at -3dB) across two collimated optic paths stemming from a tungsten 100 W lamp, filtered by an interferential heat filter Calflex-A (Balzers) with bandpass ranging from 400 nm to 800 nm. The unattenuated irradiance of the light source at the surface of specimen was 4.95×10^2 μ W/ cm^2 .

Electrophysiological recordings. Pineal organ. Glass ultrafine microelectrodes (Clark Electronic Instr., 1.2 mm/0.5 mm, outer and inner diameter) with a tip resistance of 100-200

MΩ were used for intracellular recordings. Electrodes were filled with a solution of either 2M KCl or 3M K-acetate in bidistilled water, occasionally replaced by 6% Lucifer Yellow in 0.1M LiCl. Signals were fed into a single electrode current-voltage amplifier (VF-1800, Biologic). Amplified signals were stored on the digital tape recorder and analyzed by DATAC, a dedicated data analysis software, generously provided by Dr. D. Bertrand.

After recording intracellular responses Lucifer Yellow was injected by applying negative current steps (500 ms at 1 s^{-1}, -1 nA, 1 min). The specimen was then fixed in a 4% *para*-formaldehyde solution in 0.1 M phosphate buffer (pH 7.35) at 4°C for 2 hr, rinsed in buffer, dehydrated, clarified and mounted on a coverslip for the subsequent observation at a Leizt fluorescent microscope. Photoreceptors were identified by their short outer segment and one or more ending varicosities at the other cell extreme. The coarse morphology of injected pineal cells resembles that of pineal photoreceptors previously identified by electron microscopic observation (TEM and SEM) of disks in the outer segment.

Retinal photoreceptors. Patch-pipette current recordings were made in the "whole-cell" configuration from isolated single cones. The pipettes were pulled using a two-stage horizontal puller (Mecanex BB-CH). They had a tip diameter of 0.5-1 μm and when filled with artificial intracellular solution had a resistance of 4-10 MΩ. The basic patch-pipette solution contained (in mM): KCl 110; $MgCl_2$ 3; $CaCl_2$ 1; EGTA 9; HEPES 10; Na_2-ATP 1; Na_2-GTP 1; pH 7.25. Pipette positioning was observed under infrared illumination on the TV monitor. The patch pipette was centered on the cone and then carefully lowered on the inner segment while monitoring its conductance (using rectangular pulse of 0.2 mV). Upon the increase of pipette resistance and the visual evidence of contact, a gentle negative pressure was applied until the membrane sealed. Sometimes the patch ruptured spontaneously, but often an additional negative pressure was required. The whole cell current measured with tight-sealed electrodes at -40 mV (estimated dark resting potential) was nearly zero, photocurrent were outwardly directed and were displayed as positive in polarity.

The patch clamp amplifier was a DAGAN Model 8900. The pipette current signals were stored on a digital tape recorder (DTR-1200, Biologic) and subsequently analyzed by AXOTAPE and pCLAMP (Axon Instr.) data analysis programs installed on a PC computer.

Results Pineal photoreceptors share many morphological, physiological and biochemical analogies with retinal rods and cones. They are ciliary in origin, and are endowed with a typical outer segment for their sensory function as well as with an inner segment containing the nucleus and all the organelles necessary for cellular metabolism and extending into a synaptic pedicle. Their inner segment is divided into two portions connected by a neck, while the outer segment is inserted on the apical border of the distal portion of the inner segment The outer segment is about 1.5-3 μm long and consists of about 100-120 membrane invaginations, which contain the photopigments as the disks of retinal cones.

The average membrane potential recorded from 51 pineal photoreceptors, dark adapted for about 20 min. is -23 mV. Brief light flashes induce hyperpolarizing response increasing in amplitude for increasing intensities within a restricted range (figure 1).

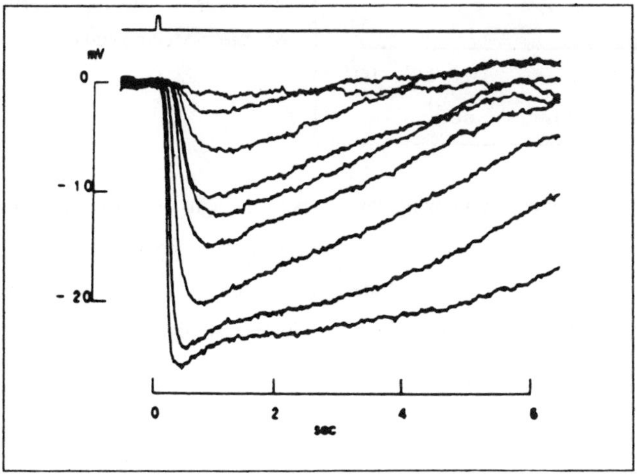

Figure 1. Series of responses of an identified pineal photoreceptor to 50 ms flashes of increasing intensity. Zero in the ordinate indicates the value of the dark membrane potential. The upper line indicates the time of illumination.

The relation between amplitude response and light intensity is well described by a hyperbolic function, a modified form of the Michaelis-Menten equation

$$\frac{V}{V\max} = \frac{I^n}{I^n + I_o^n}$$

where V is the amplitude in mV with maximum V_{max}, , I the flash intensity and I_0 the half saturating intensity. The exponent n is related with the cell sensitivity, i.e. the intensity range over which pineal photoreceptor responds with increasing amplitude responses. Data are best fitted by adding a value of n ranging from 0.7 to 0.85 to the intensity. The modified Michaelis-Menten equation fitting the data shown in figure 2 indicates that the dynamic range of pineal photoreceptors spreads over more than four logarithmic units of light intensity. This value is about twice as wide as that of retinal cones.

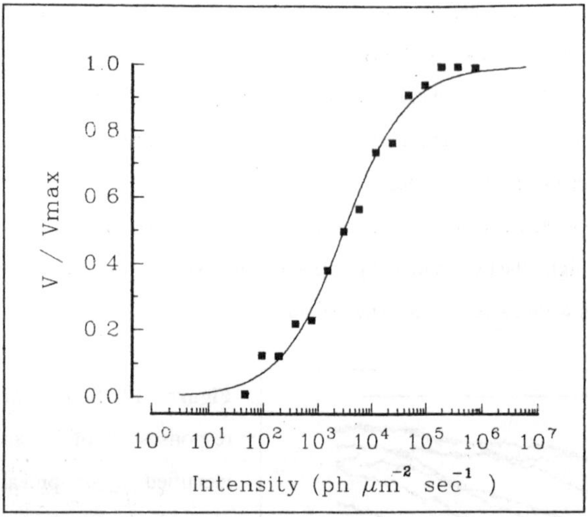

Figure 2. Amplitude-intensity relation of pineal photoreceptors. Data were obtained by averaging the normalized peak response amplitude from 31 cells. Data points are indicated by filled squares and the continuous line represents the modified Michaelis-Menten equation.

Despite of the similarity of the hyperpolarizing responses to the light of pineal and retinal receptor cells, the differences in their time course are evident when comparing the time elapsing from the stimulus onset to the peak amplitude of the response (i.e. the time to peak) and the time for membrane recovery following the flash stimulus as well (figure 3).

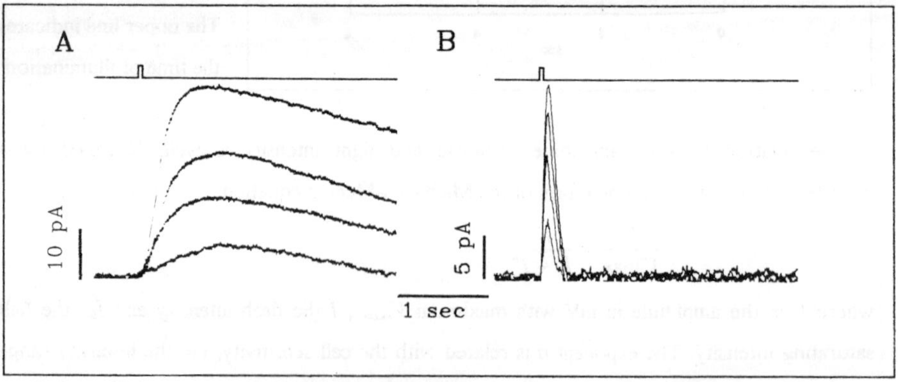

Figure 3. Pineal (A) and retinal (B) series of photocurrents to 50 ms flashes of increasing intensity. Lines above the traces indicate the time of illumination.

The pineal response to near threshold light intensities shows time to peak up to 1.2 sec, decreasing to about 0.25 sec for saturating responses (figure 4).

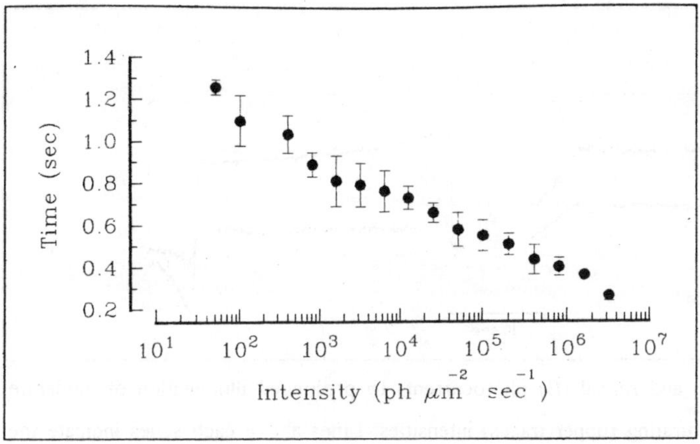

Figure 4. Relation between the time to peak of pineal responses and the intensity of 50 ms flashes measured averaged from 11 photoreceptors. Bars represent s.d. values.

These values are slower than those recorded in retinal cones, in which they vary between about 200 ms to80 ms from threshold to saturating intensities, respectively.

When dark adapted pineal photoreceptor is exposed to prolonged illumination the photoresponse maintains the same amplitude throughout the whole period of stimulation, independently of the light intensity used (figure 5).

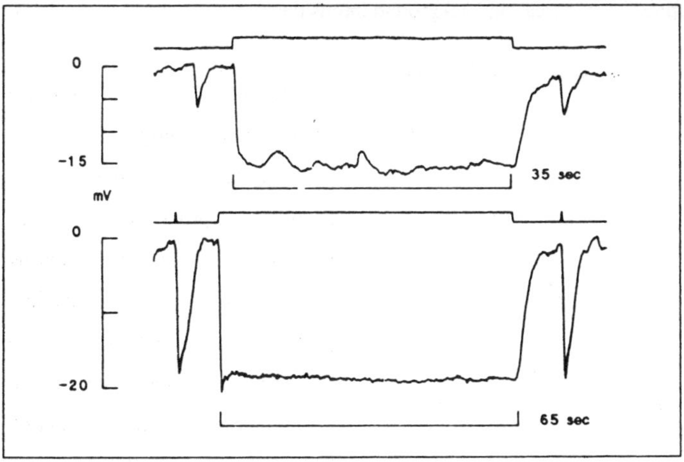

Figure 5. Pineal photoresponses to 50 ms and prolonged light stimulation. Lines above each trace indicate the time of illumination. Light intensities are 1.5×10^3 and 2.5×10^4 ph μm^{-2} sec^{-1}, respectively.

Such steady response represents a specific property of pineal photoreceptors and indicates that, unlike retinal photoreceptors, there are no time-dependent changes of the cell sensitivity to light other than those occurring at the onset of the stimulus (figure 6).

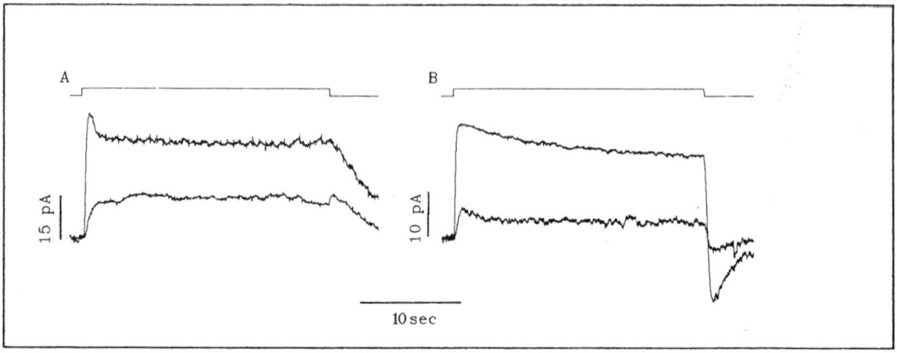

Figure 6. Pineal (A) and retinal (B) photocurrents to prolonged illumination of moderate (lower traces) and saturating (upper traces) intensities. Lines above each series indicate the time of illumination.

Superimposed flash responses are depressed as function of the background intensity. Their amplitude, however, is not modified by the duration of the background illumination. The amplitude-intensity relations plotted in figure 7 show that the background-induced incremental threshold is not associated with any variation of the saturating intensity, whereby the dynamic range is gradually compressed at increasing background intensities.

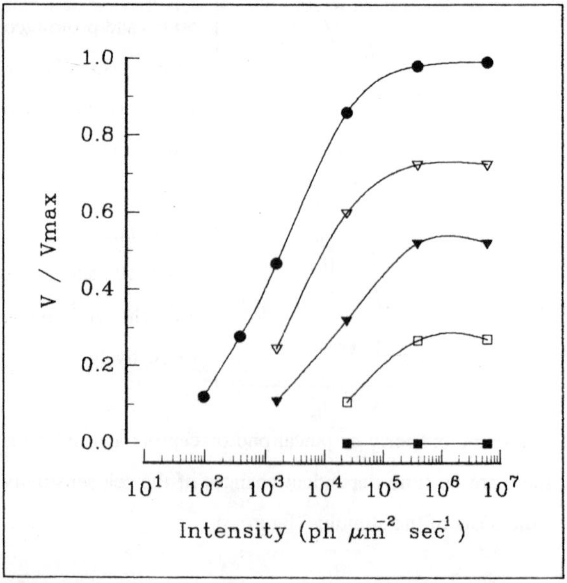

Figure 7. The average amplitude-intensity relations obtained from two pineal photoreceptors in the dark (filled circle) and during background illumination. Background intensities: 4.5 (empty triangle), 45 (filled triangle), 711 (empty square), and 1.11×10^4 (filled square) ph μm^{-2} sec^{-1}. The responses of each cell were normalized to their peak amplitude.

Discussion. Pineal photoreceptors produce hyperpolarizing flash responses similar to those obtained in retinal cones of the same species [23,25,26].

A specific property of pineal photoresponses is the slow time course which is about six times and three times as long as in retinal cones from very dim to saturating intensities, respectively. The slow kinetics does not allow pineal photoreceptors to discriminate quickly changing light intensities as retinal rods and cones do. Indeed, pineal photoreceptors are provided with other peculiar properties which make them highly specialized for detection of slow and gradual changes in the surrounding illumination.

Pineal photoreceptors are clearly distinguishable from retinal receptor cells by their steady amplitude response to prolonged illumination and by the properties of superimposed flash responses. One of the consequences deriving from such a behavior is that the saturating light intensity does not shift towards higher values with time of stimulation, but it remains locked to its dark value, whereby superimposed flash responses are depressed as a hyperbolic function of the background intensity. As a result the pineal photoreceptor dynamic range is gradually compressed with background illumination.

This latter evidence does not agree with the observed recovery of either membrane potential and amplitude of superimposed flash responses occurring in light adapted retinal photoreceptors [27-29] by which they can adjust over an additional 3 logarithmic units with respect to the dark value.

Pineal photoreceptors partly supply for their peculiar behavior by showing a dark dynamic range about 2 logarithmic units wider than that measured in retinal photoreceptors [30-33].

The constant relationship between the membrane potential and the actual level of light intensity observed during a prolonged illumination, - the most physiological condition for these cells -, allows the pineal photoreceptors to lock the cellular metabolism of indolamines and the seric level of melatonin in a steady and fixed proportion with the absolute level of diurnal light intensity.

References

1. K., von Frisch (1911) Beitrage zur Physiologie der Pigmentenzellen in der fishant. *Plugers Arch*., 138, 319-387.

2. T., van Veen, H.G., Hartwig and K., Muller (1976) Light dependent motor activity and photonegative behavior in the eel (*Anguilla anguilla* L.). Evidence for extraretinal photoreception. *J. comp. Physiol*., 111, 209-219.

3. J., Aschoff (1981) *Handbook of Behavioral Neurobiology, Vol IV, Biological Rythms.* Plenum Press, New York
4. H., Underwood and G., Groos (1982) Vertebrate circadian rythms: retinal and extraretinal photoreception. *Experientia*, 38, 1013-1021
5. E., Scharrer (1964) Photoneuroendocrine systems: general concepts. *Ann. N.Y. Acad. Sci.*, 117, 13-22.
6. W.A., Gern, D.W., Owens and C.L. Ralph (1978) Persistence of thr nycthermal rhythm of melatonin in pinealectomized or optic tract sectioned trout (*Salmo gairdneri*). *J.Exp. Zool.* 205, 371-377.
7. W.A., Gern and S.S., Greenhouse (1988) Examination of in vitro melatonin secretion from superinfused trout (*Salmo gairdneri*) pineal organs maintained under diel illumination or continuous darkness. *Gen. Comp. Endocrinol.* 71, 163-174.
8. W.A., Gern, J., Falcon, H. Meissl and M.A., Ali (1992) Melatonin synthesis by teleost pineal: an overview. In: *Rhythms in fishes*, M.A. Ali Ed. New York, Plenum Press, 219-222.
9. H-S., Yu and R.J., Reiter (1993) *Melatonin. Biosynthesis, physiological effects and clinical applications.* CRC Press, Boca Raton, 1-548..
10. J.P., Collin (1971) Differentiation and regression of the cells of the sensory line in the epiphysis cerebri. In: *The pineal gland.* G.E.W. Wolstenholme and J. Knight Eds. London, Churchill-Livingstone, 79-125.
11. A., Oksche (1971) Sensory and glandular elements of the pineal organ. In: *The pineal gland.* G.E.W. Wolstenholme and J. Knight Eds. London, Churchill-Livingstone, 127-146.
12. R.Y., Moore (1978) The innervation of mammalian pineal gland. In: *The Pineal and Reproduction. Progress in Reproductive Biology*, R.Y. Reiter Ed. Base, Karger, 1-29.
13. E. Dodt (1973) The parietal eye (pineal and parapineal organs) of lower vertebrates. In : *Handbook of Sensory Physiology*, R. Jung Ed. Vol VII/3B, Berlin-New York, Springer, 113-140.
14. H. Meissl and E. Dodt (1981) Comparative physiology of pineal photoreceptors organs. In: *The Pineal Organ: Photobiology - Biochronomretry - Endocrinology*, A. Oksche & P. Pevet Eds., Elsevier, 61-80.
15. K., Nakatani and K.W., Yau (1988) Calcium and light adaptation in retinal rods and cones. *Nature*, 334, 69-71.
16. P.A., McNaughton (1990) Light response of vertebrate photorecptors. *Physiol. Rev.*, 70, 847-883.
17. P.J., Morton and H.J. Forbes (1988) Pineal gland N-acetyltransferase and hydroxindole-O-methyltransferase activity in the rainbow trout (*Salmo gairdneri*): seasonal variation linked to photoperiod. *Neurosc. Letters*, 94, 333-337.
18. G.A., Pu and J.E., Dowling (1981) Anatomical and physiological characteristics of pineal photoreceptor cell in the larval lamprey, *Petromyzon marinus. J. Neurophysiol.*, 46, 1018-1038.
19. T., Nakamura, G., Thiele and H., Meissl (1986) Intracellular responses fron the photosensitive pineal organ of the teleost, *Phoxinus phoxinus. J. Comp. Physiol.*, 159, 325-330.
20. P. Ekstrom and H., Meissl (1988) Intracellular staining of physiologically identified photoreceptor cells and hyperpolarizing interneurons in the teleost pineal organ. *Neurosci.*, 25, 1061-1070.
21. H., Meissl and P., Ekstrom (1988) Photoreceptor responses to light in isolated pineal organ of the trout, *Salmo gairdneri. Neurosci*, 24, 1071-1076.
22. C., Kusmic and P.L., Marchiafava (1989) Membrane resistence properties of pineal photoreceptors in the trout. *Exp. Biol.*, 48, 183-186.

23. C. Kusmic, P.L., Marchiafava and E., Strettoi (1992) Photoresponse and light adaptation of pineal photoreceptors in the trout. *Proc. R: Soc: London* B, 248, 149-157.
24. C., Kusmic, L., Barsanti, V., Passarelli and P., Gualtieri (1993) Photoreceptor morphology and visual pigment content in the pineal and in the retina of juvenile and adult trout, *Salmo irideus*. *Micron*, 24, 279-286.
25. P.L., Marchiafava, E., Strettoi and V., Alpigiani (1985) Intracellular recording from single and double cone cells isolated from the fish retina (*Tinca tinca*). *Exp. Biol.*, 44, 173-180.
26. P.L., Marchiafava (1985) Cell coupling in double cones of the fish retina. *Proc: R. Soc. London* B, 226, 211-215.
27. D.A., Baylor and A.L., Hodgkin (1974) Changes in time scale and sensitivity in turtle photoreceptors. *J. Physiol. London*, 242, 729-758.
28. G.L., Fain (1976) Sensitivity of toad rods: dependence on wavelenght and background illumination. *J. Physiol. London*, 261, 71-101.
29. L., Cervetto, V., Torre, G., Rispoli and P., Marroni (1985) Mechanism of light adaptation in toad rods. *Exp. Biol.*, 44, 147-157.
30. D.A, Baylor and M.G.F., Fuortes (1970) Electrical responses of single cones in the retina of turtle. *J. Physiol. London*, 207, 77-92.
31. D.A., Baylor and A.L., Hodgkin (1973) Detection and resolution of visual stimuli by turtle photoreceptors. *J. Physiol. London*, 234, 163-198.
32. G.L., Fain and J.E., Dowling (1973) Intracellular recordings from single rods and cones in the mudpuppy retina. *Science*, 180, 1178-1181.
33.L., Cervetto, E., Pasino and V., Torre (1977) Electrical responses of rods in the retina of *Bufo*. *J. Physiol London*, 267, 17-51.

The retina, a network from the beginning

M. Piccolino, A. Pignatelli and F. Sappia, *Istituto di Fisiologia Generale, Università di Ferrara, Via Borsari 46, 44100 Ferrara (Italy)*

In 12th century, Gerard of Cremona, an Italian scholar working at the court of Toledo in Spain, introduced the word "retina", literally "small net", to render in Latin the Arabian term "rescheth" ("net-like") used by Avicenna to denote the tissue which, inside the eye, invests the vitreous body just "as a net invests the catch" ("*sicut rete comprehendit venationem*"; see ref. 1). The sense of network was absent in the original Greek term used in the Alexandrian school. Even if "retina" is related, at its origin, more to the net of a "hunting net", rather than to that of a network, it appears now a felicitous word to designate a complex network of nerve circuits where important elaboration of the visual information is carried out.

The operative capability of the retina is already displayed at the very beginning of its neural chain. The responses of photoreceptors do not depend exclusively on the light absorbed in their photosensitive region, but are influenced by the illumination of neighbouring retinal areas. In particular, one class of photoreceptors, the cones, are nodal points of a feedback circuit which intervenes in the elaboration of the spatial, chromatic and temporal information at the first synaptic stage of the retina. In this article we will present a rapid review of the experimental studies on the feedback circuit carried out since its discover made in 1971 by Baylor, Fuortes and O'Bryan [(2)], and, moreover, we will report some results of our recent work in this field.

Fig. 1 illustrates, in a simplified way, the pattern of connections existing, in the vertebrate retina, between cones and second order neurons, i.e. bipolar and horizontal cells (see refs. 3,4 for review). From this scheme, three aspects emerge which are particular relevant to the theme of this article: i) photoreceptors are both pre- and postsynaptic to horizontal cells (HCs), the return connection from HCs providing the structural groundwork of the feedback mechanism, ii) HCs establish mutual lateral connections through electrical synapses which allow for a bidirectional communication among them, and iii) the connections between photoreceptors and second order neurons are organised in synaptic complexes partially invaginated in the photoreceptor synaptic terminal. In these complexes a close proximity exists between the three elements of the synapse: the bipolar and horizontal cell processes and the synaptic-transmitter release-sites of photoreceptors. Transmission is chemical and sign-preserving in the forward

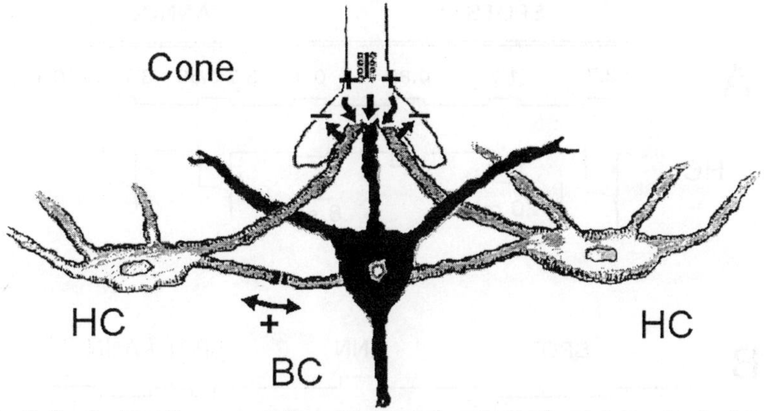

Fig. 1. Semi-schematic representation of the synaptic pathways involving cones, horizontal cells (HC) and bipolar cells (BC) in the outer plexiform layer of the retina. Single arrows indicate sites of presumptive chemical transmission, with the (+) and the (-) symbols corresponding to sign-conserving and sign-inverting transmission respectively. The double-arrow contact represents an electrical synapse (where transmission is bidirectional and sign-preserving). In the cone->bipolar cell synapse, transmission can be either sign-preserving or sign-inverting according to the bipolar cell type.

pathway from photoreceptors to HCs, whereas it is sign-inverting in the feedback synapse from HCs to cones [5-7]. Through the forward synapse the hyperpolarising response elicited in photoreceptors by light absorption leads to a hyperpolarisation of HCs. Through the feedback synapse, on the other hand, HC hyperpolarisation exerts a depolarising influence on cones (Fig. 2). Feedback efficacy depends on the HC response amplitude. This, on turn, depends in an important way on the geometrical extension of the light stimulus. As illustrated in Fig. 2 *A*, the response induced by light spots grows up by increasing stimulus diameter up to several millimetres. The large-area summation of HCs depends principally on the presence of electrical junctions among neighbouring cells. By permitting an easy spread of currents among large numbers of neighbouring HCs, these junctions transform the ensemble of these cells in true functional syncytia. Responses are small with small-size stimuli, because the synaptic current from the stimulated cells is "diluted" as it spreads laterally in the syncytium. Response amplitude grows up by increasing stimulus area since more HCs receive direct synaptic input from the illuminated photoreceptors and less current diffuses laterally to unstimulated cells. The lateral communication in the HC network allows a given cell to generate large-size responses even when the photoreceptors directly connected to it do not receive any direct light stimulus. This happens with large light annuli flashed on the receptive field periphery, and is the consequence of the current flow from the peripheral HCs toward the cell in the dark centre (see Fig. 2 *A*).

Feedback was discovered in the turtle retina as the mechanism responsible for a

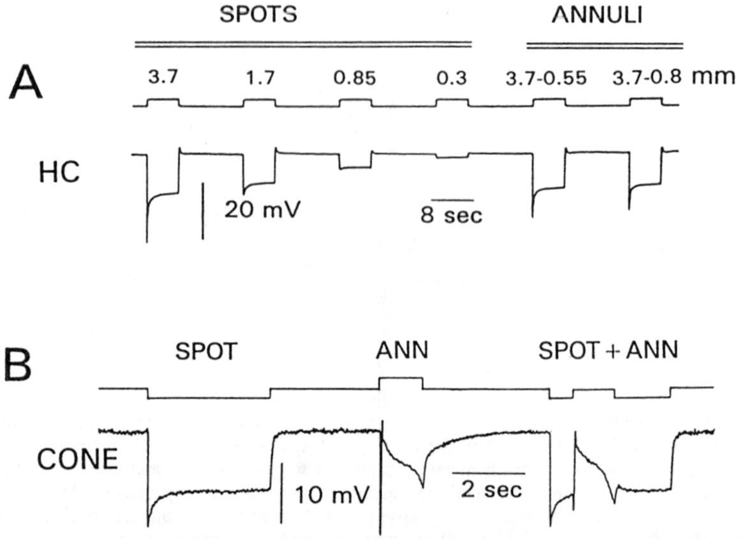

Fig. 2. *A* Intracellular recording of the light responses recorded from a horizontal cell of the turtle retina (HC). The stimuli, monitored by the upper tracing, are steps of white light of different geometrical extension: either spots of light of decreasing outer diameter, or annuli of light of fixed external diameter and decreasing inner diameter (indicated in mm above the stimulus trace). *B* Central and peripheral responses induced in a turtle cone by step illumination of its receptive field. The stimuli are either a 0.18 mm spot, or a 3.700-0.28 mm diameter annulus, or a temporal combination of the spot and the annulus. The hyperpolarising response induced by the annulus alone is largely due to light scattering to the centre of cone receptive field (the annulus was 0.6 log unit brighter than the spot). Notice the depolarising influence of the peripheral stimulation particularly evident when the peripheral stimulus is applied in the presence of the central light spot.

late depolarising deflection appearing in the recovery phase of cone response to large size light stimuli. The demonstration of the involvement of HCs at this level was provided by experiments in which a cone depolarisation was reproduced by direct hyperpolarisation of a HC, obtained through intracellular current injection, in the absence of any light stimulus. Moreover, the depolarising influence of peripheral stimulation could be isolated by flashing the large size stimulus during the maintained hyperpolarising response induced by prolonged central illumination with a small spot (see Fig. 2 *B*). Feedback appeared thus to be responsible for the presence of a centre-surround organisation in the cone receptive-field. Centre-surround antagonism is considered a fundamental aspect of the operative capability of sensory networks, particularly in relation to the analysis of the spatial contrast. Its presence at the very initial level of the visual system underlines its importance in the processing of visual information in the retina (see later).

The synaptic and ionic mechanisms underlying cone feedback responses have been investigated in subsequent studies. O'Bryan [8] demonstrated that feedback depolarisations were associated with an increase of cone membrane conductance. He also showed that feedback responses involved multiple membrane processes, and that under certain circumstances they could become regenerative and display spike-like properties. By studying these "feedback spikes", Piccolino and Gerschenfeld reached the conclusion that a fundamental mechanism underlying feedback depolarisations was an increase of the Ca^{2+} permeability of the cone membrane [9-11]. Feedback spikes were not modified by tetrodotoxin, but were blocked by Ca^{2+} channels antagonists. Their amplitude increased by increasing Ca^{2+} concentration in the perfusing medium or by adding ions capable of permeating Ca^{2+} channels such as Ba^{2+} or Sr^{2+}. As a matter of fact, in the presence of Ba^{2+} or Sr^{2+}, all cones displayed spike responses upon large-area peripheral illumination thus suggesting that feedback could increase Ca^{2+} conductance in all cones, even in those which did not show spikes in control conditions. On the basis of these results it was suggested that feedback acts as a "local" modulator of the release of the synaptic transmitter from cone synaptic pedicles. The release mechanism depends on the opening of Ca^{2+} channels situated in the active zones of the cone synaptic terminal. The processes of HCs penetrate deeply at this level, and thus they may control these channels in an effective way even though the activation of Ca^{2+} channels may not result in an important modification of cone membrane potential. According to this view, the consequence of the feedback action on the membrane potential of postsynaptic neurons could be more important than on cone themselves. For instance, it could bee then easily explained by this way why centre-surround antagonism is more developed in bipolar cells than in cones, even assuming that both depend exclusively on the feedback.

With reference to the membrane mechanisms of feedback responses in cones, it is worth referring here that, in some circumstances, peripheral stimulation can also lead to long duration depolarisations of the cone membrane. These "prolonged depolarisations", which can attain large amplitudes in some cones, are due to the activation of a Ca^{2+}-dependent Cl^- conductance, and are thus secondary to the feedback-induced Ca^{2+} inflow [12,13].

Although the ionic mechanisms associated with feedback responses in cones are relatively well understood, some uncertainness exists as to the synaptic processes underlying feedback responses. The transmission at the feedback synapse is sign-inverting, and thus inhibitory according to the current neurophysiological terminology. The classical view assumes that transmission is chemical at the feedback and that GABA is the neurotransmitter at this level. That GABA could be involved in the genesis of depolarising responses should not be considered surprising, since, as a matter of fact, in the feedback scheme, cone depolarisations would result from a reduction of the release of the putative HC transmitter. There is substantial, although not conclusive, experimental evidence to support the involvement of GABA in the feedback synapse. First, numerous studies indicate that HCs in many species synthesise GABA and release it when their membrane is depolarised, although the release may occur, at least in part, through an unconventional, carried mediated process (reviewed in ref. 14). Moreover, application of GABA, or muscimol, near the synaptic terminal of dissociated turtle cones is capable of evoking a Cl^- current, through the activation of $GABA_A$ receptors [(15)]. The presence of these receptors in the synaptic terminals of cones is also suggested by immunohistochemical studies [(16)].

Against the GABA hypothesis stands the ineffectiveness of bicuculline in blocking feedback responses in some species, including the turtle, where most feedback studies have been carried out [(17)]. There is also uncertainness as to the equilibrium potential of Cl^- in turtle cones, which could be either negative or positive with respect to the resting potential [(12)]. A relatively positive E_{Cl} would be against the feedback hypothesis because, in that case, a sudden reduction of GABA release should cause a hyperpolarisation and not a depolarisation as implied by the feedback scheme [(12)]. In view of these difficulties, other possibilities have been envisioned in order to account for the feedback transmissions. In particular, Byzov and collaborators have proposed that the feedback may act via a direct extracellular field-effect [(18)], a mechanism similar to that operating in fish spinal cord between Mauthner cells and adjacent neurones [(19)]. It has also been suggested that feedback responses could be generated as a consequence of the modifications of the extracellular concentration of glutamate, brought about by a voltage-dependent uptake mechanism present in HCs [(20-21)].

Cones, feedback and second-order neurones

Feedback do not intervenes only in the genesis of the centre-surround organisation of the receptive field of cones and bipolar cells, but it plays also an important role in the analysis of chromatic contrast at the beginning of the visual chain of the retina [(22,23)]. This role emerges particularly in the retina of cold-blooded species. Typically, in many vertebrates there are three different chromatic classes of cones (red, green and blue-sensitive), and there are also different subtypes of horizontals cells. In lower vertebrates HCs may be distinguished on the basis of the colour dependence of their light responses in three main types: i) luminosity-type (or L-HCs), characterised by hyperpolarising responses to all wavelengths, with maximal sensitivity in the red region of the spectrum, ii) red/green chromaticity HCs (RG-HCs), which are hyperpolarised by green lights and depolarised by red lights, and green-blue chromaticity horizontal cells (GB-HCs), which are hyperpolarised by blue lights and depolarised by green lights. Morphological studies indicate that there is a specificity in the connections between different cone types and HCs [(23)]. Correlation between morphological and physiological studies[(22,23)] has led

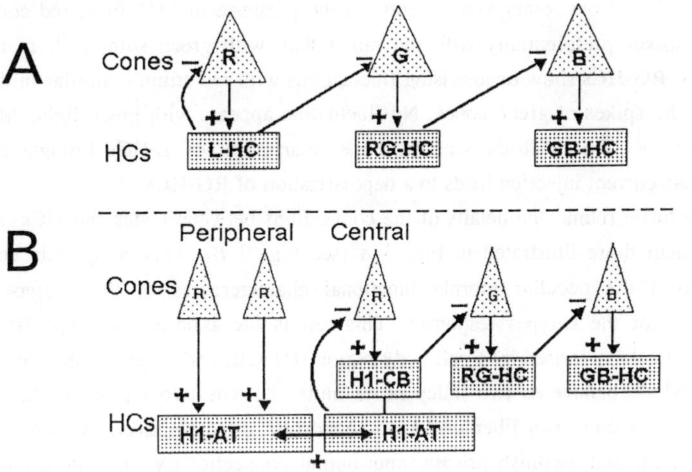

Fig. 3. Diagrams of the feedforward and feedback connections between cones and horizontal cells of the turtle retina in a simple (*A*) and a more detailed version (*B*). (+) and (-) indicate sign-preserving and sign-inverting synapses, respectively. R, G and B are the red, green and blue sensitive cones, while L-HC, RG-HC and GB-HC indicate the luminosity-type, the red-green and the green-blue chromaticity-type horizontals respectively. In *B*, H1-AT and H1-CB stand for the axon-terminal and the cell-body of the axon-bearing horizontal cell (or H1-type) responsible for the luminosity-type response (L-HC). Moreover, "Central" or "Peripheral" refer to the pool of cones inside or outside a circular region of the receptive field of about 1 mm-diameter.

to the model of functional connections illustrated in Fig. 3 *A*. This model assumes that any subtype of HCs receives its dominant input, in the forward pathway, from one chromatic type of cones, and that this input is responsible for the main hyperpolarising responses of each HC: L-HCs from red cones, RG-HCs from green cones and GB-HCs from blue cones. In the feedback pathway, L-HCs would impinge on red and green cones, while RG-HCs would act on blue cones. The depolarisation induced by red lights in RG-HCs would be due to the activation of the loop red cones -> L-HCs -> green cones -> RG-HCs, while the loop green cones -> RG-HCs -> blue cones -> GB-HCs would be responsible for the green-light induced depolarisation of GB-HCs. According to the model, also the responses of green and blue cones should exhibit chromatic opponency. Clear evidence in favour of this possibility has been provided in the turtle retina at least for green cones [24,25]. Changing stimulus wavelength from the green to the red region of the spectrum can lead in these cones to a change of the response polarity, from hyperpolarising to depolarising. However, consisting with the feedback scheme, this happens, only for relatively large-area illumination, whereas small spots evoke hyperpolarising responses if they are green, and fail to evoke any response at all if they are red. Support to the feedback hypothesis of the chromatic opponency of RG-HCs comes out also from other experiments. In the presence of Sr^{2+} ions, red cones show feedback spikes preferentially with red rather than with green stimuli[25]. In the same conditions, RG-HCs show depolarising fluctuations with red stimuli, similar in their time course to the spikes of green cones. No fluctuation appears with green light. Moreover, as implied by the feedback scheme, hyperpolarisation of L-HCs brought about by intracellular current injection leads to a depolarisation of RG-HCs.

In the turtle retina the details of the connections between cones and HCs are more complex than those illustrated in Fig. 3 *A* (see Fig. 3 *B*). This complexity comes out mainly from the peculiar morpho-functional characteristics of the subtype of HC responsible for the L-types responses. This cell is the axon-bearing type HC (or H1 cells), and is characterised by a cell body region (H1-CB) and an expanded axon terminal (H1-AT) which behave as two independent units [26]. The two regions of the cells are connected by a thin axon fiber unable to support an effective electrical communication between them, and establish private input-output connections with photoreceptors. The axon terminals are connected only to red sensitive types of cones, while the cell body contact both red and green sensitive cones. Functionally, H1-ATs are characterised by very large receptive fields, while H1-CBs have relatively small summation areas, the difference being the result of the different degree of electrical coupling existing in the network of H1-ATs compared to the network of H1-CBs[27]. Both H1-ATs and H1-CBs receive their main input from red cones, but, while H1-ATs feedback to red cones, H1-CBs establish a feedback connection with green cones. A surprising property of the H1-CBs is that they exhibit a weak but clearcut centre-surround antagonism in their receptive field, whereby the response induced by diffuse light may be smaller than that elicited by intermediate size spot (about 1 mm-diameter, see ref. 28). This antagonism is not present in the H1-AT receptive field. The experimental evidence suggests that the spatial antagonism of H1-CBs depends on the feedback action exerted by H1-ATs on the red cones situated in their receptive field centre. According to this scheme, the two regions of the axon-bearing HC are capable of indirect, and sign-inverting, communication through the red cones, even though they do not communicate directly through the axon fiber. Moreover, since green cones receive the feedback input from H1-CBs, red-light induced depolarisations of these photoreceptors are brought about preferentially by 1 mm-diameter spots compared to diffuse red lights. Large annuli of red light may actually reduce the feedback influence exerted by relatively small red spots in green cones[29]. This means that the receptive field of green cones is arranged in three concentric areas: i) a small (about 100 μm-diameter) central area, sensitive to green lights, responsible for the hyperpolarising response, ii) a near surround (1 mm-diameter) sensitive to red light and responsible for the feedback depolarisations, and iii) an outer fringe capable of an indirect hyperpolarising influence. This last influence arrives to the green cones through a double-loop circuit (peripheral red cones -> H1-ATs -> central red cones -> H1-CBs -> green cones (see Fig. 3 *B*). On these grounds the notion that cones are the nodal points of a complex elaborations of the visual messages appears to be fully justified.

Oscillations in horizontal cells

We have recently found experimental evidence suggesting that also H1-AT light-responsess may be modified by the feedback(21). The interesting point about these units is that their light responses probably reflect their own feedback influence on the cones from which they receive their main input. The basic observation was that the H1-AT responses induced by long-duration peripheral stimulation with large light annuli could show small oscillatory fluctuations, which never appeared with small light spots, even though these stimuli elicited the same level of maintained hyperpolarisation (Fig. 4). The oscillations consisted of a main frequency typically around 45 Hz, usually modulated in amplitude by slower frequencies. Many of the trigger features of these oscillations correspond to those of feedback responses in cones. First, as already mentioned, oscillations were evoked only by large-area illumination of the receptive field periphery. Moreover, as with feedback cone responses, the oscillations could be blocked by illumination of the receptive field centre with small light spots. Furthermore, perfusion of the retina with Ba^{2+} or Sr^{2+} usually resulted in an increase of the amplitude of the H1-AT oscillations. The frequency of the main component of the oscillation is compatible with the delay of 23 ms estimated in fish for the conduction time in the feedback

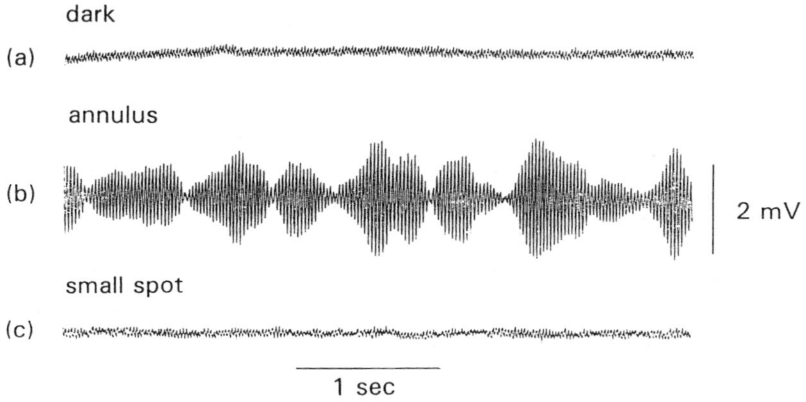

Fig. 4. High-gain intracellular recordings of the membrane potential of a H1-type horizontal cell axon terminal (H1-AT) obtained respectively in the darkness (*a*), after 22 minutes from the onset of fixed intensity illumination with a light annulus (5-1.8 mm outer and inner diameter) (*b*), or after changing the stimulus from the annulus to a small light spot (0.8 mm diameter) (*c*), 28 minutes after the onset of light stimulation. The mean level of maintained hyperpolarization was 17 mV in both (*b*) and (*c*).

loop[30]. On these grounds we suggest that the oscillations are expression of the signal reverberating in the feedback circuit between red cones and H1-ATs. Unfortunately, there is no specific blocker of the transmission at the feedback synapse and thus our hypothesis cannot be submitted to a direct experimental test. It must be noted, however, that there is no report that isolated HCs in culture could display high-frequency oscillations, and thus it is highly probable that these oscillations arise from interactions in the retinal network. If future studies will confirm the feedback origin of the H1-AT oscillations, the study of these responses could provide important information on the kinetics of feedback responses and, more in general, on the mechanism originating oscillatory potentials in the nervous system.

Concluding remarks

A fundamental advance in the study of the retina in modern times has been the recognition of its complicated network organization, appropriate for a tissue, which carries out complex neural operations on the visual input. These operations are apparently aimed to extract from the visual image the elements more rich of information and adaptive value, and to convey them in an optimised form toward the visual centres. They imply the presence of lateral and recurrent interactions between visual channels, whereby the response evoked by the illumination of a point in the retina depends in a more or less complex way on the illumination of neighbouring retinal regions. Electrophysiological evidence for these interactions was first provided in 1953 by Kuffler with the demonstration of the existence of a centre-surround antagonism in the receptive field of ganglion cells in the cat retina[31]. Ganglion cells are the output elements of the retina and their response is the final result of the neural operations carried out by the retinal circuits. The work reviewed in this article shows that important processes of the elaboration of visual signal are carried out already at the initial stage of the visual system, i.e. at the level of photoreceptors. It may be asked why this is so, and, in particular, why these processes are not postponed to later stages of the retinal network. A first hint comes from the consideration that, in most species, ganglion cells are much less numerous than photoreceptors. This poses a great constraint to eliminate the redundance in the visual signal at the initial stages of retinal processing, such as not to overcome the limited capacity of information transport of visual channels at later stages. Moreover, the visual system is faced from the outset by the problem of optimising the signal-to-noise ratio, while keeping a high sensitivity and a wide working range[32-33]. In such a situation a convenient strategy is to amplify the signal prior to the source of noise. Since in the retina, important noise arises in the chemical synaptic transmission from photoreceptors to bipolar cells, it would be thus appropriate to have large signal amplitudes in photoreceptors. However, this could cause saturation in the transmission process, a likely possibility in view of the large gain of transmission at this synapse. To face both redundance and noise problems it becomes then fundamental to remove from photoreceptors any signal poor in information content. This may be done effectively by a feedback circuit. Feedback subtracts from the response of a cone the signal which is correlated in spatial and chromatic domain with the signal of neighbouring receptors. By doing so it eliminates redundance (correlated signal is poor in information content) already at the photoreceptor level. At the same time it prevents synaptic saturation, by reducing the amplitude of the cone signal to be transmitted to postsynaptic neurons. Feedback appears thus to be one of the key mechanisms underlying the ensemble of processes referred to by Barlow as "neat packaging" of information[32]. These processes well show that the retina is basically different from a simple sensory organ aimed to generate and transmit a faithful neural replica of the optical image, and reveal how the definition of true nervous centre is appropriate for the retina, not only embryologically, but also functionally.

REFERENCES

1) S. Duke-Elder, K.C. Wybar and K. Tansley (1961) *The anatomy of the visual system.* Cap. 3/Sec. 4. In S. Duke-Elder (ed.) , *System of Ophthalmology*, London, H. Kimpton, p. 220.
2) D.A. Baylor, M.G.F. Fuortes and P.M. O'Bryan (1971) Receptive fields of cones in the retina of the turtle, *J. Physiol. (Lond.)*, 214, 265-294.
3) M. Piccolino and and P. Witkovsky, P. (1984) *Local circuits in the distal retina of vertebrates* In L. Bolis, R.D. Keynes and S.H.P. Maddrell (Eds.) *Comparative Physiology of Sensory Systems*, Cambridge, Cambridge University Press, pp. 371-404.
4) A. Kaneko (1987) The functional role of retinal horizontal cells, *Jpn.J. Physiol.*, 37, 341-358.
5) Y.A. Trifonov (1968) Study of synaptic transmission between photoreceptors and horizontal cells by means of electrical stimulation in the retina, *Biofizika*, 13, 809-817.
6) A. Kaneko and M. Tachibana (1985) Effect of L-glutamate on the anomalous rectifier potassium current in horizontal cells of *Carassius auratus* retina, *J. Physiol. (Lond.)*, 358, 169-182.
7) L. Cervetto and M. Piccolino (1974) Synaptic transmission between photoreceptors and horizontal cells in the turtle retina, *Science*, 183, 417-419.
8) P.M. O'Bryan (1973) Properties of the depolarizing synaptic potential evoked by peripheral illumination in cones of the turtle retina, *J. Physiol. (Lond.)*, 235, 207-223.
9) M. Piccolino and H.M. Gerschenfeld (1978) Activation of a regenerative calcium conductance in turtle cones by peripheral stimulation, *Proc. R. Soc. Lond. [Biol.]*, 201, 309-315.
10) H.M. Gerschenfeld and M. Piccolino (1980) Sustained feedback effects of L-horizontal cells on turtle cones, *Proc. R. Soc. Lond. [Biol.]*, 206, 465-480.
11) M. Piccolino and H.M. Gerschenfeld (1980) Characteristics and ionic processes involved in feedback spikes of turtle cones, *Proc. R. Soc. Lond. [Biol.]*, 206, 439-463.
12) W.B. Thoreson and D.A. Burkhardt (1991) Ionic influences on the prolonged depolarization of turtle cones in situ, *J. Neurophysiol.*, 65, 96-110.
13) D.A. Burkhardt, J. Gottesman and W.B. Thoreson (1988) Prolonged depolarization

in turtle cones evoked by current injection and stimulation of the receptive field surround, *J. Physiol. (Lond.)*, 407, 329-348.
14) S. Yazulla (1986) GABAergic mechanisms in the retina, *Prog. Ret. Res.*, 5, 1-52.
15) M. Tachibana and A. Kaneko (1984) γ-Aminobutyric acid acts at axon terminals of turtle photoreceptors: Difference in sensitivity among cell types, *Proc. Natl. Acad. Sci. USA*, 81, 7961-7964.
16) S. Yazulla, K.M. Studholme, J. Vitorica and A.L. de Blas (1989) Immunocytochemical localization of GABAa receptors in goldfish and chicken retinas, *J. Comp. Neurol.*, 280, 15-26.
17) W. Thoreson and D.A. Burkhardt (1991) Effect of synaptic blocking agents on the depolarizing responses of turtle cones evoked by surround illumination, *Visual Neurosci.*, 5, 571-583.
18) A.L. Byzov, K. V. Golubtzov and Y.A. Trifonov (1977) *The model of mechanism of feedback between horizontal cells and photoreceptors in the vertebrate retina* In H.B. Barlow and P. Fatt (Eds.) *Vertebrate Photoreception*, New York, Academic Press, pp. 265-274.
19) T. Furukawa and E.J. Furshpan (1963) Two inhibitory mechanisms in Mauthner neurons of goldfish, *J. Neurophysiol.*, 26, 140-176.
20) M. Schütte and E. Schlemermeyer (1993) Depolarization Elicits, While Hyperpolarization Blocks Uptake of Endogenous Glutamate by Retinal Horizontal Cells of the Turtle, *Cell Tissue Res.*, 274, 553-558.
21) M. Piccolino (1995) The feedback synapse from horizontal cells to cone photoreceptors in the vertebrate retina, *Progr. Retin. Eye Res.*, 14/I, 1-59.
22) M.G.F. Fuortes and E.J. Simon (1974) Interactions leading to horizontal cell responses in the turtle retina, *J. Physiol. (Lond.)*, 240, 177-198.
23) W.K. Stell, D.O. Lightfoot, T.G. Wheeler and H.F. Leeper (1975) Goldfish retina: functional polarization of cone horizontal cell dendrites and synapses, *Science*, 190, 989-990.
24) M.G.F. Fuortes, E.A. Schwartz and E.J. Simon (1973) Colour-dependence of cone responses in the turtle retina, *J. Physiol. (Lond.)*, 234, 199-216.
25) M. Piccolino, J. Neyton and H.M. Gerschenfeld (1980) Synaptic mechanisms involved in responses of chromaticity horizontal cells of turtle retina, *Nature*, 284, 58-60.
26) H.F. Leeper (1978) Horizontal cells of the turtle retina-I. Light microscopy of Golgi preparations, *J. Comp. Neurol.*, 182, 777-794.
27) M. Piccolino, J. Neyton, P. Witkovsky and H.M. Gerschenfeld (1982) γ-Aminobutyric acid antagonists decrease junctional communication between L-horizontal cells of the retina, *Proc. Natl. Acad. Sci. USA*, 79, 3671-3675.
28) M. Piccolino, J. Neyton and H. Gerschenfeld (1981) Center-surround antagonistic organization in small-field luminosity horizontal cells of turtle retina, *J. Neurophysiol.*, 45, 363-375.
29) J. Neyton, M. Piccolino and H. Gerschenfeld (1981) Involvement of small-field horizontal cells in feedback effects on green cones of turtle retina, *Proc. Natl. Acad. Sci. USA*, 78, 4616-4619.
30) H. Spekerijse and A.L. Norton (1970) The dynamic characteristics of color-coded S-potentials, *J. Gen. Physiol.*, 56, 1-15.
31) S.W. Kuffler (1953) Discharge patterns and functional organization of mammalian retina, *J. Neurophysiol.*, 16, 37-68.
32) H.B. Barlow (1981) The Ferrier Lecture, 1980. Critical limiting factors in the design of the eye and visual cortex, *Proc. R. Soc. Lond. Biol.*, 212, 1-34.
33) M.V. Srinivasan, S.B. Laughlin and A. Dubs (1982) Predictive coding: a fresh view of inhibition in the retina, *Proc. R. Soc. Lond. Biol.*, 216, 427-459.

600 MILLION YEARS OF VISION EVOLUTION

Silvana Vallerga
Istituto di Cibernetica e Biofisica del CNR, Via De Marini, 6 - 16149 Genova, and International Marine Centre, Lungomare Eleonora D'Arborea 22 - 09072 Oristano.

Introduction

The history of visual systems extends back in time several hundreds of millions of years [3,14,19]. Since then nature has experienced a large number of solutions to best accomplish visual tasks[1]. Two types of evolutionary processes have produced diversity: (i) the *increase in general efficiency* of the system, a slow and continuous process to raise the upper limit of efficiency from the molecular, sub-cellular, cellular and tissue grades; and (ii) *adaptive radiation* to populate the available ecological niches, a faster process from which a species adapts to a new environment. The first life forms that swam in ancient seas developed patches of light-sensitive cells, allowing them to tell light from dark. As life evolved these patches became eyes that could detect first motion, then form and finally colours. New habitats became available, as land and air, requiring performances that were satisfied by specific visual systems. Terrestrial and aquatic animals, diurnal and nocturnal creatures, preys and predators all differentiated their vision according to habitat and/or habits.

The oldest eye of which we have record belongs to the trilobite, an ancient marine creature which lived 500 millions years ago. The trilobite had a compound, faceted eye that could see side but not up. Another ancient creature, that managed to survive to the present day , the *Limulus* resembles the long extinct trilobite. Interestingly, the *Limulus* visual system has been in the early sixties one of the first model systems for studies on vision, through the pioneer work of Borsellino and Fuortes.

1. Forms of vision

Generally three different ways of forming images have evolved [Fig. 1]. The simplest eye scans the scene like a tv camera, building a picture from successive impressions. This system is thought to be the link between the primitive eye-spot and the more complex, image-forming eye[15].This form of vision, the least common, is used by the smallest creature capable of forming a visual image, the marine copepod, copilia.

The second form of vision is the compound eye, widely adopted by invertebrates. This is a fixed-focus eye with individual refractive units, ommatidia, each responsible for a portion of the visual field. The compound eye allows a close view over an almost spherical field of vision free of spherical and chromatic aberration.

A third form of vision is found in vertebrates. The vertebrate eye is a camera-like device that brings an image into focus on the retina, a continuous receptor surface, apposed onto neural layers. The photoreceptors lie in the innermost part of the eye, thus the light crosses the whole retina before being absorbed by the visual pigment molecules, in the photoreceptor outer segments. For this reason the vertebrate retina is called "reverse", in contrast with the "direct" retina of invertebrates.

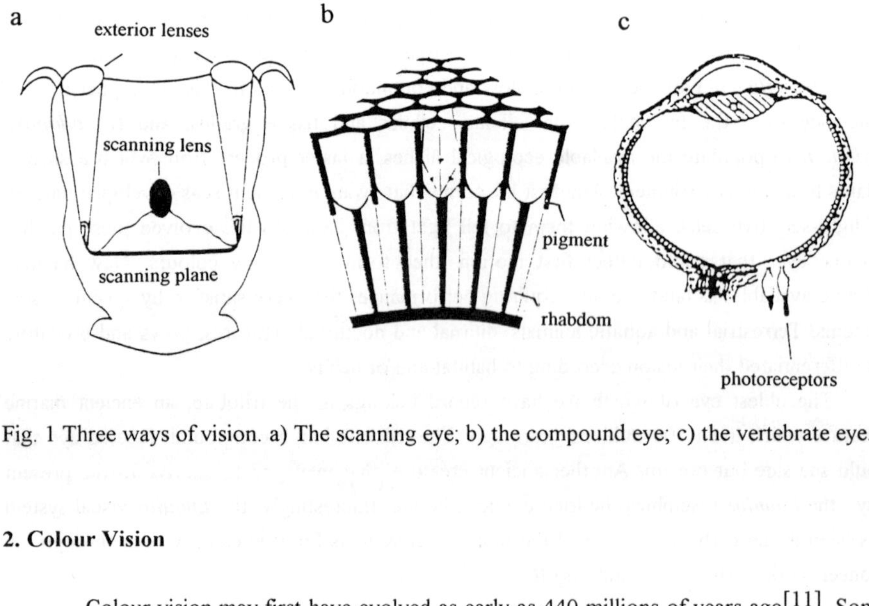

Fig. 1 Three ways of vision. a) The scanning eye; b) the compound eye; c) the vertebrate eye.

2. Colour Vision

Colour vision may first have evolved as early as 440 millions of years ago[11]. Some of the ostracoderm fish gave up their benthic habits at that time and become free-swimming. Their paired dorsally located eyes moved away from the median eye, down onto the sides of the head. No doubt vision became more important for the feeding of these animals. Away from the bottom of the sea the whole visible spectrum became available. Multiple receptor mechanisms were developed for daytime use, when light intensity permits the separate sensory pathways required for high visual acuity and for colour vision.

One can easily image that colour vision has evolved in terrestrial animals in response to the same factors. The advantage of colour vision in providing better photocontrast is so evident that it is difficult to image its gradual development in an evolutionary sequence[13]. Strongest contrast would be the most useful way to solve the need to recognise objects with the same size and form, but differing in important properties as ripe or unripe, flowers with nectar or not, sexes of the same of close species.

3. Invertebrates

In the visual systems of invertebrates we find the largest diversity. From the pigment depression that serves as an eye for the earthworm to the largest eye (370 cm in diameter) of any animal, that of *architeuthis*, the giant squid.

3.1 Spiders

Nearly 300 millions of years ago, during the Devonian period spiders appeared. They have simple eyes, to distinguish them by the compound eyes of other invertebrates, even is these eyes are far from being simple. Spider's eyes occur in clusters of six to eight, they have a form of colour vision, can detect UV and polarised light. Spiders, as do the majority of insects can see clearly up to 40 cm, the best image is at about 8-10 cm. The frontal pair of eyes form the visual image and detect the plane of polarisation of light, which is used as an aid for orientation. The three lateral pairs control the periphery of the visual field, estimate distance and motion. Accommodation is performed shifting the retina back and forth to bring the image in focus. The retina can be also moved sideways to increase the field of vision. Some species are diurnal, as the jumping spider, which is considered as having the best vision of all arachnids. The net casting spiders, a nocturnal species, has the lens of the frontal eyes totally free of any spherical and chromatic aberration, with f-stop of 0.58. This eye can absorb 2000 times as many photons than a human eye.

3.2 Insects

Insects have been very successful in evolution. The number of known insects species is over 700,000. The other groups, except arachnids, 30,000, do not exceed 5,000 species. Insects have compound eyes with a number of units that can reach 28,000 as in the dragonfly giving the animal an almost spherical field of view. Colour vision is typical of most insects. Their photoreceptors are tuned to biologically important signals such as feeding and mating and to the secondary emission of the sky, which shows pronounced peaks in the blue and UV because of the scattering effect of the atmosphere. The distribution of UV photoreceptors allows for the orientation of insects with the so called sky compass, though the detection of polarised light [5].

3.3 Cephalopods

The eye of cephalopods has features common to both invertebrate and vertebrate eyes. The photoreceptors are rhabdom facing the incoming light as in invertebrates, but they are organised in a retina layered on an eye-cup where the image is focused through a single lens as in vertebrates.

4. Vertebrates

The eyes of most vertebrates conform to a similar pattern. They have transparent media, the cornea, lens and vitreous, through which the light passes before it reaches the retina [Fig. 2]. Photoreceptors input signals to the neural network in which information is processed through horizontal, bipolar, amacrine, interplexiform and ganglion cells. From the ganglion cells visual information is sent to the brain via the optic fibres. The human eye as well as all other vertebrates' eyes work on this principle

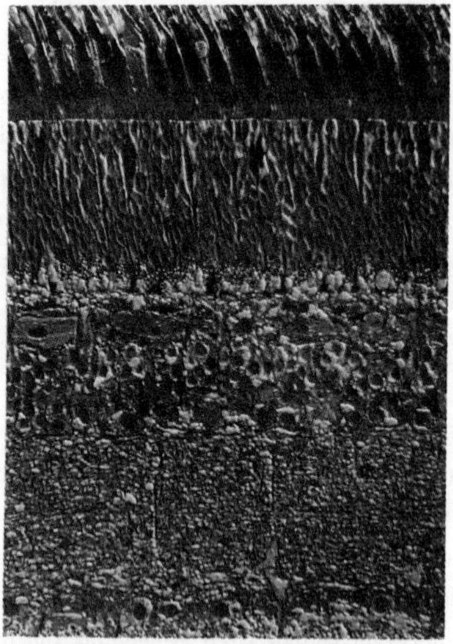

Fig. 2. Photomicrograph of a cross section of a vertebrate retina. C, cones. R, rods, H, horizontal cells. B, bipolar cells. G, ganglion cells. The total width of retina is about 250 μm.

Horizontal cells control the general sensitivity of the retina, mediate laterally the signal, are colour coded, in species with colour vision, and provide surround information to bipolar cells. Bipolar cells begin the coding of forms, which adds to the coding of light intensity and wavelength. They feed such signals directly to ganglion cells and through feedback circuits, to amacrine cells. Amacrine cells provide the surround information to ganglion cells, are colour coded and initiate the detection of movement. Ganglion cells code intensity, movement, colours and forms and feed the visual information to the brain through their long axons. In high vertebrates the more complex visual tasks, as motion detection, are performed by the brain.

4.1 Fishes

The first vertebrates, fish, appeared 600 million years ago. From them originated all the other vertebrate classes. Some ancestral fish are still dwelling in present oceans to prove the success of this group of vertebrates. Vision is undoubtedly the most important sense for fish. They have good perception of forms and colours, their eyes are well developed, and the optic lobe, which also controls the learned behaviour, is the larger part of fish's brain. The eye of fish grows throughout life. Photoreceptors and retinal neurones are continuously added as peripheral rings. The adult fish retina maintains the capacity to adapt to changing environment that other groups have only in juveniles.

All fish have a flattened cornea, in contrast with the curved cornea of land creatures. An important reason for that is that the refractive index of the cornea is almost exactly the same as water's. The fish lens is the most important light-gather. Large and spherical, the fish lens has the highest refractive index of any vertebrate lens. Due to the round shape of the lens fish, with the exception of sharks, are nearsighted, but their eye is partly able to accommodate for far vision by pushing the lens away from the retina.

The evolutionary adaptation of fish visual system is of particular interest since fishes adapted to the widest variety of photic environments, from the sparkling waters of coral reefs, to turbid waters, to oceanic depths. Visual pigments, the sensitive elements in photoreceptors, are strongly correlated with the photic environment, deep-sea species have rod pigments maximally sensitive shorter wavelength than surface-living species, and freshwater fish have their vision shifted toward the longer wavelengths[9]. The neural retina also adapts to habitat. Bottom dwelling fish have the inner retina well developed, indicating a high degree of neural processing, while clear water animals rely more on colour vision for contrast discrimination than on inner retina processing, therefore their outer retina is more developed[18].

Sharks, rays and mantas have the same ancestor as other fish, but their eyes have distinct variations. They are farsighted, their lenses are so perfectly graded that they suffer from very little astigmatism. Their vision is probably inferior for they have fewer photoreceptors and hardly any cones, and most see only in black and white. Their visual world is a grainy and indistinct picture, though they see form at a distance more clearly than do the other fish.

4.2 Amphibians

The amphibians developed from ancestral fish that reached shores to spend on land part of their life. The word "amphibian" means with a double life, since these animals are aquatic as juveniles and terrestrial as adults. Their visual system adjusts during metamorphosis to adapt to the change of habitat. Amphibian visual neurophysiology has provided a great deal of research on the visual system, starting with the work of Lettvin, Maturana, McCulloch and Pitts in 1959[8], that showed that the retina of anurans can process great deal of visual analysis: a) sharp edges and contrast, b) movement of edges, c) local dimming produced by movements, d) the curvature edge of a dark object. Vision, by means of the lateral eyes, is intimately involved in prey catching behaviour, in prey-avoidance and with the extra optic receptors in orientation[6,7]. Further processing of the visual information is accomplished in the optic tectum and elsewhere.

In general the visual system of many amphibians is characterised by a high degree of peripheral (retinal) coding. It is supposed that as we ascend the phylogenetic scale the central nervous system becomes better developed, behaviour more flexible, and the need for peripheral coding decreases[13]. While there may be some truth in this view, it must be remembered that even in the primates' retina there is some degree of peripheral coding.

4.3 Reptiles

Reptiles were the first vertebrates to live on land, in many ways their visual problem became much simpler. To them air offered a clear medium for vision with great advantages secured by the rectilinear passage of light. The difference in the refractive indexes of air and water resulted in having the cornea becoming the major focusing element and the lens a fine focusing device. Furthermore since water absorbs heavily in both the ultraviolet and the infrared end of the spectrum, terrestrial vertebrates had a much broader band of electromagnetic radiation available for visual information.[12].

The higher standards of vision required by life on land, allowed for new anatomy and function to develop. The environmental adaptation of the reptile visual system, especially of the retina, is an example of the evolutionary process. Walls[19] believed that the primitive reptiles were all diurnal and he supposed that they had inherited the cones of their ancestors, and lacked rods. As competition on land increased, many reptiles were forced to become partially or completely nocturnal. This process was accompanied by an increase in the size of the outer segments of some cones, so they gradually became rod-like both in function and appearance. Walls summarises this hypothesis as "*the first rods in the world were produced by the transmutation of cones, and the process has been occasionally repeated, whenever needed, ever since the vertebrate came on land*".

Many lizards still have on top of their head the ancestral third eye or parietal eye, which contains photoreceptors that resemble the cone and rods of the lateral eye and is connected through a nerve to the diencephalon. The development of the parietal eye differs from the lateral eye: the retina is not cupped by invagination and lens is formed from the same outgrowth of the brain as the retina, as a result the photoreceptors point towards the entering light, unlike the situation in the lateral eyes. The function of the parietal eye is to detect solar radiation, and is probably specialised to detect changes in ambient light at sunrise and sunset, which could be important for the control of the diurnal rhythms.

The snakes probably evolved from an ancestral lizards, that became fossorial. As a result the lateral eyes developed a protective spectacle, beneath which the eye became slowly atrophied. When later the snake came back to the surface the eye was reconstructed. The eye of the present day diurnal snakes, is very different from those of other reptiles showing an example of convergent evolution.

4.4 Birds

Vision is a major sensory modality in birds. The avian eye, in general, combines a remarkably large size with a superb accommodation mechanism and a thick, neatly stratified retina devoid of blood vessels capable of high visual acuity, colour discrimination, and movement detection. Characteristics features of the avian retina include a large retinal image, a fascinating array of visual cells (rods, single and double cones), an efferent centrifugal system, elaborate and diverse areas of acute vision (with one or two foveae) and an additional nutritional source, the pecten.

Both diurnal and nocturnal birds have a duplex retina, but the proportion of cones is low in nocturnal forms. Many birds have a deep foveal pit, the convexiclivate fovea, whose function is thought to magnify the image due to the fact that the refractive index of the retina is slightly greater than that of the vitreous (ratio 1.006). This refraction would magnify the image. Walls[19] suggested that this magnification effect could increase the foveal acuity of some hawks and eagles to at least eight times that of man.

The photoreceptors of birds contain coloured oil droplets, which role is probably associated with colour vision. Bowmaker and Knowles[2] studying the chicken retina pointed out that hue discrimination within the broad spectral band dominated by double cones, which are nearly half of the cones in the chicken retina, would be very poor were it not for the presence of two types of single cones, one combining a red oil droplet with yellow visual pigment and the other combining a yellowish droplet with a green pigment. This results in having a narrow band spectral sensitivity with only a little overlap.

Birds, as a number of invertebrate and some fish, are able to detect UV light. UV vision might be important in bird's navigation since UV light penetrates haze better than longer wavelengths and sky light away from the sun peaks in the UV. The polarisation pattern of the sky is thus more pronounced at these wavelengths, and if a bird can perceive the plane of polarisation of light at very short wavelengths it will have the best opportunity to use this information as an aid in direction finding[16].

4.5 Mammals

Two completely new developments occurred in the visual system of mammals: consensual pupillary movements and co-ordinated eye movements. Both are the result of the increased visuomotor organisation commanded by the brain. In all sub-mammalian species, including marsupials and monotremes, the visual information from one eye crosses over to the opposite hemisphere. In primitive mammals starts a new development and optic fibres of each eye split at the optic chiasm to reach both hemispheres. The result of the split of information is responsible for a peculiarity of primates, three dimensional stereoscopic vision. The brain compares the two retinal images of each eye and judges distances comparing the angle between the sighted object and the eye. The movement of the oculomotor muscles bring the object into focus to help in judging the exact distance. As the distance increase the pupil widens to let in more light in.

Mammals exhibit a variety of sensory adaptation for specialised ecological niches. The specialisation for nocturnal and diurnal vision implies that whereas diurnal species can sacrifice visual sensitivity to improve acuity, nocturnal animals are limited in their visual acuity by the necessity of achieving maximum sensitivity. Nocturnal mammal were the first to appear. Most creatures active at night are either the descendants of creatures who fled to the relative safety of the dark or pursued them. The dark-adapted eye has wide apertures. The best resolution for cat's eye has a f-stop of 0.09, whilst for humans is at f-2.4. The cornea and lens are huge in comparison to those of diurnal animals. The retina consists of all, or nearly all, rods, which have higher sensitivity than cones and are frequently pooled to connect to the second order neurones. The sensitivity is shifted towards shorter wavelength. For diurnal mammals the peak of sensitivity is in the yellow region whilst for nocturnal animals green objects appears brighter.

When mammals returned to daylight they had lost the coloured oil droplets found in more primitive vertebrates to protect photoreceptors from UV damage and reduce chromatic aberration. Their lens adapted becoming opaque to UV and chromatic aberration was reduced filtering out shorter wavelength. The diurnal mammalian eyes retains the advantage of the bird's eye. The lens is pushed toward the cornea increasing the size of the image falling on the retina. The cornea provides the first focusing on the retina, and some mammals rely entirely on the cornea for focusing.

5. Adaptation to photic environment

Visual systems are the result of a dynamic process of adaptation to visual tasks to be performed that has occurred for the past 600 million years. The extraordinary adaptability of the eye and its ability to respond in the most varied conditions are without question, not only in geological periods, but within the life span of an individual. This ability is observed in many vertebrate classes, including mammals. Ungulates that spend their life scanning the horizon for carnivorous predator become horizontally astigmatic by the third year. Domesticated birds become nearsighted even if birds are naturally farsighted. The most interesting adaptation occurs in fish. Visual systems of fish are highly dynamic in their functional organisation. They may change their visual pigments within a few weeks, and change cones for rods, and vice versa, whenever needed by the photic environment[9,10]. The synaptic organisation of the neural retina also undergoes significant changes during light adaptation. In particular, external horizontal cells, which are intimately involved in colour processing, go through major structural changes, extending and retracting finger-like processes called "spinules"[4]. Recent evidence suggests that similarly plastic mechanisms are active in the inner retina, which we must remember is part of the central nervous system.

Acknowledgements

This paper is mainly derived from: Silvana Vallerga (1994) "The phylogenetic evolution of the visual system" in *Human and Machine Vision, Analogies and Divergencies,* edited by Virginio Cantoni for the Plenum Press, New York and London.

References

[1] M.A. Ali.(1977). In: *Sensory physiology: Review and Perspectives*, Ed. M.A. Ali.. NATO ASI, Plenum Press, New York and London, Preface.

[2] J.K. Bowmaker and A. Knowles.(1977) The visual pigments and oil droplets of the chicken retina, *Vision Res.* 17, 755-764.

[3] F. Crescitelli.(1977) The visual pigments of geckos and other vertebrates. In: *The Visual System of Vertebrates.* Handbook of Sensory Physiology, VII/5, Ed. F. Crescitelli, Springer-Verlag, Berlin, Heidelberg, New York, 391-450.

[4] M.B.A. Djamgoz and M. Yamada.(1990) *The visual system of fish.*, Chapman & Hall, London, 159-210.

[5] M. Gogala.(1977) Ecosensory functions in insects. In: *Sensory physiology: Review and Perspectives.* NATO ASI Ed. M.A. Ali, Plenum Press, New York and London, 123-154.

[6] R.G. Jaeger (1977) Sensory functions in amphibians. In: *Sensory physiology: Review and Perspectives.* NATO ASI Ed. M.A. Ali, Plenum Press, New York and London, 169-196.

[7] E. Kicliter (1973) Flux, wavelength and movement discrimination in frogs: forebrain and midbrain contributions. *Brain Behav. Evol.*. 8, 340-365.

[8] J.Y. Lettvin, W.S. Maturana, W.S. McCulloch and W.H. Pitts (1959). What the frog's eye tells the frog's brain. *Brain. Proc. Inst. Radio Eng.* 47, 1940-1951

[9] J.N. Lythgoe. (1979) The adaptation of visual pigments to the photic environment. In: *Handbook of Sensory Physiology. Vol. VII/1.* Springer-Verlag, Berlin, Heidelberg, New York.

[10] J.N. Lythgoe, (1977) *The Ecology of Vision,* Academic Press, London.

[11] W.N. McFarland and W.R.A. Munz (1975) Part III: The evolution of photopic visual pigments in fishes *Vision Res.* 15, 1071-1080.

[12] W.R.A. Munz. (1977) The reptile sensory system. In: *Sensory physiology: Review and Perspectives.* NATO ASI Ed M.A. Ali, Plenum Press, New York and London, 197-216.

[13] W.R.A. Munz. and W.N. McFarland. (1977) Evolutionary adaptation of fishes to photic environment. In: *The Visual System of Vertebrates.* Handbook of Sensory Physiology, VII/5, Ed. F. Crescitelli, Springer-Verlag, Berlin, Heidelberg, New York, 194-274.

[14] E.C. Olson (1977) The history of the vertebrates. In: *The Visual System of Vertebrates.* Handbook of Sensory Physiology, VII/5, Ed. F. Crescitelli, Springer-Verlag, Berlin, Heidelberg, New York, 1-46.

[15] S. Sinclair (1985) *How Animals See. Other Visions of Our World* (Crom Helm, London & Sidney.

[16] R.A. Suthers (1977) Sensory ecology of birds. In: *Sensory physiology: Review and Perspectives.* NATO ASI Ed M.A. Ali, Plenum Press, New York and London, 217-251.

[17] S. Vallerga (1994) The phylogenetic evolution of the visual system. In *Human and Machine Vision,* Ed. V. Cantoni, Plenum Press, New York & London, 1-12.

[18] S. Vallerga, S., Arru, A. and D. Lupi (1990) Retinal Plasticity during eye migration in flatfish. *Invest. Ophthal. Visual Sci.* 31, 159

[19] G.L. Walls (1942). *The Vertebrate Eye and Its Adaptive Radiation.* (Cranbrook Inst. of Sci. Bull. 19. Bloomfield Hills, Mich.

AN ADAPTIVE MODEL WITH SOME IMPORTANT PROPERTIES OF VERTEBRATE PHOTORECEPTORS

William H.A. BEAUDOT *

Laboratoire de Traitement d'Images et Reconnaissance de Formes

Institut National Polytechnique de Grenoble

46 avenue Félix Viallet, 38031 Grenoble Cedex, France

1 Introduction

The photoreceptor is the locus of complex chemical reactions which have not been completely analyzed and understood: Indeed all the steps of the visual transduction has been only recently brought to light, and it still remains sometimes mysterious, particularly regarding the mechanisms of adaptation. The topic of this paper is the investigation of the adaptation at the photoreceptor level and its implications on further retinal processing. We propose a model of photosensor (§2) which shares some important properties with vertebrate photoreceptors. This engineering model demonstrates the advantages of these basic principles in visual adaptation (§3), and finally lies within a coherent structural model of the retina whose function is the control of sensitivity, and in which rather than functioning independently, each photoreceptor is an integral part of the elaborate retinal synaptic network (§4).

2 A Biophysical Model of Phototransduction

Phototransduction deals with the detection of photons, the amplification of the resulting signal and its coding into a neural signal. All the mechanisms involved in the phototransduction in vertebrate photoreceptors begin to be well known [3]. The phototransduction can be analyzed at a molecular level and modelled with enzymatic reactions. The law of mass action allows to consider the following enzymatic reaction and its mathematical expression:

$$M \xrightarrow{I,\alpha} M^* \xrightarrow{\beta} M \qquad \text{and} \qquad \beta . M^* + \frac{dM^*}{dt} = \alpha . I . (M_o - M^*)$$

They express the transformation of the molecule M to another M^* (e.g., an activated form) under the influence of a catalyst I at a rate α, and its free reformation at a rate β, with the hypothesis of a constant total amount of the molecule (its active and inactive forms), that is $[M]+[M^*]=[M_o]$. If β is related to a conductance, C to a membrane capacity, M_o to a potential generator, and $\alpha.I$ to a conductance modulated by the input signal I, the first order differential equation expresses the dynamics of our photosensor (figure 1a) in which M^* takes the place of the receptor potential V, and also expresses an overall property of photoreceptors: their nonlinear and compressive transfer function.

---

* Present Address: C.S.E.M. SA, Centre Suisse d'Electronique et de Microtechnique, Maladière 71, Case postale 41, CH-2007, Neuchâtel, Switzerland.

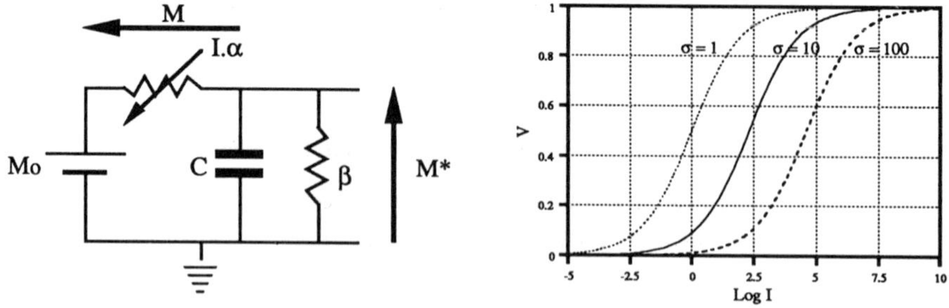

a) b)

Fig. 1: a) Electrical Model of the Photoreceptor (see text for explanation).
b) The *shift property* of the *Fechner* law V = f(log I) for 3 values of σ.

3 Some Outstanding Properties

3.1 Michaelis-Menten Law and Weber Law

The response of the model at steady state (just as well the peak response as the plateau potential) follows the *Michaelis-Menten* law:

$$V(I) = \frac{I.V_{max}}{I+\sigma} \quad \text{with} \quad \sigma = \frac{\beta}{\alpha},$$

where I is the steady light stimulation, V_{max} the saturation value of the potential and σ a dissociation constant which acts on the compression effect. Such an intensity-response relation has been observed in direct photocurrent measurement as well as in voltage recording in cone outer segments of salamander and turtle [6]. If the stimulus I is now composed of the superposition of a background I_0 and an increment ΔI, a response to I will be elicited if V(I) exceeds a threshold Γ: ΔI is the just noticeable increment above the background intensity I_0. Using the model's response (i.e. *Michaelis-Menten* law) and expressing the constancy of the differential ratio $\Delta I/I_0$ (i.e. *Weber law*), we find that σ must be proportional to the background intensity I_0 ($\sigma = K.I_0$, where K is a constant).

3.2 An Optimal Sensory Coding

An optimal coding of sensory information must take place as early as the photoreceptor level. The optimality of the coding implies an optimal representation of the input range of light intensity onto the response range, that is to maximize the sensitivity S of the system around the input mean level. Let us consider a light stimulation I composed of a mean level I_0 and a standard deviation ΔI. Using *Michaelis-Menten* law and maximizing its sensitivity around I_0 according to the parameter σ, we obtain:

$$\sigma_{opt} = \sqrt{I_o^2 - \Delta I^2} \quad \text{and} \quad \sigma_{opt} \cong I_o \text{ when } \Delta I \ll I_o .$$

Thus, an optimal sensory coding can be realized by adjusting the photoreceptor transfer function (via σ) according to the mean I_0 and the standard deviation ΔI of the light intensity.

3.3 Fechner Law and Shift Property

When we consider a *Michaelis-Menten* law which satisfies *Weber* law (i.e. $\sigma = K.I_o$), we can express V(L) according to the logarithm L of the light intensity I: $V(L) = V_{max} /(1+K.e^{-(L-L_o)})$. This expression denotes a sigmoid curve which leads to the *Fechner* law: there exists a medium range of intensity, above the very small values and below the very large values, for which the response curve according to the logarithm of the intensity is almost a straight line, that is V(L)=A.log L (figure 1b). According to the criterion of optimal sensory coding and for $\Delta I << I_o$ (i.e. $K \approx 1$), we finally obtain a set of sigmoid curves where L_o only acts on the translation of this sigmoid on the log(I) axis (figure 1b). The adaptation of the parameter σ with the mean intensity provides the shift of the sigmoid curve. This sigmoid curve and its shift property are also in agreement with some measurements of the cone intracellular responses and in a more general way with the sensory responses [4].

3.4 The Consequences for a Collective Behaviour

A photoreceptor is actually always integrated into a set of photoreceptors, and it happens that in some species of vertebrates there exists an electrical coupling via gap-junctions between neighbouring cones [1]. This coupling takes place at the level of the basal terminal of photoreceptors where is located the membrane potential V that we have already considered in our photoreceptor model. The addition of a coupling in our model simply leads to connect the nodes V of the neighbouring photoreceptors by a resistance. Then beyond the addition of a nonlinearity, the multiplicative effect of the input signal I onto a conductance implies that the spatiotemporal low-pass filtering performed by the photoreceptor layer is modulated in a coherent way by the light intensity: the signal is integrated on a spatiotemporal window which size is inversely proportional to the intensity. When one reminds that the signal-to-noise ratio (SNR) decreases with the light intensity, the situation becomes clearer: the dependence of the space and time constants with the light intensity is such that it allows to thwart the influence of spatiotemporal noise. In that way, as early as the first neural layer of the retina, the notion of *adaptive spatiotemporal regularization* is naturally introduced.

4 The Photoreceptors in the Retinal Circuit

The properties of *Michaelis-Menten* law are simple but become fundamental as soon as the parameter σ is hold adaptable: even though its function can be considered as a compression function which would have the disadvantage to limit the range of operating signal, it becomes essential when it evolves with the mean signal, allowing an optimal coding. Towards the electrical analogue of *Michaelis-Menten* law (figure 1a), the relation between σ and I_o generates an outstanding result: some parameters of this circuit (e.g. the conductance β if $\alpha = 1$) might be modulated by some statistical properties of the light intensity such its mean and its standard deviation. There is some anatomical and physiological evidence for the presence of numerous feedbacks in the retinal network. Firstly, there is a chemical feed-back synapse from the horizontal cells onto the photoreceptors which is sign inverting (through a GABA release gating Cl^- channels). Even if single photoreceptor already shows temporal adaptation, we think that the need of adaptation at the photoreceptor level can explain the existence of such a *synaptic feedback*: indeed, the horizontal cells perform a spatiotemporal integration (a kind of local mean estimation) which can be used as a reference point for the fast dynamic adjustment of the photoreceptor operation onto the range of useful signal. And secondly, there is also a feedback from the inner plexiform layer onto the outer plexiform layer through a dopaminergic *neuromodulation* [5]. We also think that this neuromodulation of the outer retina processing (just as well at the photoreceptor level as the horizontal or/and bipolar ones) is performed according to the estimation of the variance of the input signal by a spatiotemporal integration of the bipolar signal in the inner retina. Thus, the control of the retinal sensitivity could be performed through an adaptive photoreceptor whose parameters are locally estimated by the subsequent retinal circuit [7].

5 Conclusion

We have proposed a model of the photoreceptor which links *Michaelis-Menten*, *Weber* and *Fechner* laws, and which provides an optimal sensory coding (through a curve shifting) according to the statistical properties of the input signal (e.g., mean and standard deviation). These statistical characteristics are estimated by the subsequent retinal network and fed back onto the photoreceptor in order to adjust its nonlinear transfer function. Moreover, its nonlinearity also induces a spatiotemporal filtering, locally adapted to the characteristics of the luminous signal. This consistent model clarifies the role of retina in visual adaptation and points out some of the neural strategies involved in this mechanism: fast local synaptic feedbacks to adjust the photoreceptor response around the local spatiotemporal mean, and slow neurohormonal feedbacks to enhance the sensitivity of bipolar cells' response according to the slow (light/dark) change from a nonstationary environment. Moreover, this work emphasizes the fact, as claimed in [2], that rather than functioning independently, each photoreceptor is an integral part of an elaborate synaptic network.

References

[1] P.B. Detweiler and A.L. Hodgkin (1979) Electrical coupling between cones in turtle retina. *J. Physiol. (Lond.)*, 291:75-100

[2] D.G. Green (1986) The search for the site of visual adaptation. *Vision Res.*, 26:1417-1429

[3] E.N. Pugh and T.D. Lamb (1993) Amplification and kinetics of the activation steps in phototransduction. *Biochimica et Biophysica Acta*, 1141:111-149

[4] F.S. Werblin (1973) The control of sensitivity in the retina. *Scientific American*, 1:71-79

[5] P.Witkovsky and A. Dearry (1992) Functional roles of dopamine in the vertebrate retina. *Progress in Retinal Research*, 10:247-292

[6] S. Yasui (1992) On the square root intensity coding at the level of cone photoreceptors. *Vision Res.*, 32(1):199-202

[7] W. Beaudot (1994) Le traitement neuronal de l'information dans la rétine des vertébrés: un creuset d'idées pour la vision artificielle. PhD Thesis, I.N.P. Grenoble, France

IDENTIFIED CUTANEOUS PATHWAY IN THE FRONTAL ORGAN OF THE FROG *RANA ESCULENTA* SUGGESTS A POSSIBLE RELATIONSHIP WITH PHOTORECEPTIVE ELEMENTS

V. Guglielmotti, L. Fiorino and E. Sada
Istituto di Cibernetica del CNR, 80072 Arco Felice, Naples, Italy

The frontal organ (FO) of amphibians (Fig. 1) and the epiphysis (EPI) or pineal are referred to in the literature as pineal complex. The FO, also called "stirnorgan", is equivalent to the parapineal organ of fishes and to the parietal eye of reptilians. It is located within the skin in a medial position between the lateral eyes. The EPI is the intracranial component of the complex and lies over the habenular commissure like a cover of the IIIrd ventricle of the diencephalon. The frontal nerve (FN), also known as pineal nerve, connects the FO to the EPI and the brain. The pineal complex of the frog is a photosensitive structure that contains photoreceptive pinealocytes interconnected with pineal nerve cells [1]. This has been demonstrated both ultrastructurally [2, 3] and electrophysiologically [4]. The structural organization of the pineal complex components suggests that this complex is specialized to utilize light-dependent impulses, arising in the pineal complex, and that these are conducted by pinealofugal fibers to di- and mes-encephalic areas, such as the habenular regions, the area of anterior commissure, the neuropil adjacent to the ventrolateral nucleus of the thalamus and the dorsolateral part of the posterior preoptic area [5, 6]. Some of these areas and the EPI are involved in the entrainment of circadian and circannual pacemakers for the synchronization of rhythmic body functions with the photoperiod of the environment. Another peculiar characteristic of the pineal complex is the electrophysiological difference of the chromaticity of neurons, that appears to be more common in the extracranial portion than in pineal organ [7, 8].

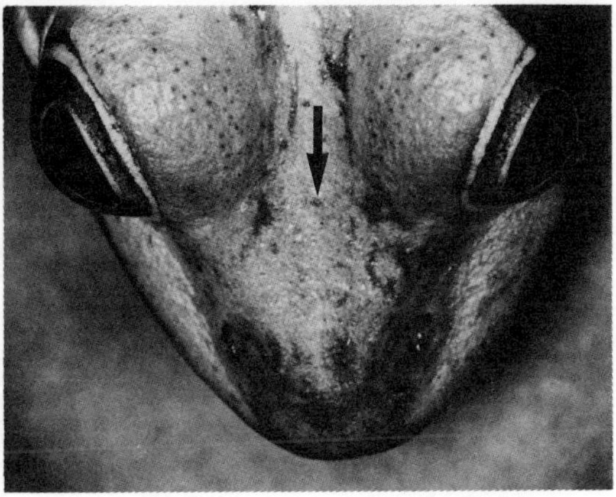

Fig. 1 Frontal organ (arrow) of the frog Rana esculenta

It is known that the pineal complex of the frog contains afferent and efferent fibers for the interconnections between the elements themselves and the brain and studies performed up to now have considered a system constituted by the FO, the FN, the EPI and the brain. Recently

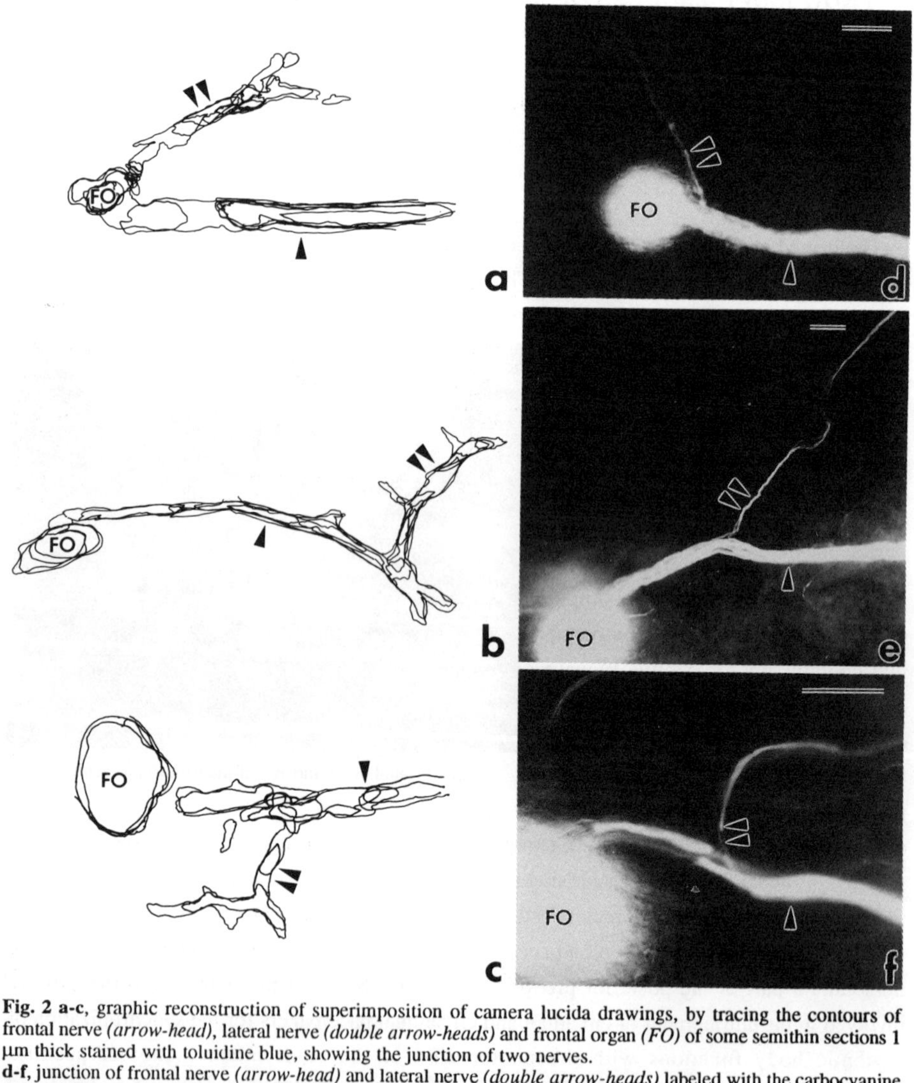

Fig. 2 a-c, graphic reconstruction of superimposition of camera lucida drawings, by tracing the contours of frontal nerve *(arrow-head)*, lateral nerve *(double arrow-heads)* and frontal organ *(FO)* of some semithin sections 1 μm thick stained with toluidine blue, showing the junction of two nerves.
d-f, junction of frontal nerve *(arrow-head)* and lateral nerve *(double arrow-heads)* labeled with the carbocyanine DiI application. In **d** , the junction occurs in close proximity of the frontal organ *(FO)*, whereas in **e** and **f** the junction occurs more distally. Epifluorescence images using filter G1 on Polyvar photomicroscope. Calibration bars 100 μm.

we described for the first time a nerve branch of the epithelial neural net of the frog *Rana esculenta,* which joins the FN and which we named the "lateral nerve" (LN) [9]. In order to show these structures, maximal care is required during the dissection when removing the zone containing the FO and cutting the FN anchored to the cranial cap. The fragility of this area, the risk of microlacerations and detachment of the LF from the FN in the part where their joining occurs can easily account for the previous lack of reports in the literature of the LN. The occurrence of the LN and its junction with the FN was verified by means of different methodological approaches: the method of methylene blue intravital staining, the Landau-Ignesti method for myelinated fibers, the study on semithin 1 μm thick toluidine blue-stained sections. Some significant results that reveal the morphological presence of the LN are illustrated in Fig. 2 a-c. These figures show the graphic reconstructions of superimposed drawings of separate histological semithin sections obtained by tracing the contours of structures with camera lucida. In this study we have used the *in vitro* tracing method with the carbocyanine DiI (1,1'-dioctadecyl-3,3,3',3'-tetramethylindocarbocyanine perchlorate; Molecular Probes, Eugene, OR). The carbocyanine DiI was selected for our study since it has been proven to be a useful tracer in anura amphibia [10, 11] and has been reported in some systems to be more sensitive than horseradish peroxidase [12]. In addition, the efficacy of DiI in fixed tissue allowed us to avoid the extremely difficult manipulation necessary for the administration of tracers *in vivo* in such a delicate structure.

The results obtained (Fig. 2 d-f) fully confirmed all previous observations. In addition, the LN seems to be divided into two branches when it joins the FN: one reaches the FO and the other is directed to the brain. We do not know the direction of these fibers, but our anatomical results open the way to different hypotheses, one of which is the possibility that the LN could have relations also with photoreceptive cells of the FO. In this case, the LN could represent a new pathway connecting the skin to the FO and/or the brain: thus, the well known current model of the pineal complex of the frog *Rana esculenta* might need to be revised if the involvement of the above LN with the processes of light transduction in the FO could be confirmed by further neuroanatomical and by electrophysiological studies.

References

[1] J.P. Collin and A. Oksche, (1981), In: *The Pineal Gland: Anatomy and Biochemistry.* Ed. by R.J. Reiter. Boca Raton: CRC Press, 27-67.

[2] D.E. Kelly and S.W. Smith, (1964), Fine structure of the pineal organ of the adult frog *Rana pipiens. J. Cell Biol.,* 22: 653-674.

[3] A. Oksche and M. Vaupel-Von Harnack, (1963), Elektronenmikroskopische Untersuchungen am Stirnorgan der Anuren (zur Frage der Lichtrezeptoren). *Z. Zellforsch.,* 59: 239-288.

[4] E. Dodt, (1973), Thc parietal eye (pineal and parietal organs) of lower vertebrates. In: *Handbook of Physiology*. Vol. VII/3 B. Central processing of visual information. Part B. Ed.: R. Jung. Berlin, Heidelberg, New York: Springer, 114-140.

[5] W.D. Eldred, T.E. Finger and J. Nolte, (1980), Central projections of the frontal organ of *Rana pipiens*, as demonstrated by the anterograde transport of horseradish peroxidase. *Cell Tissue Res.*, 211: 215-220.

[6] M. Kemali and A. De Santis, (1983), The extracranial portion of the pineal complex of the frog (frontal organ) is connected to the pineal, the hypothalamus, the brain stem and the retina. *Exp. Brain Res.*, 53: 193-196.

[7] H. Meissl and E. Dodt, (1981), Comparative physiology of pineal photoreceptor organs. In: *The pineal organ: photobiology-biochronometry-endocrinology*. Eds A. Oksche and P. Pévet. Amsterdam: Elsevier-North Holland, 61-80.

[8] M. Ueck, K. Wake and H. Kobayashi, (1989), Nervous organization of the pineal complex in lower vertebrates. *Zoological Science*, 6: 817-831.

[9] V. Guglielmotti, L. Fiorino and E. Sada, (1994), A new neural connection traced in the frontal nerve of the frog *Rana esculenta*. *European Journal of Neuroscience*, S7: 147.

[10] C. Artero, A. Fasolo and M.F. Franzoni, (1994), Multiple sources of pituitary pars intermedia innervation in amphibians: a DiI retrograde tract-tracing study. *Neuroscience Letters*, 169: 163-166.

[11] H. Tan and V. Miletic, (1990), Bulbospinal serotoninergic pathways in the frog *Rana pipiens*. *J. Comp. Neurol.*, 292: 291-302.

[12] H. Wicht and R.G. Northcutt, (1990), Retinofugal and retinopetal projections in the pacific hagfish *Eptatretus stanti* (myxinoidea). *Brain Behav. Evol.*, 36: 315-328.

INTRACELLULAR K^+ BINDING SITE PROPERTIES OF THE Na^+:Ca^{2+},K^+ EXCHANGER OF VERTEBRATE PHOTORECEPTORS

Anacleto Navangione, Vittorio Vellani and Giorgio Rispoli
Istituto di Fisiologia Generale v. Borsari 46 - 44100 Ferrara

Summary. The properties of the intracellular potassium binding site of the Na^+:Ca^{2+},K^+ exchanger were investigated in detached rod outer segments (OS) perfused with high Ca^{2+}_i and recorded under whole-cell voltage-clamp conditions. The OS were loaded with known amounts of K^+ with the aid of a few Ca^{2+}-activated K^+ channels while perfusing with high external K^+ (K^+_o). The substitution of K^+_o with Na^+_o elicited an inward, transient exchange current (forward mode: Na^+ intruded per Ca^{2+} and K^+ extruded) whose time integral was equal to the amount of K^+ charges loaded in the OS, thus demonstrating that K^+ is transported by the exchanger with a stoichiometry consistent with one K^+ ion extruded per charge imported. The activation of forward exchange by intracellular K^+ (K^+_i) was fitted by first-order kinetics with a Michaelis constant, $K^{app}(K^+_i) \approx 1.6$ mM and a maximal current, $I_{max} \approx -65$ pA. $[Na^+]_i$ larger than 100 mM had little effect on $K^{app}(K^+_i)$ and I_{max}, indicating that Na^+_i does not compete with K^+_i for exchange sites under physiological conditions, and that Na^+ release at the intracellular side is not a rate limiting step for the exchange process.

Methods. OS were mechanically isolated from rods of the nocturnal lizard Gekko gecko (Rettili srl, Varese, Italy). The methods are described in detail elsewhere.[1] Electrical recordings were carried out in room light with the whole-cell configuration of the patch recording technique using an Axopatch 1D amplifier (Axon Instruments, Burlingame, CA, USA). Pipettes were fabricated from 100 µl Drummond glass capillaries in the conventional manner[2] and fire polished to a pipette resistance of 7-11 MΩ. All solutions were buffered to pH≈7.4 with N-2-hydroxyethyl-piperazine-N'-2-ethanesulphonic acid (Hepes) and NaOH, KOH or tetraethylammonium hydroxide. The composition of Ringer solution was: 160 mM Na^+, 3.3 mM K^+, 1 mM Ca^{2+}, 1.7 mM Mg^{2+}, 165 mM Cl^-, 1.7 mM SO_4^{2-}, 2.8 mM Hepes and 10 mM dextrose. Isotonic solutions contained 166 mM of the monovalent cation indicated in the text and 2 mM EGTA. Rapid changes (t≈50 msec) of the external solution were performed by moving horizontally a multibarreled microperfusing pipette placed in front of the recorded OS. All chemicals were obtained from Sigma (St. Louis, MO, USA) except Charybdotoxin (Alomone labs, Jerusalem, Israel).

Ca^{2+}-activated K^+ channels are occasionally present in OS. In order to demonstrate that K^+ is transported by the Na^+:Ca^{2+},K^+ exchanger and to study the dependence of the exchange current from $[K^+]_i$, we controlled intracellular potassium with the aid of some Ca^{2+}-activated channels, occasionally present in the OS, that we found to be highly selective for K^+.[1] This novel approach was necessary also because ion diffusion between patch pipette and OS cytoplasm was found to be hindered (see following), thus making the patch pipette ineffective on controlling $[K^+]_i$. These Ca^{2+}-activated channels are probably located in a small piece of the inner segment that remains attached to the outer one.[3,4] Single channel activity (present in about 60% of the cells) was detected upon depolarizing the OS perfused with 30 mM Ca^{2+}_i and 124 mM K^+_i: the unitary current size was outward (K^+ flowing out of the OS) for holding potentials larger than -30 mV, and its amplitude, together with the probability and duration of channel opening, increased with depolarization (Fig.1). Under the above recording conditions, the only conductances present were the Na^+:Ca^{2+},K^+ exchanger and the Ca^{2+}-activated channels, while the cGMP-gated channels were closed. Consequently, the OS input resistance was as high as 10 GΩ, thus allowing the recording of single channel activity in whole cell configuration. The Ca^{2+}-activated channels were Charybdotoxin (100 nM) and TEA (20 mM) sensitive but Apamin (1 μM) insensitive: all these features are typical of the BK channels.[5,6]

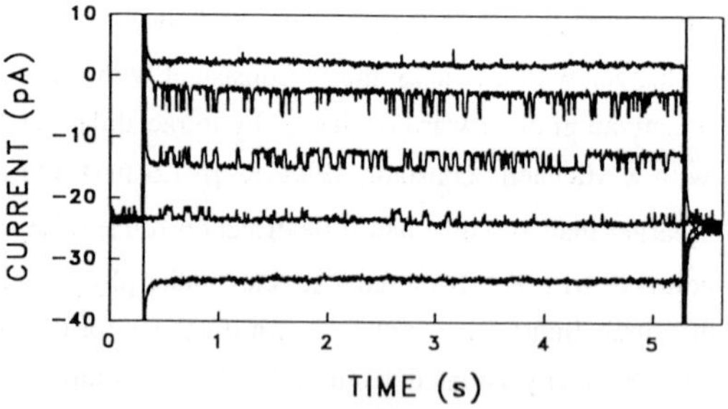

Fig. 1. Current-voltage characteristics of single Ca^{2+}-activated channel activity superimposed to exchange and leak current in OS perfused with 124 mM K^+_i + 30 mM Ca^{2+}_i and bathed in Ringer. The capacitive transients at the onset and closing of the voltage pulse (20 mV steps from +40 to -40 mV) display the whole cell configuration of this experiment.

$K^{app}(K^+_i)$ estimate. In the OS where the Ca^{2+}-activated channels were present, the relationship between exchange current amplitude and $[K^+]_i$ was estimated with the protocol illustrated in Fig. 2. To attain precise control over $[K^+]_i$, endogenous K^+_i must be depleted first. Thus the OS, perfused with 30 mM Ca^{2+}_i and 124 mM Li^+_i, was initially bathed in Ringer, so that any endogenous K^+_i was extruded through forward exchange activation and/or washed out by the patch pipette (Fig. 2, *a*). The time integral of this exchange current gave a lower limit for the number of endogenous K^+_i ions present in the OS preparation at the beginning of whole-cell recording (endogenous $[K^+]_i$ = 3 - 22 mM: such a low value may be due to the contamination of the OS cytoplasm by extracellular solution during its isolation). Once depleted of endogenous K^+_i, the 0 current level was checked by switching the OS to Isotonic Li^+_o, an ion that does not operate the exchanger[7] and does not permeate the Ca^{2+}-activated channels.[8] A following switch to Isotonic K^+_o produced an inward current due to K^+ flowing through the Ca^{2+}-activated channels and the leak (Fig. 2, *b*). Finally, a switch to Isotonic Na^+_o elicited a transient inward current, due to exchange activity produced by the pipette Ca^{2+} and by the K^+ ions imported during Isotonic K^+_o perfusion (Fig 2, *c*). The current declined to a small steady level, representing essentially the inward leak driven by the junction potential between Isotonic Na^+_o and Isotonic Li^+_o (-3 mV). The time integral of the exchange current recorded in Isotonic Na^+_o (leak subtracted) was equal to the time integral of the K^+ current in Isotonic K^+_o (the ratio between the two integrals was 1.0±0.1: n=10, 6 cells: see Table I). This is the first direct demonstration that K^+ is transported by the exchanger with a stoichiometry consistent with one K^+ ion extruded per charge imported, according to previous indirect evidences by Cervetto et al.[9] Furthermore, according to this stoichiometry, the K^+ imported through the Ca^{2+}-activated channels was entirely extruded by the exchanger, thus the K^+ washout by the patch pipette and the channels within the period of Isotonic Na^+_o perfusion was negligible. In fact, the time integral of the exchange current was not appreciably reduced by an intermediate switch to Isotonic Li^+_o for several tens of seconds between the Isotonic K^+_o and Isotonic Na^+_o

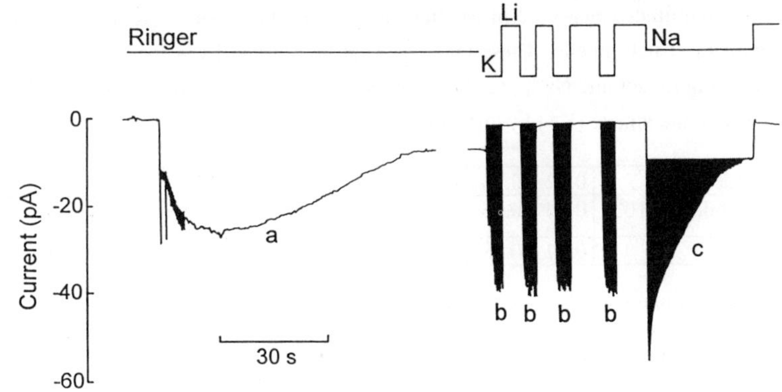

Fig. 2. Chart record of whole-cell current from an OS (retaining Ca^{2+}-activated channels) intracellularly perfused with 30 mM Ca^{2+}_i + 124 mM Li^+_i and exposed to Isotonic K^+_o, Isotonic Li^+_o and Isotonic Na^+_o. The OS used in the present work contained 7±1.4 channels (range: 3-14; n=13 OS).

perfusion (or up to a minute for access resistances larger than 40 MΩ), showing that K^+ diffusion was truly hindered. Thus, the time integral of the exchange current or the time integral of the K^+ current gave a reasonable measure of the actual number of K^+ ions trapped in the cytoplasm. However, the former time integral was smaller than the latter one for very large K^+ loads, since the K^+_i washout became significant for large K^+ gradients between cytoplasm and patch pipette and for long integration times, as expected. Poor diffusion between patch pipette and the cytoplasm was found experimentally[10] and predicted theoretically, showing that the pipette tip is a significant barrier to diffusion in whole-cell configuration;[11] in addition, the OS is not an empty bag, and the disk stack is the main cytoplasmic barrier opposing free intracellular solute diffusion. Olson and Pugh[12] found that it takes more than 3 minutes for cGMP to diffuse from one end of the OS to the other. K^+ diffusion is certainly much faster, since it is not expected that K^+ is buffered as strongly as cGMP within the cytoplasm.[13] The K^+ load occurs presumably at one end of the OS (that is, where a channel belonging to the inner segment should be located), thus K^+_i may diffuse longitudinally along the OS through the 15 nm wide pathway delimited by the plasma membrane and the disk stack of the OS.[14] K^+ loads larger than $K^{app}(K^+_i)$ (Fig. 3) gave exchange current peaks near I_{max} (which was determined controlling $[K^+_i]$ with the patch pipette: see below), showing that within the loading time $[K^+]_i$ attained a reasonably uniform saturating level along the OS. The plot of the exchange current peak in isotonic Na^+_o perfusion versus the $[K^+]_i$ produced by K^+_o perfusion was well fitted by a Michaelis-Menten relationship with $K^{app}(K^+_i) \approx 2$mM (assuming $v \approx 1$ pl[15]) and I_{max} about -60 pA (Fig. 3, **Δ**). This sets a lower limit for exchange stoichiometry of $n \geq 1$ K^+ ions extruded per exchange cycle (or charge imported[8,15]). Since the number of charges imported during exchange activity in Na^+_o following a K^+ load is equal to the number of K^+ ions loaded (see Table I), then the only possibility is $n=1$. I_{max} and $K^{app}(K^+_i)$ were in good agreement with the current peak amplitudes measured in Isotonic Na^+_o from OS in which $[K^+]_i$ was controlled directly with the patch pipette. These OS were kept in Isotonic Li^+_o for at least 2 minutes before switching to Isotonic Na^+_o, to allow the $[K^+]_i$ at the level of the plasma membrane to equilibrate with the pipette $[K^+]$ by diffusion.

K^+ ions loaded (x 10^9)	0.62	0.75	0.84	1.52	1.66	2.06	2.35	2.87	3.14	3.35
K^+ ions extruded (x 10^9)	0.64	0.85	0.77	1.51	1.52	2.24	1.97	2.47	3.82	3.97
Ratio	0.97	0.88	1.09	1.01	1.09	0.92	1.19	1.16	0.82	0.84

Table I. Ratios (third row) between the number of K^+ ions loaded through the Ca^{2+}-activated channels (first row) and the number of K^+ ions extruded by the Na^+:Ca^{2+},K^+ exchanger (second row). 10 measurements in 6 OS.

Influence of Na^+_i on forward ion exchange. Both $K^{app}(K^+_i)$ and I_{max} were little affected upon substituting Li^+_i (124 mM) with Na^+_i in the presence of saturating $[Ca^{2+}]_i$ (Fig. 3, □). This shows that Na^+_i does not compete with K^+_i for exchange sites under physiological conditions. This is quite different with respect to the reversed exchange, where $K_m{}^{app}(K^+_o)$ rises when $[Na^+]_o$ is increased in the presence of saturating $[Ca^{2+}]_o$,[16] suggesting that the

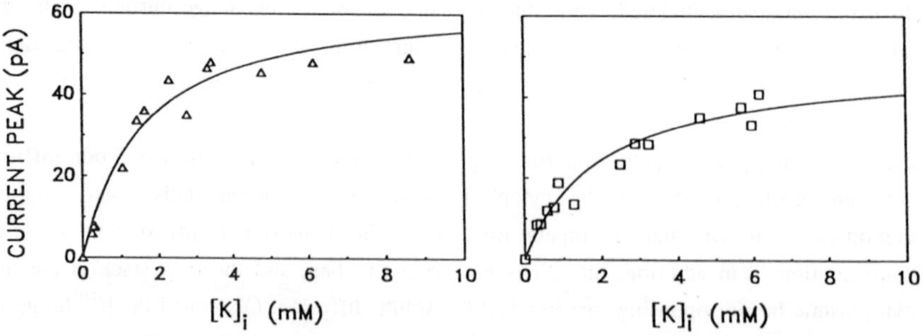

Fig 3. Plot of the intracellular potassium concentration versus the exchange current peak after a switch to Isotonic Na^+_o. Intracellular solutions: **Δ**=30 Ca^{2+}_i+124 Li^+_i (I_{max}= -64.5 pA; $K^{app}(K^+_i)$= 1.6 mM; 16 measurements in 8 OS); □=30 Ca^{2+}_i+124 Na^+_i (I_{max}= -50 pA; $K^{app}(K^+_i)$= 2.0 mM; 15 measurements in 5 OS)

exchange process is asymmetrical. Consistent with this view, the largest current amplitudes of reversed exchange recorded under saturating conditions (that is, in the presence of 124 mM K^+_i, 30 mM Ca^{2+}_i and 166 mM Na^+_o, data not shown) were at least twice as large as those recorded in the forward mode. The insensitivity of forward exchange to Na^+_i (in the presence of saturating $[Ca^{2+}]_i$) shows also that Na^+ release at the intracellular side is not a rate limiting step in the exchange process[17], suggesting that $K^{app}(Na^+_i)$ has the same order of magnitude as $K^{app}(Na^+_o)$ (93 mM[18]). In conclusion, Na^+ unbinding from exchange sites can be regarded as diffusion limited under physiological conditions.

REFERENCES

1. G. Rispoli, W. A. Sather, and P. B. Detwiler (1993). Visual transduction in dialyzed detached rod outer segments from lizard retina. *J. Physiol.* 465,513-537.
2. O. P. Hamill, A. Marty, E. Neher, B. Sakmann, and F. J. Sigworth (1981). Improved patch-clamp techniques for high resolution current recording from cells and cell-free membrane patches. *Pflügers Arch.* 391,85-100.
3. C. R. Bader, D. Bertrand, and E. A. Schwartz (1982). Voltage-activated and calcium-activated currents studied in solitary rod inner segments from salamander retina. *J. Physiol.* 331,253-284.
4. S. Barnes and B. Hille (1989). Ionic channels of the inner segment of tiger salamander cone photoreceptors. *J. Gen. Physiol.* 94,719-743.
5. B. S. Pallotta, K. L. Magleby, and J. N. Barrett (1981). Single channel recordings of Ca^{2+}-activated K^+ currents in rat muscle cell culture. *Nature* 293,471-474.
6. M. L. Garcia, A. Galvez, M. Garcia-Calvo, V. F. King, J. Vazquez and G. J. Kaczorowski (1991). Use of toxins to study potassium channels. *J. Bioenerg. Biomem.* 23,615-646.
7. A. L. Hodgkin, P. A. McNaughton, and B. J. Nunn (1987). Measurement of Sodium-Calcium exchanger in salamander rods. *J. Physiol.* 391,347-370.
8. A. L. Blatz and K. L. Magleby. (1987). Calcium-activated potassium channels. *Trends Neurosci.* 10,463-467.
9. L. Cervetto, L. Lagnado, R. J. Perry, D. W. Robinson, and P. A. McNaughton (1989). Extrusion of Calcium from rod outer segment is driven by both Sodium and Potassium gradients. *Nature* 337,740-743.
10. M. Push and E. Neher (1988) Rates of diffusional exchange between small cells and a measuring patch pipette. *Pflügers Arch.* 411,204-211.
11. C. Oliva, I. S. Cohen, and R. T. Mathias (1988). Calculation of time constants for intracellular diffusion in whole cell patch clamp configuration. *Biophys. J.* 54,791-799.
12. A. Olson and E. N. Pugh (1993). Diffusion coefficient of cyclic GMP in salamander rod outer segments estimated with two fluorescent probes. *Biophys. J.* 65,1335-1352.
13. A. Caretta, A. Cavaggioni, and R. T. Sorbi (1985). Binding stoichiometry of a fluorescent cGMP analogue to membranes of retinal rod outer segments. *Eur. J. Biochem.* 153, 49-53.
14. T. D. Lamb, P. A. McNaughton, and K.-W. Yau (1981). Spatial spread of activation and background desensitization in toad rod outer segments. *J. Physiol.* 319,463-496.
15. K.-W. Yau and Nakatani (1984). Electrogenic Na-Ca exchange in retinal rod outer segments. *Nature* 311,661-663.
16. R. J. Perry and P. A. McNaughton (1993). The mechanism of ion transport by the Na^+-Ca^{2+},K^+ exchanger in rods isolated from the salamander retina. *J. Physiol.* 466,443-480.
17. A. Navangione (1994). Lo scambiatore Na,Ca,K dei fotorecettori retinici di vertebrato. PhD Thesis, University of Ferrara, Ferrara, Italy.
18. L. Lagnado, L. Cervetto, and P. A. McNaughton (1988). Ion transport by the Na:Ca exchanger in isolated rod outer segments. *Proc. Natl. Acad. Sci. USA* 85,4548-4552.

PARTICIPANTS

List of Participants

Members of the Scientific Committee, Lecturers and Members of the Organizing Committee are indicated respectively by three, two and one asterisks.

Angelini Nicola
Istituto di Biofisica, CNR
Via San Lorenzo, 26
I-56127 Pisa
Italy
tel: 0039 50 513211
fax: 0039 50 553501

Beaudot William
Laboratoire de Traitment d'Image et
Reconnaissance de Formes
Institut National Polytechnique di Grenoble
46 Avenue Félix Viallet
F-38031 Grenoble
France
tel: 033 76 574822
fax: 033 76 574790
e-mail: beaudot@tirf.Inpg.fr

**Caretta Antonio
Istituto di Fisiologia Umana
Via A. Gramsci, 14
I-43100 Parma
Italy
tel: 0039 521 290380/82
fax: 0039 521 291304
e-mail: fisioumi@ipruniv

Checcucci Giovanni
Istituto di Biofisica, CNR
Via San Lorenzo, 26
I-56127 Pisa
Italy
tel: 0039 50 513211
fax: 0039 50 553501
e-mail: checcucc@ib.pi.cnr.it

Chien Chiu-Yuan
National Taiwan Normal University
Department of Biology
Roosevelt Road 88
Section 5,
Taipei
11718 Taiwan
Republic of China
tel: 00886 2 9326234
fax: 00886 2 9312904

**Colombetti Giuliano
Istituto di Biofisica, CNR
Via San Lorenzo, 26
I-56127 Pisa
Italy
tel: 0039 50 513237
fax: 0039 50 553501
e-mail: giuliano@ib.pi.cnr.it

*Cotugno Antonio
Istituto di Cibernetica
CNR
Via Toiano, 6
I-80072 Arco Felice (Napoli)
Italy
tel: 0039 81 8534113
fax: 0039 81 5267654
e-mail: neuros@mail.irtemp.na.cnr.it

Di Maio Vito
Istituto di Cibernetica,
CNR
Via Toiano 6
I-80072 Arco Felice (Napoli)
Italy
tel: 0039 81 8534107
fax: 0039 81 5267654
e-mail:vdm@biocib.cib.na.cnr.it

**Frediani Carlo
Istituto di Biofisica, CNR
Via San Lorenzo, 26
I-56127 Pisa
Italy
tel: 0039 50 513233
fax: 0039 50 553501

Gerster Uli
Institute for Biochemical Pharmacology
University of Innsbruck
Peter-Mayr-Str. 1
A-6020 Innsbruck
Austria
tel: 0043-512-507-3166
fax: 0043-512-507-2858
e-mail: uli.gerster@uibk.ac.at

**Ghetti Francesco
Istituto di Biofisica, CNR
Via San Lorenzo, 26
I-56127 Pisa
Italy
tel: 0039 10 513214
fax: 0039 10 553501
e-mail: ghetti@vm.cnuce.cnr.it

**Giulio Ludovico Filippo
Università di Torino
Dipartimento di Morfofisiologia Veterinaria
Settore Fisiologia ed Etologia
Viale Mattioli, 25
I-10125 Torino
Italy
tel e fax: 0039 11 6687887

Guglielmotti Vittorio
Istituto di Cibernetica, CNR
Via Toiano, 6
I-80072 Arco Felice (Napoli)
Italy
tel: 0039 81 8534114/117
fax: 0039 81 5267654

***Häder Donat-Peter
Friedrich-Alexander Universität
Institut für Botanik und Pharmazeutische
Biologie Leurstuhl für Botanik I
D-Erlangen, Germany
tel: 0049 9131 858216
fax:0049 9131 858215
e-mail: haeder@botanik1-pc.biologie.uni-erlangen.dbp.de

***Hardie Roger C.
Cambridge University
Department of Zoology
Downing Street
CB2 3EJ Cambridge
UK
tel: 0044 223 336623/6622
fax: 0044 223 336676
e-mail: RCH14@hermes.cam.ac.uk

**Hengstenberg Roland
Max-Planck-Institut für biologische Kybernetik
Spemannstrasse, 38
D-72076 Tübingen, Germany
tel: 0049 7071 601566
fax: 0049 7071 601575
e-mail: hbg@sunwan.mpik-tueb.mpg.de

Herrmann Heike
Institut für Botanik und Pharmazeutische
Biologie
Universität of Erlangen-Nürnberg
Staudtstrasse, 5
D-91058 Erlangen, Germany
Mailing Address:
Istituto di Biofisica, CNR
Via San Lorenzo, 26
I-56127 Pisa
Italy
tel: 0039 50 513111
fax: 0039 50 553501

*Izzo Marcella
Istituto di Cibernetica, CNR
Via Toiano, 6
I-80072 Arco Felice (Napoli)
Italy
tel: 0039 81 8534124
fax: 0039 81 5267654

***,**Kaupp U.B.
Institut für biologische Informationsverarbeitung
Forschungzentrum
Postfach 1913, D-Jülich, Germany
tel: 0049 2461 614041
fax: 0049 2461 614216

**Kusmic Claudia
Dipartimento di Fisiologia e Biochimica
Università di Pisa
Via San Zeno, 31
I-56127 Pisa
Italy
tel: 0039 50 5555335
fax: 0039 50 552183

***,**Lenci Francesco
Istituto di Biofisica, CNR
Via San Lorenzo, 26
I-56127 Pisa
Italy
tel: 0039 50 513271
fax:0039 50 553501
e-mail: lenci@ib.pi.cnr.it

**Liquori Alfonso Maria
Università di Roma "Tor Vergata"
I-00133 Roma
Italy
tel: 0039 6 5037787
fax: 0039 6 5037787

Lucia Sabina
Università di Sassari
Piazza Università, 1
I-07100 Sassari, Italy
Mailing Address:
Istituto di Biofisica, CNR
Via San Lorenzo, 26
I-56127 Pisa, Italy
tel: 0039 50 513111
fax: 0039 50 553501

Marangoni Roberto
Dipartimento di Fisiologia e Biochimica
Università di Pisa
I-56127 Pisa, Italy
Mailing Address:
Istituto di Biofisica, CNR
Via San Lorenzo, 26
I-56127 Pisa
Italy
tel: 0039 50 513205/207
fax: 0039 50 553501
e-mail: roberto@ib.pi.cnr.it

**Marchiafava Pierlorenzo
Dipartimento di Fisiologia e Biochimica
Università di Pisa
Via San Zeno, 31
I-56127 Pisa
Italy
tel: 0039 50 5555335
fax: 0039 50 552183
e-mail: fisiolog@vm.cnuce.cnr.it

*Mazzarella Annamaria
Istituto di Cibernetica, CNR
Via Toiano, 6
I-80072 Arco Felice (Napoli)
Italy
tel: 0039 81 8534123
fax: 0039 81 5267654
e-mail: cib@secyAnn.na.cnr.it

***,**Minke Baruch
Department of Physiology
The Hebrew University
Hadassah Medical School
P.O.B. 1172
IL-91010 Jerusalem
Israel
tel: 00972 2 758407/585111
fax: 00972 2 439736
e-mail: minke@il.ac.huji.md2

**,*Musio Carlo
Istituto di Cibernetica, CNR
Via Toiano, 6
I-80072 Arco Felice (Napoli)
Italy
tel: 0039 81 8534131
fax: 0039 81 5267654
e-mail: camus@biocib.cib.na.cnr.it

Neugebauer Anna
IIGB-CNR
Via Marconi, 10
I-80100 Napoli
Italy
tel: 0039 81 7257111/268
fax: 0039 81 5936123
e-mail: neugebauer@iigbna.iigb.na.cnr.it

***Packard Andrew
Stazione Zoologica "A. Dohrn"
Villa Comunale
I-80100 Napoli
Italy
tel: 0039 81 5833230
fax: 0039 81 7641355

Park Young Mok
Korea Basic Science Center
Yeoeun-Dong 224-1 Yusung-Ku
305-333 Taejeon
Korea
tel: 0082 42 8653420
fax: 0082 42 8653419

**Payne Richard
Department of Zoology
University of Maryland
College Park, MD 20742
USA
tel: 001 301 4056924
fax: 001 301 3149358
e-mail: rp12@umail.umd.edu

***,**Pepe Isidoro
Istituto di Biofisica
Università di Genova, Genova, Italy
Mailing Address:
Istituto di Cibernetica e Biofisica, CNR
Via Dodecaneso, 33
I-16146 Genova
Italy
tel: 0039 10 3538349/3536426
fax: 0039 10 6475800

**Petracchi Donatella
Istituto di Biofisica, CNR
Via San Lorenzo, 26
I-56127 Pisa
Italy
tel: 0039 50 513205/207
fax: 0030 50 553501
e-mail: petracch@ib.pi.cnr.it

*Piantedosi Salvatore
Istituto di Cibernetica, CNR
Via Toiano, 6
I-80072 Arco Felice (Napoli)
Italy
tel: 0039 81 8534212
fax: 0039 81 5267654
e-mail: sp@imagm.na.cnr.it

**Piccolino Marco
Dipartimento di Fisiologia Generale
Università di Ferrara
Via Borsari, 46
I-44100 Ferrara
Italy
tel: 0039 532 291470
fax: 0039 532 207143
e-mail: neficlm@vm.cnuce.cnr.it

**Ratto Gian Michele
Istituto di Neurofisiologia, CNR
Via San Zeno, 51
I-56127 Pisa
Italy
tel: 0039 50 559723
fax: 0039 50 559725
e-mail: fengari@vm.cnuce.cnr.it

**Rispoli Giorgio
Dipartimento di Fisiologia Generale
Università di Ferrara
Via Borsari, 46
I-44100 Ferrara
Italy
tel: 0039 532 291462/72
e-mail: mk5feqe3@vm.cineca.it

*Santillo Silvia
Istituto di Cibernetica, CNR
Via Toiano, 6
I-80072 Arco Felice (Napoli), Italy
tel: 0039 81 8534113
fax: 0039 81 5267654
e-mail: neuros@mail.irtemp.na.cnr.it

Schettino Giuseppe
Università di Napoli, Dipartimento di Fisica
Mostra d'Oltremare
I-80100 Napoli
Italy
tel: 0039 81 7253111
fax: 0039 81 7253449

Seifert Reinhard
Institut für biologische Informationsverarbeitung
Forschungzentrum Jülich
D-Postfach 1913, Jülich
Germany
tel: 0049 2461 614215
fax: 0049 2461 614216

**Selinger Zvi
Department of Biological Chemistry
Institute of Life Science
The Hebrew University, Givat Ram
IL-91010 Jerusalem
Israel
fax: 00972 2 527427
e-mail: selinger@vms.huji.ac.il

**Song Pill-Soon
University of Nebraska-Lincoln
Department. of Chemistry
Hamilton Hall
Lincoln
NE 68588-0304
USA
tel: 001 402 4723501
fax: 001 402 4722044
e-mail: pssong@unl.edu

***,**Stieve Henning
Institut für Biologie II RWTH
Kopernikusstrasse 16
D-5100 Aachen,
Germany
tel: 0049 241 804835
fax: 0049 241 8888133
e-mail:stieve@RWTH-Aachen.DE

**,*Taddei-Ferretti Cloe
Istituto di Cibernetica, CNR
Via Toiano, 6
I-80072 Arco Felice (Napoli)
Italy
tel: 0039 81 8534113
fax: 0039 81 5267654
e-mail: neuros@mail.irtemp.na.cnr.it

Ukhanov Kyrill
Sechenov's Institute
M. Torez Ave., 44
CSE-194223 St-Petersburg
Russia
tel: 007 812 552 3090
fax: 007 812 552 3012
e-mail: ukhanov@sens.ief.spb.su

**Vallerga Silvana
Istituto di Cibernetica e Biofisica, CNR
Via Dodecaneso, 33
I-16146 Genova
Italy
tel: 0039 10 335626
fax: 0039 10 6475800
e-mail: vallerga@ced.ice.ge.cnr.it
International Marine Center
Lungomare d'Arborea, 22
09072 Torre Grande (Oristano)
Italy
tel 0039 783 22027/32
fax 0039 783 22002

Vellani Vittorio
Istituto di Fisiologia Generale
Università di Ferrara
Via L. Borsari,46
I-44100 Ferrara
Italy
tel: 0039 532 291472
fax: 0039 532 307143
e-mail: mk5feqe3@vm.cineca.it

**Vorobyev Misha
CSE-St. Petersburg
Russia
and
Freie Universität Berlin
Fachberaich Biologie
Institut für Tierphysiologie (WE 5)
Neurobiologie
Königin-Luise-Str. 28-30
D-14195 Berlin
Germany
tel: 0049 30 838 6537
fax: 0049 30 838 5455
e-mail (c/o Prof. Menzel):
menzel@castor.zedat.fu-berlin.de